BIOTECHNOLOGY IN AGRICULTURE, INDUSTRY AND MEDICINE

TRENDS IN LIFE SCIENCE RESEARCH

Biotechnology in Agriculture, Industry and Medicine

Additional books in this series can be found on Nova's website under the Series tab.

Additional e-books in this series can be found on Nova's website under the eBooks tab.

BIOTECHNOLOGY IN AGRICULTURE, INDUSTRY AND MEDICINE

TRENDS IN LIFE SCIENCE RESEARCH

RAJESHWAR P. SINHA
AND
UMESH P. SHRIVASTAVA
EDITORS

nova
biomedical
New York

Additional color graphics may be available in the e-book version of this book.

Library of Congress Cataloging-in-Publication Data

ISBN: 978-1-53613-241-0

Published by Nova Science Publishers, Inc. † New York

Dedicated to our PhD supervisor
Professor Ashok Kumar
School of Biotechnology, Institute of Science, Banaras Hindu University, Varanasi, India

On the occasion of his 65th birthday

CONTENTS

PREFACE

The modern age life science research is more inclined towards interdisciplinary studies. Recent developments in the technologies have led to a better understanding of living systems, and this has removed the demarcations between various disciplines of life sciences. A new trend in life science incorporates biological research involving a merger of diverse disciplines such as microbiology, biotechnology, molecular biology, biochemistry, ecology and agriculture. This book is primarily written with the objective of providing standard information on recent topics in the fields of stress biology, molecular biology, ecology, agriculture, bioremediation, human diseases and over all, several biotechnological approaches towards the studies ranging from green photosynthesizing plants to human pathogens. The book encompasses topics on cyanobacteria, bacteria, fungi, higher plants, ecology and human diseases, distributed throughout eleven chapters for the benefit of graduate and postgraduate students as well as young researchers and scientists. The ever-increasing population has resulted in an increasing demand of food, medicine and other natural products. Researchers are continuously working to resolve and develop a better understanding of life and other biological processes; they need to conserve genes, plants and ecosystems, and the apprehension of land, water and environment have led to the strengthening of traditional disciplines as well as the emergence of diverse disciplines such as stress biology, molecular biology, biotechnology, agriculture and human health. This book imparts the latest developments and the state-of-the art knowledge in the specific fields of life sciences.

The subject matter is presented in a logical sequence so that it is easily understandable to the students and researchers. The recent information on the subject is integrated throughout the chapters. In addition, this book provides newer techniques and the use of modern tools in achieving the potential of microbiology, ecology, herbal medicine, biotechnology and stress biology; this is all used to understand the challenges found in life sciences. This book covers a wide range of topics, and the authors anticipate that the contents may appeal to a wider range of readers. Therefore, the authors present this book with great pleasure and express their most sincere gratitude to all the contributors and research scholars, namely Mr. Jainendra Pathak, Mr. Rajneesh, Ms. Vidya Singh, Mr. Deepak Kumar, Mr. Haseen Ahmed, Mr. Deepak K. Singh and Ms. Abha Pandey. They are also particularly thankful to Nova Science Publishers, Inc., for their continued patience and for the unfailing attention they have given to the publication of this book.

Rajeshwar P. Sinha
Umesh P. Shrivastava

November, 2017

In: Trends in Life Science Research
Editors: R. P. Sinha and U. P. Shrivastava

ISBN: 978-1-53613-241-0

Chapter 1

ENZYMATIC AND NON-ENZYMATIC PROTECTION STRATEGIES OF CYANOBACTERIA AGAINST ULTRAVIOLET RADIATION

Jainendra Pathak[1], Rajneesh[1], Vidya Singh[1], Deepak Kumar[1], Haseen Ahmed[1], Deepak K. Singh[1], Vinod K. Kannaujiya[2], Richa[1], Shailendra P. Singh[3] and Rajeshwar P. Sinha[1,*]

[1]Laboratory of Photobiology and Molecular Microbiology, Center of Advanced Study in Botany, Institute of Science, Banaras Hindu University, Varanasi, India
[2]Department of Botany (MMV), Banaras Hindu University, Varanasi, India
[3]Centre of Advanced Study in Botany, Institute of Science, Banaras Hindu University, Varanasi, India

ABSTRACT

Climate change and enhanced ultraviolet radiation (UVR; 280-400 nm) acts synergistically and strongly affects terrestrial and aquatic ecosystems. Phytoplankton productivity and species composition is expected to change due to altered external conditions such as temperature, nutrient accessibility, pH and exposure to photosynthetically active radiation (PAR; 400-700 nm) and UVR. Cyanobacteria are found in almost every habitat; ranging from oceans to fresh water and from bare rock to soil. They are major biomass producers in aquatic ecosystems representing more than 50% of the biomass in many aquatic ecosystems. These photosynthetic organisms are simultaneously exposed to PAR and UVR in their natural habitats. UVR is one of the most potent genotoxic agents which adversely affect the living organisms by altering their genomic stability. UVR also inhibit number of photochemical and photobiological processes in cyanobacteria. However, cyanobacteria have evolved a number of defense

[*]Corresponding Author Email: r.p.sinha@gmx.net.

strategies to reduce the direct and indirect damaging effects of UVR. These strategies include scavenging of ROS by non-enzymatic and enzymatic antioxidant molecules, synthesis of UV screening compounds such as scytonemin and mycosporine-like amino acids (MAAs), repair of UV-induced damage of DNA and resynthesis of damaged proteins. Each of these strategies involves enzymatic activity either directly or indirectly. This chapter presents an update on various enzymatic and non-enzymatic defense mechanisms employed by cyanobacteria to withstand UVR stress.

Keywords: cyanobacteria, photosynthetically active radiation (PAR), ultraviolet radiation (UVR), enzymes, scytonemin, mycosporine-like amino acids (MAAs)

INTRODUCTION

Ultraviolet radiation (UVR), an important component of solar radiation, imposes number of deleterious effects on biological systems. The increment in atmospheric pollutants such as chlorofluorocarbons (CFCs), fluorocarbons (FCs) and organobromides (OBs) have depleted the ozone layer which has resulted in increased dose of ultraviolet-B radiation (UV-B; 280-320 nm) on the Earth's surface (Crutzen 1992; Stolarski et al. 1992). In addition to the Antarctic ozone hole, ozone depletion has been reported to increase and spread in a broader range of altitudes and latitudes throughout the world since the late 1970s (Hoffman and Deshler 1991). The ozone layer is not likely to recover to 1980s levels in coming decades (Weatherhead and Andersen 2006). Natural production of reactive nitrogen species (RNS) such as nitric oxide (NO^-), peroxynitrite ($ONOO^-$) and nitrous oxide (N_2O) from unpolluted terrestrial and aquatic ecosystems or from anthropogenic sources also contribute to the depletion of the ozone layer (Kramlich and Linak 1994). The UV-B radiation, which is$<$1% of the total solar irradiance, may negatively affect the living organisms by damaging the DNA, RNA and protein molecules or *via* producing reactive oxygen species (ROS) (Peak and Peak 1982; Quaite et al. 1992; Pakker et al. 2000; Vincent and Neale 2000; Buma et al. 2001; Sinha et al. 2001; He and Häder 2002; Rastogi et al. 2010; Singh et al. 2014). In contrast, UV-A (315-400 nm) radiation is not absorbed directly by the nucleic acids, but still induce damage either by producing a secondary photoreaction of existing photoproducts or *via* indirect photosensitizing reactions (Hargreaves et al. 2007). UV-B radiation can reach at different depths in a water column which is determined by the optical properties of the water column (Jerlov 1950; Fleischmann 1989; Smith et al. 1992). The variables of the atmosphere and water that affect the quality and quantity of UVR are known to determine the transmission of solar UVR into the water column (Häder et al. 2007; Smith et al. 2007; Llabrés et al. 2013; Andrady et al. 2016). UV-B radiation can reach higher depths in clear oceanic waters where its levels are sufficient to cause mortality of photosynthetic plankton. UV-B radiation has been reported to reach upto 60 m depth in the subtropical Atlantic Ocean (Llabrés and Agustí 2006), and upto 26 m depth in the Mediterranean Sea (Llabrés et al. 2010). The basis of marine food web is constituted by primary producers, e.g., phytoplanktons, cyanobacteria and macroalgae, which are food source for primary and secondary consumers (Ali et al. 2007; Häder et al. 2007).

Investigations have shown that UV-B radiation adversely affects the key physiological and biochemical life processes, such as morphology, cell differentiation, survival, growth, pigmentation, motility and orientation, N_2 metabolism, phycobiliprotein composition, protein profile, DNA and $^{14}CO_2$ uptake and membrane permeability, in various photosynthetic organisms (Gao et al. 2007; Lesser 2008; Sinha et al. 2008; Kannaujiya et al. 2015; Richa et al. 2015; Kannaujiya et al. 2016; Kannaujiya and Sinha 2017; Singh et al. 2017a; Singh et al. 2017b). Solar UV-B radiation damages cellular DNA in freshwater phytoplankton by inducing mainly CPDs and to a smaller extent pyrimidine (6-4) pyrimidone photoproducts (6-4 PP) and their Dewar isomers (Sinha and Häder 2002; Richa et al. 2015). To counteract the deleterious effects of UVR, cyanobacteria have evolved a number of defense mechanisms.

These defense strategies include avoidance, scavenging of reactive oxygen species (ROS) by non-enzymatic and enzymatic antioxidants like catalase (CAT), superoxide dismutase (SOD), peroxidase (POD), carotenoids, etc., and synthesis of UV-absorbing compounds (scytonemin and mycosporine-like amino acids (MAAs) (Sinha et al. 2007; Xie et al. 2009; Pathak et al. 2015a; 2015b; Richa and Sinha 2015a; 2015b; Pathak et al. 2017a). Additionally, repair of UV-induced damage of DNA by photoreactivation, excision repair and resynthesis of proteins are also important mechanisms to repair the UV-induced photodamages (Britt 1995; Kim and Sancar 1995). This chapter provides an overview of various enzymatic and non-enzymatic defense mechanisms developed by cyanobacteria which helps them to neutralize the damaging effects of UVR.

Mitigation Strategies Evolved by Cyanobacteria against UVR

To counteract the damaging effects of UVR, cyanobacteria have evolved several lines of defense mechanisms (Figure 1) (Vincent and Roy 1993; Sinha et al. 1998; Cockell and Knowland 1999; Singh et al. 2010; Singh et al. 2017a).

The various enzymatic and non-enzymatic protective mechanisms employed by cyanobacteria against UVR have been discussed below in detail.

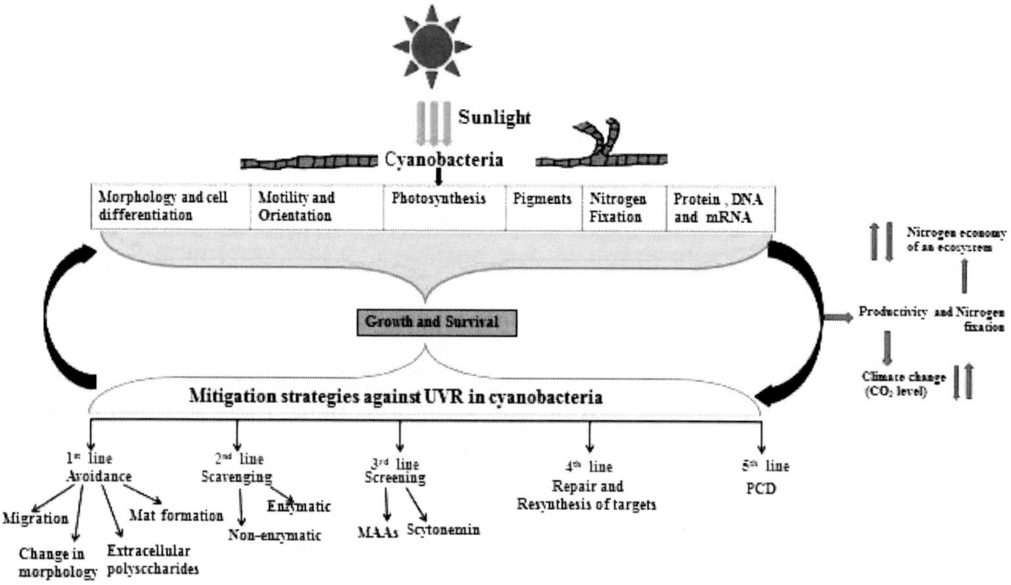

Figure 1. The possible effects of UV radiation and mitigation strategies employed by cyanobacteria.

SYNTHESIS OF ANTIOXIDATIVE ENZYMES

First line of defense mechanism adopted by cyanobacteria is avoidance by migrating into the water column at level where minimum exposure to the UVR occurs. However, once UVR reaches inside the cell, it interacts with oxygen and other organic compounds to produce toxic ROS such as superoxide (O_2^-), hydroxyl radical (OH^-) or hydrogen peroxide (H_2O_2) which results in oxidative stress. To overcome the oxidative stress, cyanobacteria have developed antioxidant systems as a second line of defense mechanism against UVR. This system includes enzymatic and non-enzymatic antioxidants. The enzymatic antioxidants include superoxide dismutase (SOD), catalase, glutathione peroxidase (GSH-Px) and the enzymes involved in the ascorbate-glutathione cycle to detoxify ROS such as ascorbate peroxidase (APX), monodehydroascorbate reductase (MDHAR), dehydroascorbate reductase (DHAR) and glutathione reductase (GR) (Figure 2) (He and Häder 2002b; Richa and Sinha 2015b).

Non-enzymatic antioxidants comprise ascorbate (vitamin C), α-tocopherol (vitamin E), carotenoids and reduced glutathione (GSH). The detrimental effects of UV-B radiation on cyanobacteria are mostly mediated through free radicals which induces multilevel oxidative stress (Wang et al. 2007; Janknegt et al. 2008). In cyanobacteria, ROS are inevitably generated by photosynthetic electron transport; especially when the intensity of light-driven electron transport outpaces the rate of electron consumption during CO_2 fixation (Latifi et al. 2009). Therefore, the generation of ROS during photosynthesis is a normal phenomenon. However, many stress condition provokes increased production of toxic oxygen derivatives (Halliwell and Gutteridge 1989; Pospíšil 2009; Demidchik 2010). ROS are generated as intermediates of O_2 reduction, or by its energization as well as during photorespiration and respiration. ROS, including singlet oxygen ($^1O_2$), the superoxide anion (O^{2-}), hydrogen peroxide (H_2O_2) and the hydroxyl radical (OH^-), are powerful oxidizing agents. Singlet oxygen ($^1O_2$) is produced by energy input to oxygen. It is highly reactive, has a short half life in cells and reacts with target molecules (proteins, pigments and lipids) in the immediate neighborhood. This byproduct of biological redox reactions can inactivate enzymes, damage important cellular components and initiate lipid peroxidation, thereby producing lipid peroxyl radicals and lipid hydroperoxides (Halliwell and Gutteridge 1989). The overproduction of ROS and resultant oxidative stress can negatively affect biomolecules such as, DNA, proteins and lipids, however, DNA is considered to have a more critical influence on the survival and growth of living organisms (Rastogi et al. 2010; Singh et al. 2014; Rajneesh et al. 2017a).

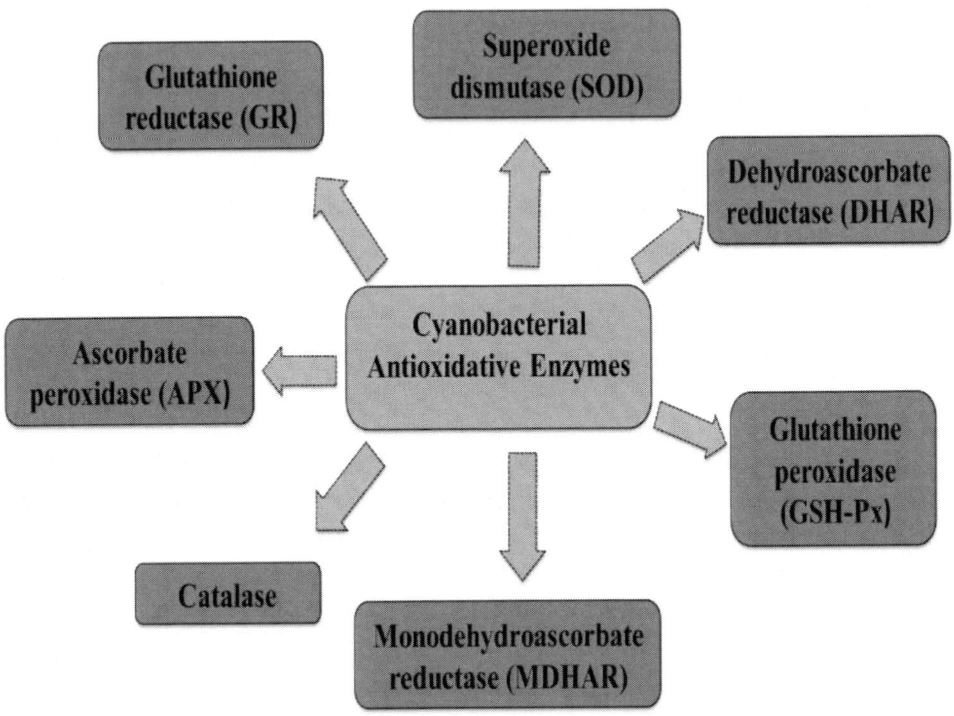

Figure 2.Various antioxidative enzymes found in cyanobacteria.

Antioxidative enzymes such as superoxide-dismutase (SOD), catalase (CAT), peroxidases (PODs) and ascorbate peroxidases (APXs) can scavenge reactive oxygen species (Zhang et al. 2005; Janknegt et al. 2008; Richa and Sinha 2015b; Ahmed et al. 2017a; Singh et al. 2017b).

Superoxide-Dismutase

SOD might be the first antioxidants which function against nascent oxygen species. It plays a critical role in mitigating the toxic effect of superoxide ion. The first implication on the protective role of cyanobacterial SOD in photo-oxidative damage was shown in *Anacystis nidulans* (Herbert et al. 1992). Subsequently, several studies on protective role of SOD against various physiological processes/stresses have been reported in cyanobacteria (Saha et al. 2003; Rajendran et al. 2007). SOD was first isolated as a copper storage protein (Mann and Kleilin 1938). Shirkey et al. (2000) reported that an accumulation of active iron superoxide dismutase (FeSOD) in the cyanobacterium *N. commune* reverse the effects of oxidative stress which is imposed by multiple cycles of desiccation and rehydration during the UV-A or UV-B irradiation. Zhao et al. (2007) reported that Mn SOD protects the nitrogenase enzyme from inactivation by ROS generated in heterocysts under aerobic conditions. Depending upon the nature of the catalytic metals, four forms of SODs namely, FeSOD, MnSOD, Cu/ZnSOD and NiSOD have been reported. Comparative genomic analysis by Priya et al. (2007) revealed that NiSOD is the only SOD found in primitive cyanobacteria, however, Fe and Mn occupy the higher orders of cyanobacteria, and Cu/ZnSOD is rarely found in cyanobacteria. These workers also revealed that primitive unicellular *Prochlorococcus* possesses only NiSOD while the more evolved middle order forms of cyanobacteria posses a combination of Fe and Ni or Fe and Mn SODs. The most evolved filamentous, heterotrichous and heterocystous forms predominantly have only Fe and Mn metalloforms (Priya et al. 2007). SOD catalyses the dismutation of highly reactive superoxide to hydrogen peroxide and oxygen *via* the reaction:

$$2O_2^{-} + 2H^{+} \rightarrow H_2O_2 + O_2$$

Catalase

CAT is one of the main tetrameric enzymes which are responsible for the dismutation of hydrogen peroxide (H_2O_2) into water and oxygen *via* the reaction:

$$2H_2O_2 \rightarrow 2H_2O + O_2$$

CAT is also known to prevent dehydration related oxidative damage inbiological membranes and helps in complete recovery of cells (Wai et al. 2000). Three groups of catalases, i.e., monofunctional haem-containing catalases, bifunctional haem-containing catalase-peroxidases and nonhaem-manganese catalases have been reported till date in 20 cyanobacterial genomes (Latifi et al. 2009).

Peroxidases

Peroxidases detoxify harmful H_2O_2 present in almost every organism (Passardi et al. 2004). Recently, another type of peroxidase called rubrerythrin-homologue (RbrA) was identified in *Anabaena* PCC 7120 and was found to protect the nitrogenase enzyme from negative effects of oxidative stress (Zhao et al. 2007).

Ascorbate Peroxidases

APX catalyzes the reduction of H_2O_2 to water using the reducing power of ascorbate (Noctor and Foyer 1998). It was found that the cells of *Synechocystis* 6803 containing-ascorbate peroxidase supported the quenching of chlorophyll fluorescence induced by hydrogen peroxide whereas it was completely absent in case of the cells of *Anacystis nidulans* which were devoid of ascorbate peroxidase (Miyake et al. 1990). Ascorbate peroxidase was found to be the major enzyme involved in the removal of hydrogen peroxide under oxidative stresses in *Synechococcus* PCC 9742 cells (Mittler and Tel-or 1991). The increased activities of APX, glutathione reductase and SOD in pea buds were observed under UV-B and high light irradiation which is known to induce ROS production. 3D structures of SOD, CAT and APX are shown in Figure 3.

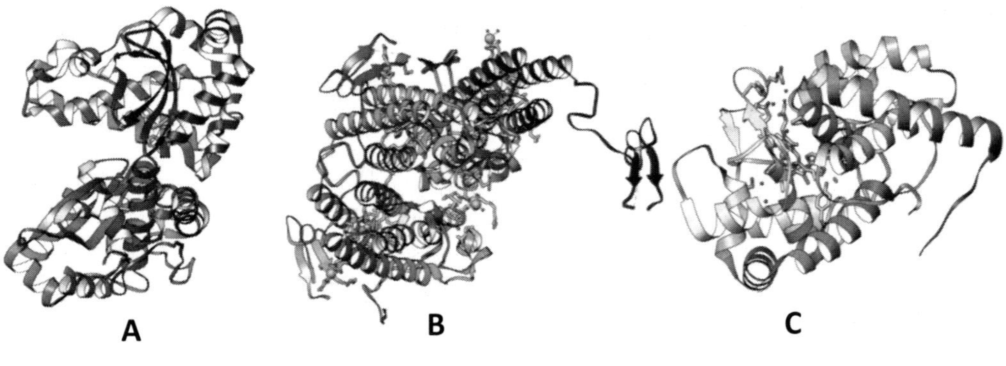

Figure 3. 3D structures of superoxide-dismutase (A); catalase (B); and ascorbate peroxidase (C).

These antioxidative enzymes counteract the damages caused by UV-B stress and thus contribute to the survival of cyanobacteria under UV-B radiation. The work on UV-B-induced alteration in antioxidant enzymes of cyanobacteria is limited to few studies (Zeeshan and Prasad 2009; Singh et al. 2013). Richa and Sinha (2015b) studied the enzymatic defense mechanisms and the alteration in the activity of antioxidant enzymes such as SOD, CAT, APX and POD in two *Nostoc* species isolated from diverse habitats. After exposure to UV-B radiation (1.0 Wm^{-2}) for varying periods of time, the antioxidative enzymes such as SOD, CAT, POD and APX showed a multifold induction upto 10 h of UV-B exposure. The increased levels of antioxidative enzymes were found to be associated with survival of organisms even after 48 h of continuous UV-B exposure. Richa and Sinha (2015b) reported that *Nostoc* sp. strain HKAR-2, a hot-spring isolate, possess high antioxidative enzymatic activity which confer better survival in comparison to the rice-field cyanobacterium *Nostoc* sp. strain HKAR-6 which posses comparatively low level of enzymatic activity. However, different species may show varying degree of induction in the level of abovementioned enzymes under UV-B stress which could be associated with different degree of survival (Singh et al. 2013; Ahmed et al. 2017a; Singh et al. 2017b). Kesheri et al. 2014 showed that the Antarctic cyanobacterium *Nostoc commune* possess two antioxidative enzymes *viz.*, superoxide dismutase (SOD) and catalase that jointly cope with environmental stresses prevailing at its natural habitat. Carotenoids protect cells against photooxidative damage by absorbing energy from excited chlorophyll molecule and quench singlet state oxygen, while α-tocopherol prevents lipid peroxidation by scavenging ROS (Niyogi 1999). However, ascorbate plays an important role in direct quenching of ROS, regeneration of α-tocopherol and acts as a substrate in both violaxanthin de-epoxidase and APX reactions. Glutathione protects thiol groups in various enzymes and it is also involved in α-tocopherol and ascorbate regeneration through the glutathione-ascorbate cycle (Niyogi 1999). Ehling-Schulz et al. (1997) reported changes in the carotenoids pattern of the cyanobacterium *N. commune* in response to UV-B irradiation, and myxoxanthophyll and echinenone were suggested to act as outer membrane-bound UV-B photoprotectors. An increase in the activity of SOD and APX was observed in *N. spongiaeforme* and *Phormidium corium* as a result of high-light treatment (Bhandari and Sharma 2006). The exogenous addition of antioxidants such as ascorbic acid and N-acetylcysteine (NAC) is associated with higher survival rate of *Anabaena* sp. due to reduction in chlorophyll bleaching, damage to photosynthetic apparatus, lipid peroxidation and DNA strand breaks under UVR-induced oxidative stress (He and Häder 2002a). Two NADPH-dependent glutathione peroxidase-like proteins have been characterized in *Synechocystis* PCC 6803 which are essential for the protection of membranes against lipid peroxidation (Gaber et al. 2004).

Synthesis of UV Protective Compounds

Screening of damaging UVR by UV-absorbing compounds has been developed by several cyanobacteria as a third line of defense mechanism under prolonged UVR exposure (Cockell and Knowland 1999; Pathak et al. 2017a). MAAs and scytonemin are well known UV-screening compounds that provide photoprotection against UV-B and UV-A radiation (Sinha et al. 1998; Singh et al. 2010). Photoprotective role of these UV-screening compounds is briefly described below.

Mycosporine-Like Amino Acids

MAAs are a diverse group of UV-absorbing, small (<400 Da), colorless, water-soluble, secondary metabolites composed of a cyclohexenone or cyclohexenimine chromophore conjugated with the nitrogen substituent of an amino acid or its imino alcohol (Singh et al. 2008; Richa and Sinha 2015a; Pathak et al. 2017a). Generally, the ring system contains a glycine subunit at the third carbon atom. Differences between the absorption spectra of MAAs are due to variations in the attached side groups and nitrogen substituents. Several characteristics of MAAs, such as strong UV absorption maxima between 309 and 362 nm (Figure 4) and high molar extinction coefficients (ε = 28,100 - 50,000 M^{-1} cm^{-1}) suggest their role in UV-screening.

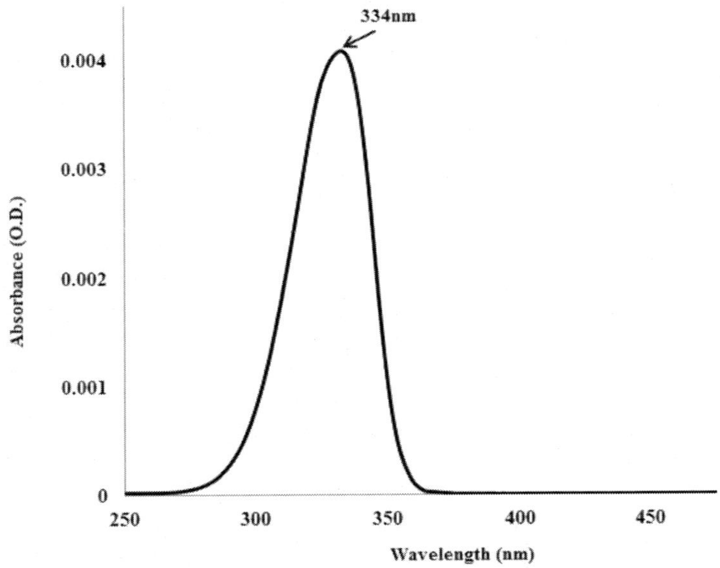

Figure 4. Absorption spectrum of MAAs (Porphyra-334; RT 3.1 min) having absorption peak at 334 nm.

Genomics and Proteomics of MAAs

The precise mechanisms of MAAs biosynthesis are still controversial. However, it was earlier proposed that 3-dehydroquinate formed during the early stages of the shikimate pathway serves as a precursor for the synthesis of fungal mycosporines (Favre-Bonvin et al. 1987) and MAAs *via* gadusols (Bandaranayake 1998; Shick and Dunlap 2002; Singh et al. 2008; Singh et al. 2017a; Richa et al. 2018). Genome mining of certain strains identified a combination of genes, YP_324358 (predicted DHQ synthase) and YP_324357 (O-methyltransferase), which were present only in MAAs biosynthesizing *A. variabilis* PCC 7937 and missing in the other non-MAAs biosynthesizing cyanobacteria

(Singh et al. 2010b; Ahmed et al. 2017b). Phylogenetic analysis revealed that these two genes are transferred from a cyanobacterial donor to dinoflagellates and finally to metazoa by a lateral gene transfer event (Singh et al. 2010b).

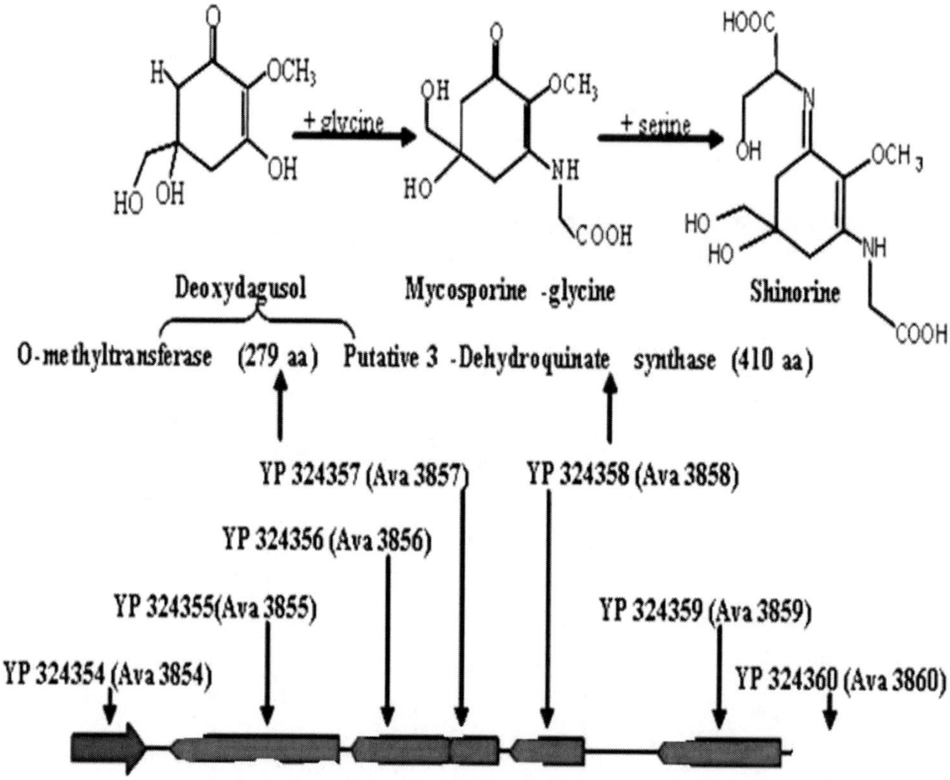

Figure 5. Genes involved in MAAs biosynthesis (Source: Singh et al. 2010b).

All other cyanobacteria, which have these two genes, also had another copy of the DHQ synthase gene. The study provides first insight into the genes of cyanobacteria involved in MAA biosynthesis (Figure 5). Singh et al. (2010b) also suggested that the protein structure of YP_324358 is different from the chemically characterized DHQ synthase of *Aspergillus nidulans*, however, YP_324879 was predicted to be similar to the DHQ synthase of *A. nidulans* (Wu et al. 2007). Based on the results, it was proposed that YP_324358 and YP_324357 genes products are involved in the biosynthesis of the common precursor (deoxygadusol) of all MAAs (Singh et al. 2010a, 2010b; Ahmed et al. 2017b). Thus, YP_324879 gene product is exclusively involved in the shikimate pathway catalyzing the formation of dehydroquinate, while the YP_324358 gene product in conjunction with the YP_324357 gene product (O-methyltransferase) catalyzes the formation of deoxygadusol, which is the core of all MAAs. Balskus and Walsh (2010) challenged the established assumption that MAA biosynthesis involved a shikimate pathway intermediate, as their group failed to observe any 6-deoxygadusol production when the putative substrate 3-dehydroquinate was incubated with thedehydroquinate synthase (DHQS) and O-methyltransferase (O-MT) homologues (NpR 5600 and NpR 5599 respectively) from *Nostoc punctiforme* (ATCC 29133) and the typical cofactors S-adenosylmethionine (SAM), nicotinamide adenine dinucleotide (NAD^+) and Co^{2+}. Molecular structure of some of the common MAAs present in cyanbacteria has been depicted in Figure 6.

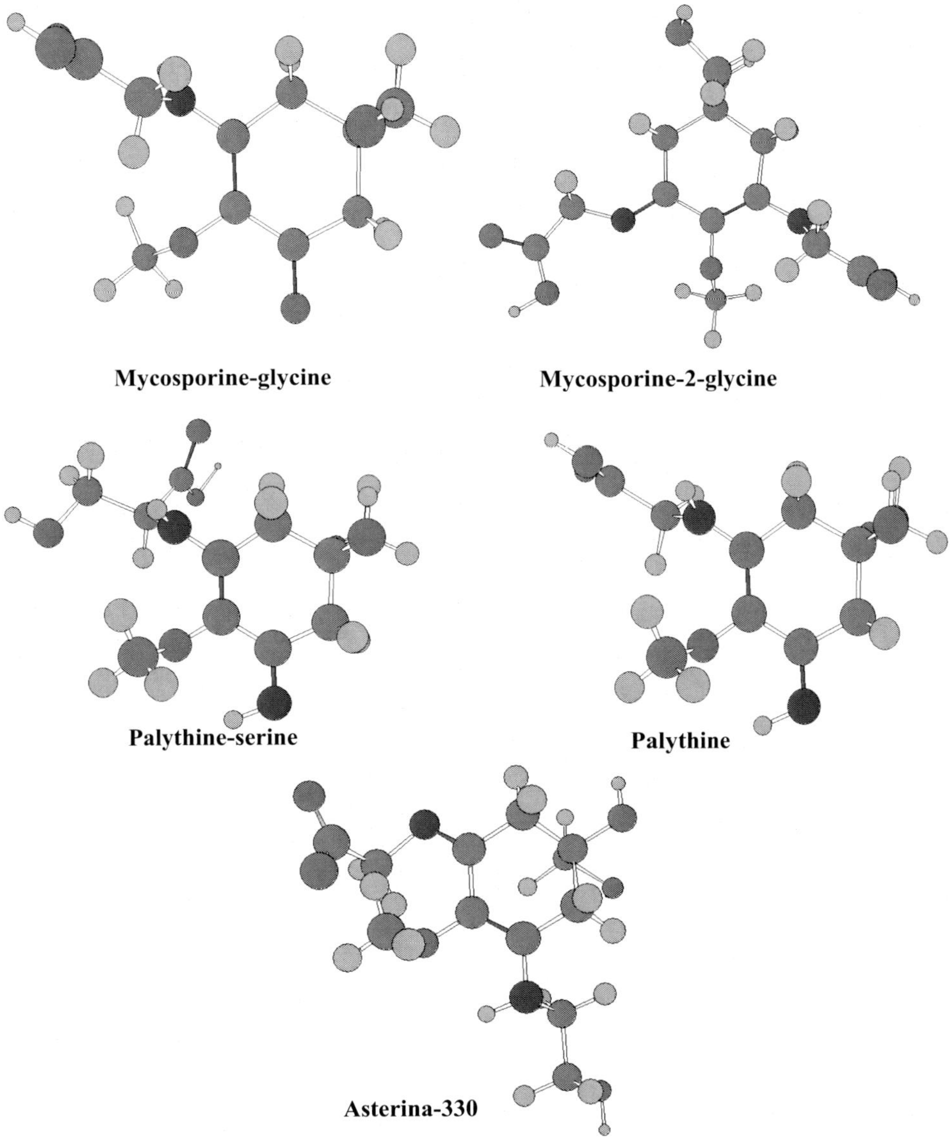

Figure 6. The molecular structures of different MAAs.

Scytonemin

Scytonemin is a highly stable, lipid soluble, yellow-brown hydrophobic pigment which has been reported from extracellular polysaccharide sheaths of about 300 cyanobacteria (Figure 7) (Pathak et al. 2015b; 2017b).

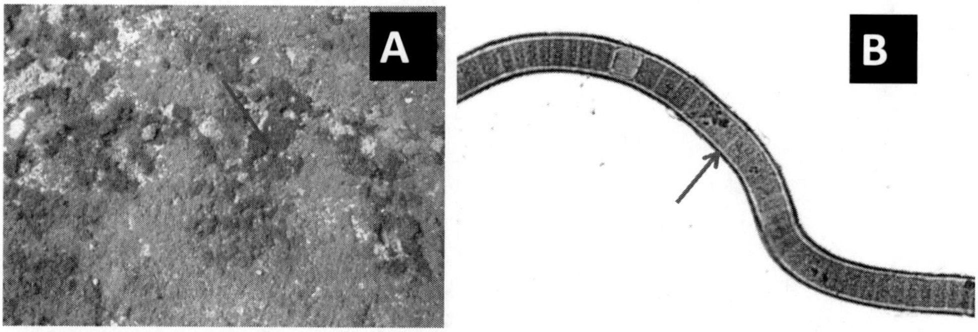

Figure 7. Cyanobacterium growing at the roof top as black-brown leathery mats (A); a filament of *Scytonema geitleri* HKAR-12 with sheath containing yellow-brown UV-protective pigment, scytonemin (B).

Scytonemin has an *in vivo* absorption maximum at 370 nm whereas purified scytonemin shows maximum absorption at 386 nm, but it also absorbs significantly at 252, 278 and 300 nm (Pathak et al. 2017b; 2017c, Rajneesh et al. 2017b). Absorption spectrum of scytonemin extracted from *Scytonema geitleri* HKAR-12 has been shown in Figure 8 which show absorption peaks at 252, 300 and 386 nm.

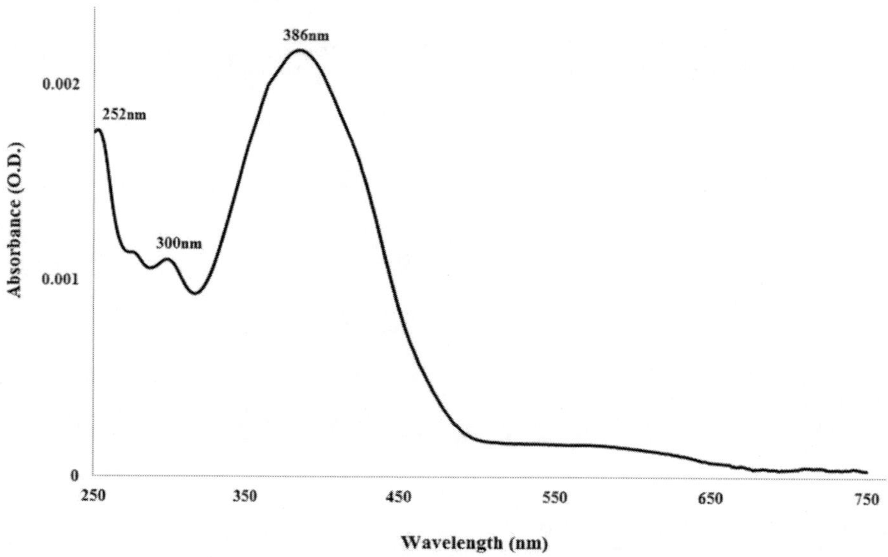

Figure 8. Absorption spectrum of scytonemin extracted from *Scytonema geitleri* HKAR-12 showing peaks at 252, 300 and 386 nm.

Scytonemin is a dimer of indole and phenolic subunits having a molecular mass of 544 Da. The two subunits in scytonemin are linked through an olefinic carbon atom which is unique among natural products. Scytonemin exists in two forms, i.e., oxidized (green) and reduced (red) forms (Garcia-Pichel and Castenholz 1991) which were named as fuscochlorin and fuscorhodin, respectively (Kylin 1927, 1937). Figure 9 shows structures of different forms of scytonemin.

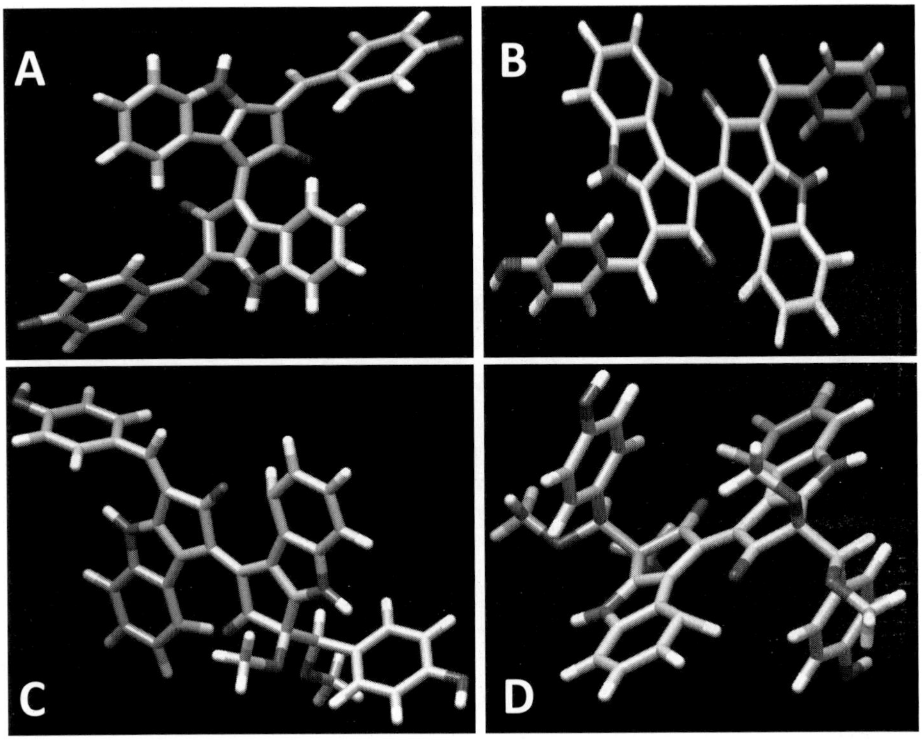

Figure 9. Structures of different forms of scytonemins: Scytonemin (A); Reduced scytonemin (B); Dimethoxy scytonemin (C); and Tetramethoxy scytonemin (D).

Genetics of Scytonemin

Gene cluster associated with scytonemin biosynthesis has been studied in *N. punctiforme* by Soule et al. (2007). The genomic region flanking the mutation revealed an 18-gene cluster (NpR1276 to NpR1259) among which the gene clusters NpR1274-NpR1271 plays central role in scytonemin biosynthesis. NpR1273 was found to be directly involved in scytonemin biosynthesis in *N. punctiforme.* All 18 genes are induced by UV-A radiation and relative transcription levels generally peak after 48 h of continuous UV-A exposure. All of the genes in the 18-gene region are co-transcribed as part of a single transcriptional unit (Soule et al. 2009a). Balskus and Walsh (2008) have proposed possible biosynthetic route (Figure 10) for the scytonemin biosynthesis and the acyloin reaction was found to be the key step in constructing the carbon framework of scytonemin.

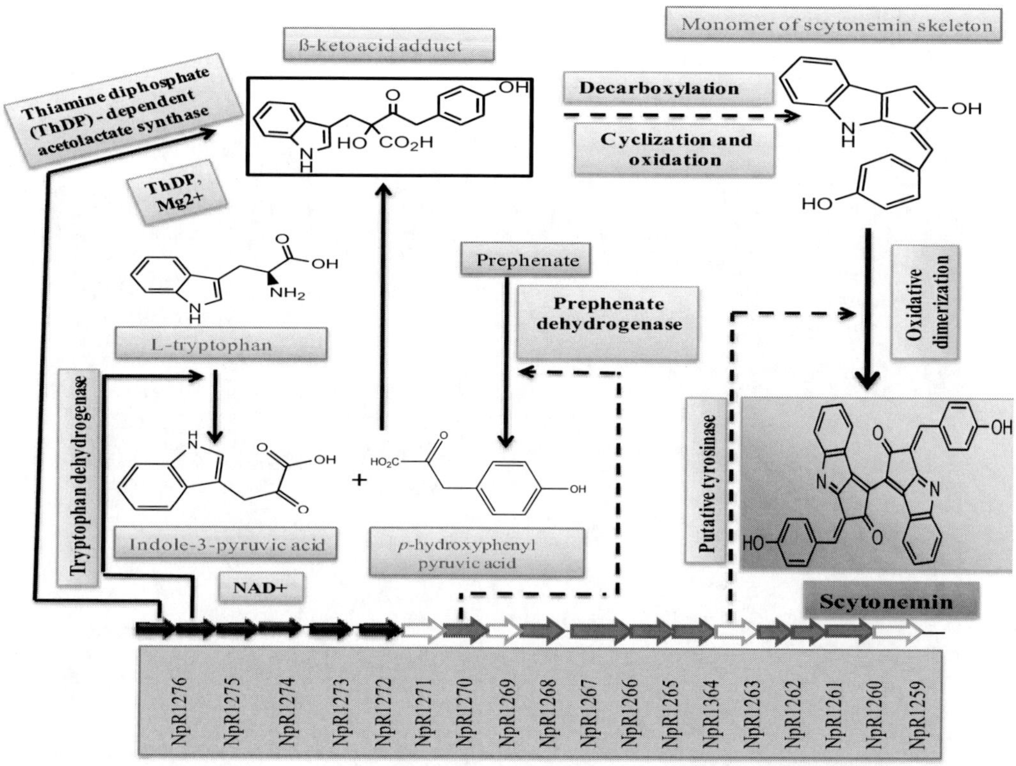

Figure 10. Biosynthetic route for the scytonemin, and corresponding gene products involved in each step. Functionally characterized gene products are represented by continuous arrows while gene products indicated by broken arrows are still to be functionally characterized (Source: Soule et al. (2007; Balskus and Walsh 2008).

Genes and corresponding enzymes associated with scytonemin biosynthesis have been given in Table 1. Two enzymes encoded by ORF NpR1275 and NpR1276 from the gene cluster have been functionally characterized and are involved in the initial step of scytonemin biosynthesis. However, the products of NpR1263 and NpR1269 ORFs are still to be functionally characterized. On the basis of conservation and location, two additional conserved clusters i.e., NpF5232 to NpF5236 and a putative two-component regulatory system (NpF1277 and NpF1278) were proposed to be likely involved in the scytonemin biosynthesis and regulation. (Soule et al. 2009b). Probable use of scytonemin and MAAs as sun screening compounds has been depicted in Figure 11.

Table 1. Genes and corresponding enzymes associated with scytonemin biosynthesis in *Nostoc punctiforme*

S.N.	Gene	Description	Annotation
1	NpR 1259	Transmembrane	Unknown
2	NpR 1260	Shikimate pathway enzyme	aroG
3	NpR1261	Tryptophan synthesis enzyme	trpD
4	NpR1262	Tryptophan synthesis enzyme	trpB
5	NpR1263	Transmembrane	tyrP
6	NpR1264	Tryptophan synthesis enzyme	trpA
7	NpR1265	Tryptophan synthesis enzyme	trpC
8	NpR1266	Tryptophan synthesis enzyme	trpE
9	NpR1267	Shikimate pathway enzyme	aroB
10	NpR1268	Disulfide isomerase	dsbA
11	NpR1269	Prephenate dehydrogenase	tyrA
12	NpR1270	Glycosyltransferase	Glycosyltransferase
13	NpR1271	Unknown, scytonemin biosynthesis enzyme, signal peptide	scyF
14	NpR1272	Unknown, scytonemin biosynthesis enzyme, signal peptide	scyE
15	NpR1273	Unknown, scytonemin biosynthesis enzyme, signal peptide	scyD
16	NpR1274	Unknown, scytonemin biosynthesis enzyme	scyC
17	NpR1275	Leucine dehydrogenase	scyB
18	NpR1276	Acetolactate synthase (TPP binding) transmembrane	scyA
19	NpR1277	Regulatory protein	Sensor kinase
20	NpR1278	Regulatory protein	Response regulator

Source: Soule et al. 2009a, 2009b.

Repair and Resynthesis

The repair and resynthesis of sensitive targets in cyanobacteria becomes critically important as a fourth line of defense mechanism when UVR escape from first, second and third line of defense mechanisms; and damage of biomolecules such as DNA and proteins occurs. Although the presence of multiple copies of genomic DNA in cyanobacteria may nullify the effect of single mutations caused by UVR, the existence of several DNA repair mechanisms strengthen cyanobacteria to cope with the UV radiation. These mechanisms include photoreactivation by photolyases which converts UV-induced dimers into monomers, the dark or excision repair and the recombinational repair.

Figure 11. Defense strategies adapted by human against harmful solar UV radiation. UV protective compounds such as scytonemins and MAAs have shown good potentials for their possible use as sunscreens.

Photolyase

The process of photoreactivation is executed by means of a photoreactivating enzyme known as "photolyase," which is well conserved and found throughout the three domains of life. The enzyme binds specifically to the CPDs (CPD photolyase) or 6-4PPs (6-4 photolyase) and directly monomerizes the cyclobutane ring of the pyr <>pyr, using the energy of visible/blue-light and protects the genome from deleterious effects of UVR (Kim et al. 1992; Essen and Klar 2006; Rajneesh et al. 2017c).

The absorption of every blue-light photon may split approximately one dimer (Britt 1996). CPD photolyases have been reported in diverse groups such as archaea, bacteria, fungi, virus, plants, invertebrates and many vertebrates including aplacental mammals (Rajneesh et al. 2017c). Photolyases seem to be absent or nonfunctional in placental mammals like human (Yasui et al. 1994; Sancar 1996; Todo 1999). However, Sutherland (1974), Sutherland and Bennett (1995), and Harm (1980) have demonstrated photolyase activity in cells and tissues, including white blood cells (WBCs) of several placental mammals, such as humans, ox, cat, and mouse. DNA photolyases (45-66 kDa) having 420-616 amino acid residues (Essen and Klar 2006) are monomeric flavin-dependent repair enzymes which contain two known cofactors, a catalytic cofactor and a light-harvesting cofactor.

Till date, 5,10-methenyltetrahydrofolate (MTHF) (Johnson et al. 1988), 8-hydroxy-5-deaza-riboflavin (8-HDF) (Eker et al. 1990) and FMN (Ueda et al. 2005) are known as light-harvesting cofactors, which absorb light energy efficiently and transfer them to $FADH^-$ (Saxena et al. 2004). Deprotonated reduced flavin adenine dinucleotide ($FADH^-$) is found in all known photolyases as a catalytic cofactor, which transfers energy in the form of an electron to the CPD, splitting the cyclobutane ring with the generation of two monomeric bases (Kim et al. 1992; Kao et al. 2005). In comparison to other eukaryotic systems, reports on the repair of UV-induced DNA damage in plants are still very limited. To avoid the deleterious effects of UVR, plants have acquired two main protective strategies; shielding by flavonoids and phenolic compounds (Stapleton and Walbot 1994; Schmitz-Hoerner and Weissenbock 2003) and DNA repair by photoreactivation.

Photoreactivation mediated by the enzyme photolyases is thought to be the major DNA repair pathway in several higher plants such as rice, *Arabidopsis*, wheat, and maize (Kimura 2004; Kaiser et al. 2009). Studies on *Arabidopsis* seedling, rice and alfalfa indicate that photoreactivation greatly enhances the rate of removal of dimers, although, in the absence of photoreactivating (blue) light, dimers are slowly eliminated from bulk DNA while 6-4PPs are generally observed to be repaired more quickly than CPDs (Chen et al. 1994; Kang et al. 1998). The structural information about the interaction between CPD lesions and photolyases became clear with the help of X-ray crystallography (Mees et al. 2004) and nuclear magnetic resonance (NMR) spectroscopy (Torizawa et al. 2004). However, how DNA photolyases find lesions in the DNA molecule is still not clear (Essen and Klar 2006). MacFarlane and Stanley (2003) have suggested that photolyase enzyme is indeed left in the semiquinonid state after accomplishment of repair of the CPD lesion. The presence of a UV-inducible photoreactivation system has been reported in strains of *Anabaena* sp. such as *Anabaena* sp. PCC7120, *A. variabilis* PCC 7937, *Anabaena* sp. M-131 and *A. variabilis* sp. PCC 7118 (Levine and Thiel 1987). The genes for photolyase homologs have been identified in *Synechocystis* sp. PCC 6803 and these homologs were functionally characterized for their role in photoreactivation (Hitomi et al. 2000; Ng and Pakrasi 2001). Han et al. (2001) have also reported photoreactivation of UV-B-induced inhibition of photosynthesis in *Anabaena* sp. An increase in the transcript level of *recA* gene and the concomitant increase in the abundance of the corresponding 37-38-kDa polypeptide were reported in *A. variabilis* after UV exposure (Owttrim and Coleman 1989). The gene for the DNA repair enzyme Fpg (formamydopyrimidine-DNA glycosylase) has been reported from *Synechococcus elongatus* and was suggested to be involved in photoprotection against oxidative damage (Mühlenhoff 2000).

Thus, cyanobacteria are capable of repairing DNA lesions by both photoreactivation and excision repair.

Recently, photorepair of UV-B-induced damaged DNA has been reported in the cyanobacterium *A. variabilis* PCC 7937 (Rastogi et al. 2011). In additions to DNA repair, cyanobacteria are also capable of synthesizing new proteins to replace damaged copies. The UV-B shock and acclimation response of UV-B stress was found to be a completely different and complex phenomenon which influence accumulation of 493 proteins comprising of an early shock response (214 proteins) and a late acclimation response (279 proteins) (Ehling-Schulz et al. 2002). An increased turnover of D1 and D2 proteins of the photosystem II reaction center was found to be responsible for UV-B resistance in *Synechocystis* sp. (Sass et al. 1997). Instant turnover of photosystem II reaction center proteins helps in acclimatization and provides resistance against UVR in cyanobacteria (Campbell et al. 1998). In addition, recent studies show that cyanobacteria may also undergo apoptosis or programmed cell death (PCD) when a cell is damaged beyond repair. An autocatalytic PCD has been shown to operate in the nitrogen-fixing cyanobacterium (*Trichodesmium* sp.) and was found to be induced by high irradiance, iron starvation and oxidative stress (Berman-Frank et al. 2004). The caspase activities as well as proteins reacting to human caspase-3 antibodies were reported in *Trichodesmium* sp. (Berman-Frank et al. 2004). The freshwater cyanobacterium *Anabaena* has also been found to activate PCD and increases general protease activity after exposure to univalent-cation salts (Ning et al. 2002).

Conclusion

In response to UVR stress, cyanobacteria have developed several lines of defense mechanisms that sustain their successful growth and survival in various habitats where they receive high solar UVR. A fine equilibrium between damage and defense mechanisms has the ecological importance as it maintains the productivity and nitrogen economy of an ecosystem, and thereby, regulate other climate change problems. All these defense strategies are mediated *via* fine tuning of various enzymes, of which some are characterized but majority of these enzymes needs to be studied. There are several studies reporting UVR-induced oxidative damage in cyanobacteria. ROS are known to induce the antioxidant system to remove toxic species, nevertheless, the signaling role of ROS in cyanobacteria under UVR stress can not be ruled out. There are several families of SODs and catalases found in cyanobacteria but characterization of specific SOD and catalase expressed under UVR stress needs future research. The fragmentation of cyanobacterial filaments under UVR may possibly occur due to PCD of severely damaged cells in a filament, however, degradation of proteins that maintains the structure, has been proposed for the filament breakage. The process of PCD and related proteins is found in cyanobacteria to play an important role under oxidative stress and the role of PCD under UVR needs further studies. Identification of DNA and protein repair mechanisms andtheir signalling pathways needs to be worked out. The molecular structure and reaction mechanism of DNA repair enzymes (photolyase and glycosylase), and molecular biology of MAAs as well as scytonemin in various cyanobacteria require more attention. The use of DNA microarray technology and 2-D electrophoresis would help in understanding the adaptation of cyanobacteria to UV stress at genome and proteome level which will help in establishing role of various enzymes involved in several lines of defense mechanisms in cyanobacteria against UV stress.

ACKNOWLEDGMENTS

J. Pathak, V. Singh and D.K. Singh are thankful to CSIR, New Delhi, India for the financial support in the form of SRF. Rajneesh is thankful to DBT, Govt. of India, for the research grant in the form of SRF. D. Kumar is thankful to DST for the financial support in the form of SRF. H. Ahmed acknowledges UGC, New Delhi, India for SRF. Dr. S.P. Singh acknowledges UGC and SERB for financial support in the form of Start-up Grant and Early Career Research Award, respectively.

REFERENCES

Ahmed H, Singh V, Pandey A, Kannaujiya VK, Rajneesh, Pathak J, Sinha, RP (2017a) Combined effects of UV radiation, PAR and 2,4-Dichlorophenoxyacetic acid on pigmentation, proteins and antioxidant enzymes in the cyanobacterium *Scytonema geitleri* HKAR-12. *JSM Environ Sci Ecol* 5(3): 1048.

Ahmed H, Rajneesh, Pandey A, Pathak J, Sinha RP (2017b) Comparative *In silico* analysis of different dehydroquinate synthase proteins in the cyanobacterium *Cyanothece* sp. PCC 7424. *J Bioinform Genomics Proteomics* 2(2): 1018.

Ali MM, Mageed, AA, Heikal, M (2007) Importance of aquatic macrophyte for invertebrate diversity in large subtropical reservoir. *Limnologica-Ecol. Manage. Inland Waters* 37: 155-169.

Andrady A, Aucamp PJ, Austin AT, Bais AF, Ballaré CL, Barnes PW, Bernhard GH, Björn LO, Bornman JF, Erickson DJ, De Gruijl FR, Häder DP, Ilyas M, Longstreth J, Lucas RM, Madronich S, McKenzie RL, Neale R, Norval M, Pandy KK, Paul N, Redhwi HH, Robinson SA, Rose K, Shao M, Sinha RP, Solomon KR, Sulzberger B, Takizawa Y, Torikai A, Tourpali K, Williamson CE, Wilson SR, Wängberg SÅ, Worrest RC, Young AR, Zepp RG (2016) Environmental effects of ozone depletion and its interactions with climate change: progress report 2015. *Photochem Photobiol Sci*15: 141-174.

Balskus EP, Walsh CT (2008) Investigating the initial steps in the biosynthesis of cyanobacterial sunscreen scytonemin. *J Am Chem Soc* 130: 15260-15261.

Bandaranayake WM (1998) Mycosporines: are they nature's sunscreens? *Nat Prod Rep* 15: 159-172.

Berman-Frank I, Bidle KD, Haramaty L, Falkowski PG (2004) The demise of the marine cyanobacterium, *Trichodesmium* sp., *via* an autocatalyzed cell death pathway. *Limnol Oceanogr* 49: 997-1005.

Bhandari R, Sharma PK (2006) High-light-induced changes on photosynthesis, pigments, sugars, lipids and antioxidant enzymes in freshwater (*Nostoc spongiaeforme*) and marine (*Phormidium corium*) cyanobacteria. *Photochem Photobiol* 82: 702-710.

Britt AB (1995) Repair of DNA damage induced by ultraviolet radiation. *Plant Physiol* 108: 891-896.

Britt AB (1996) DNA damage and repair in plants. *Annu Rev Plant Physiol Plant Mol Biol* 47(1): 75-100.

Buma AGJ, De Boer MK, Boelen P (2001) Depth distributions of DNA damage in antarctic marine phyto and bacterioplankton exposed to summertime UV radiation. *J Phycol* 37:200-208.

Campbell D, Eriksson MJ, Öquist G, Gustafsson P, Clarke AK (1998) The cyanobacterium *Synechococcus* resists UV-B by exchanging photosystem II reaction-center D1 proteins. *Proc Natl Acad Sci USA* 95: 364-369.

Chen JJ, Mitchell DL, Britt AB (1994) A light-dependent pathway for the elimination of UV-induced pyrimidine (6-4) pyrimidinone photoproducts in *Arabidopsis*. *Plant Cell* 6(9): 1311-1317.

Cockell CS, Knowland J (1999) Ultraviolet radiation screening compounds. *Biol Rev* 74: 311-345.

Crutzen PJ (1992) Ultraviolet on the increase. *Nature* 356: 104-105.

Demidchik V (2010) Reactive oxygen species, oxidative stress and plant ion channels. In: Demidchik V, Maathuis F, editors. *Ion Channels and Plant Stress Responses*: Springer-Verlag, Berlin/Heidelberg, pp. 207-232.

Ehling-Schulz M, Bilger W, Scherer S (1997) UV-B-induced synthesis of photoprotective pigments and extracellular polysaccharides in the terrestrial cyanobacterium *Nostoc commune*. *J Bacteriol* 179: 1940-1945.

Ehling-Schulz M, Schulz S, Wait R, Görg A, Scherer S (2002) The UV-B stimulon of the terrestrial cyanobacterium *Nostoc commune* comprises early shock proteins and late acclimation proteins. *Mol Microbiol* 46: 827-843.

Eker APM, Kooiman P, Hessels JKC, Yasui A (1990) DNA photoreactivating enzyme from the cyanobacterium *Anacystis nidulans*. *J Biol Chem* 265: 8009-8015.

Essen LO, Klar T (2006) Light-driven DNA repair by photolyases. *Cell Mol Life Sci* 63: 1266-1277.

Favre-Bonvin J, Bernillon J, Salin N, Arpin N (1987) Biosynthesis of mycosporines: mycosporine glutaminol in *Tricothecium roseum. Phytochemistry* 26: 2509-2514.

Fleischmann EM (1989) The measurement and penetration of ultraviolet radiation into tropical marine water. *Limnol Oceanogr* 34: 1623-1629.

Gaber A, Yoshimura K, Tamoi M, Takeda T, Nakano Y, Shigeoka S (2004) Induction and functional analysis of two reduced nicotinamide adenine dinucleotide phosphate-dependent glutathione peroxidase-like proteins in *Synechocystis* PCC 6803 during the progression of oxidative stress. *Plant Physiol* 136: 2855-2861.

Gao K, Yu H, Brown MT (2007) Solar PAR and UV radiation affects the physiology and morphology of the cyanobacterium *Anabaena*sp. PCC 7120. *J Photochem Photobiol B: Biol* 89: 117-124.

Garcia-Pichel F, Castenholz RW (1991) Characterization and biological implications of scytonemin, a cyanobacterial sheath pigment. *J Phycol* 27: 395-409.

Harm H (1980) Damage and repair in mammalian cells after exposure to non-ionizing radiations. III. Ultraviolet and visible light irradiation of cells of placental mammals, including humans, and determination of photorepairable damage *in vitro. Mut Res* 69(1): 167-176.

Häder DP, Kumar HD, Smith RC, Worrest RC (2007) Effects of solar UV radiation on aquatic ecosystems and interactions with climate change. *Photochem Photobiol Sci* 6: 267-285.

Halliwell B, Gutteridge JMC (1989) *Free Radicals in Biology and Medicine.* Oxford Clarendon Press, USA.

Han T, Sinha RP, Häder DP (2001) UV-A/blue light-induced reactivation of photosynthesis in UV-B irradiated cyanobacterium, *Anabaena* sp. *J Plant Physiol* 158: 1403-1413.

Hargreaves A, Taiwo FA, Duggan O, Kirk SH, Ahmad SI (2007) Near-ultraviolet photolysis of β-phenylpyruvic acid generates free radicals and results in DNA damage. *J Photochem Photobiol B Biol* 89:110-116.

He YY, Häder DP (2002a) UV-B-induced formation of reactive oxygen species and oxidative damage of the cyanobacterium *Anabaena* sp.: protective effects of ascorbic acid and N-acetyl-L-cysteine. *J Photochem Photobiol B: Biol* 66, 115-124.

He YY, Häder DP (2002b) Reactive oxygen species and UV-B: effect on cyanobacteria. *Photochem Photobiol Sci* 1: 729-736.

Herbert SK, Samson G, Fork DC, Laudenbach DE (1992) Characterization of damage to photosystems I and II in a cyanobacterium lacking detectable iron superoxide dismutase activity. *Proc Natl Acad Sci USA* 89: 8716-8720.

Hitomi K, Okamoto K, Daiyasu H, Miyashita H, Iwai S, Toh H, Ishiura M, Todo T (2000) Bacterial cryptochrome and photolyase: characterization of two photolyase-like genes of *Synechocystis* sp. PCC6803. *Nucleic Acids Res* 28: 2353-2362.

Hoffman DJ, Deshler T (1991) Evidence from ballon measurements for chemical depletion of stratospheric ozone in the arctic wind of 1989-90. *Nature* 349: 300-305.

Janknegt PJ, van de Poll WH, Visser RJW, Rijstenbil JW, Boma AGJ (2008) Oxidative stress responses in the marine Antarctic diatom *Chaetocerosbrevis* (Bacillariophyceae) during photoacclimation. *J Phycol* 44: 957-966.

Jerlov NC (1950) Ultra-violet radiation in the sea. *Nature* 11: 111-112.

Johnson JL, Hamm-Alvarez S, Payne G, Sancar GB, Rajagopalan KV, Sancar A (1988) Identification of the second chromophore of *Escherichia coli* and yeast DNA photolyases as 5,10-methenyltetrahydrofolate. *Proc Natl Acad Sci* 85: 2046-2050.

Kaiser G, Kleiner O, Beisswenger C, Batschauer A (2009) Increased DNA repair in *Arabidopsis* plants overexpressing CPD photolyase. *Planta* 230 (3) 505-515.

Kang HS, Hidema J, Kumagai T (1998) Effects of light environment during culture on UV-induced cyclobutyl pyrimidine dimers and their photorepair in rice (*Oryza sativa* L.). *Photochem Photobiol* 68(1): 71-77.

Kannaujiya, VK, Sinha RP (2015) Impacts of varying light regimes on phycobiliproteins of *Nostoc* sp. HKAR-2 and *Nostoc* sp. HKAR-11 isolated from diverse habitats. *Protoplasma* 252: 1551-1561.

Kannaujiya VK, Rahman A, Adinath, Sonker AS, Pathak J, Rajneesh, Sundaram S, Sinha RP (2016) Cumulative effects of ultraviolet radiation and photosynthetically active radiation on phycobiliproteins of a hot-spring cyanobacterium *Nostoc* sp. strain HKAR-2. *Int J Appl Sci Biotechnol* 4(3): 247-253.

Kannaujiya VK, Sinha RP (2017) Impacts of diurnal variation of ultraviolet-B and photosynthetically active radiation on phycobiliproteins of the hot-spring cyanobacterium *Nostoc* sp. strain HKAR-2. *Protoplasma* 254: 423-433.

Kao YT, Saxena C, Wang L, Sancar A, Zhong D (2005) Direct observation of thymine dimer repair in DNA by photolyase. *Proc Natl Acad Sci USA* 102: 16128-16132.

Kesheri M, Kanchan S, Richa, Sinha RP (2014) Isolation and *in silico* analysis of Fe-superoxide dismutase in the cyanobacterium *Nostoc commune*. *Gene* 553: 117-125.

Kim ST, Heelis PF, Sancar A (1992) Energy transfer (deazaflavin→$FADH_2$) and electron transfer ($FADH_2$→T<>T) kinetics in *Anacystis nidulans* photolyase. *Biochemistry* 31: 11244-11248.

Kim ST, Sancar A (1995) Photorepair of nonadjacent pyrimidine dimers by DNA photolyase. *Photochem Photobiol* 61: 171-174.

Kimura S, Tahira Y, Ishibashi T, Mori Y, Mori T, Hashimoto J, Sakaguchi K (2004) DNA repair in higher plants; photoreactivation is the major DNA repair pathway in non-proliferating cells while excision repair (nucleotide excision repair and base excision repair) is active in proliferating cells. *Nucleic Acids Res* 32: 2760-2767.

Kramlich JC, Linak WP (1994) Nitrous oxide behaviour in the atmosphere, and in combustion and industrial systems. *Prog Energy Combust Sci* 20: 149.

Kylim H (1927). Über die Karotinoide Farbstoffe der Algen. Hoppe-Scyler'szeitschr. *Physiological Chemistry*. 166: 33-77.

Kylim H (1937). Über die Farbstoffe und die Farbe der cyanophyceen. [About the dyes and the color of cyanophyceen]. *Fysiogr.Sällsk.Förhandl.* 7: 131-158.

Latifi A, Ruiz M, Zhang CC (2009) Oxidative stress in cyanobacteria. *FEMS Microbiol Rev* 33: 258-278.

Lesser MP (2008) Effects of ultraviolet radiation on productivity and nitrogen fixation in the cyanobacterium, *Anabaena*sp. (Newton's strain). *Hydrobiologia* 598: 1-9.

Levine E, Thiel T (1987) UV-inducible DNA repair in the cyanobacteria *Anabaena* spp. *J Bacteriol* 169: 3988-3993.

Llabrés M, Agustí S, Fernández Miriam, Canepa A, Maurin F, Vidal F, Duarte CM (2013) Impact of elevated UV-B radiation on marine biota: a meta-analysis. *Global Ecol Biogeogr*22: 131-144.

Llabrés M, Agustí S (2006) Picophytoplankton cell death induced by UV radiation: evidence for oceanic Atlantic communities. *Limnol Oceanogr*51: 21-29.

Llabrés M, Agustí S, Alonso-Laita P, Herndl G (2010) *Synechococcus* and *Prochlorococcus* cell death induced by UV radiation and the penetration of lethal UVR in the Mediterranean Sea. *Mar Ecol Prog Ser* 399: 27-37.

MacFarlane AW, RJ Stanley (2003) Cis-Syn thymidine dimer repair by DNA photolyase in real time. *Biochemistry* 42(28): 8558-8568.

Mann T, Kleilin D (1938) Homocuprein and heptacuprein, copper-protein compounds of blood and liver in mammals. *Proc R Soc London B* 126: 303-315.

Mees A, Klar T, Gnau, P, Hennecke U, Eker APM, Carell T, Essen LO (2004) Crystal structure of a photolyase bound to a CPD-like DNA lesion after *in situ* repair. *Science* 306 (5702): 1789-1793.

Mittler R, Tel-or E (1991) Oxidative stress responses and shock protein in unicellular cyanobacterium *Synechococcus* PCC 7942 (R2). *Arch Microbiol* 155: 125-130.

Miyake C, Michihata F, Asada K (1990) Scavenging of hydrogen peroxide in prokaryotic and eukaryotic algae: Acquisition of ascorbate peroxidase during the evolution of cyanobacteria. *Plant Cell Physiol* 32: 33-43.

Mühlenhoff U (2000) The FAPY-DNA glycosylase (Fpg) is required for survival of the cyanobacterium *Synechococcus elongatus* under high light irradiance. *FEMS Microbiol Lett* 187: 127-132.

Ng WO, Pakrasi HB (2001) DNA photolyase homologs are the major UV resistance factors in the cyanobacterium *Synechocystis* sp. PCC 6803. *Mol Gen Genet* 264: 924-930.

Ning SB, Guo HL, Wang L, Song YC (2002) Salt stress induces programmed cell death in prokaryotic organism *Anabaena. J Appl Microbiol* 93: 15-28.

Niyogi KK (1999) Photoprotection revisited: genetics and molecular approaches. *Annu Rev Plant Physiol* 50: 333-359.

Noctor G, Foyer CH (1998) A re-evaluation of the ATP: NADPH budget during C_3 photosynthesis. A contribution from nitrate assimilation and its associated respiratory activity. *J Exp Bot* 49: 1895-1908.

Owttrim GW, Coleman JR (1989) Regulation of expression and nucleotide sequence of the *Anabaena variabilis recA* gene. *J Bacteriol* 171: 5713-5719.

Pakker H, Beekman CAC, Breeman AM (2000) Efficient photoreactivation of UV-B radiation induced DNA damage in the sublittoral macroalga *Rhodymenia pseudopalmata* (Rhodophyta). *Eur J Phycol* 35:109-114.

Passardi, F, Longet D, Penel C, Dunand C (2004) The class III peroxidase multigenic in land plants family in rice and its evolution. *Phytochemistry* 65: 1879-1893.

Pathak J, Richa, Rajneesh, Sonker AS, Kannaujiya VK, Sinha RP (2015a) Isolation and partial purification of scytonemin and mycosporine-like amino acids from biological crusts. *J Chem Pharm Res* 7: 362-371.

Pathak J, Rajneesh, Richa, Kannaujiya VK, Sonker AS, Sinha RP (2015b) Diverse functions and applications of novel and unique cyanobacterial sheath pigment, scytonemin. In: Sinha RP, Richa, Rastogi RP, editors. *Biological Sciences: Innovations and Dynamics*: New India Publishing Agency, New Delhi, India. pp. 237-262.

Pathak J, Rajneesh, Ahmed H, Richa, Sinha RP (2017a) Metabolomic profiling of cyanobacterial UV-protective compounds. *Curr Metabolomics* 5(2): 138-163.

Pathak J, Rajneesh, Richa, Sonker AS, Kannaujiya VK, Sinha RP (2017b) Cyanobacterial extracellular polysaccharide sheath pigment, scytonemin: a novel multipurpose pharmacophore. In: Se-Kwon Kim, editor. *Marine Glycobiology: Principles and Applications*: Taylor and Francis Group, CRC Press, Boca Raton, Florida, USA, pp. 323-338.

Pathak J, Sonker AS, Richa, Rajneesh, Kannaujiya VK, Singh V, Ahmed H, Sinha RP (2017c) Screening and partial purification of photoprotective pigment, scytonemin from cyanobacterial crusts dwelling on the historical monuments in and around Varanasi, India. *Microbiol Res (Pavia)* 8: 6559.

Peak MJ, Peak JG (1982) Single-strand breaks induced in *Bacillus subtilis* DNA by ultraviolet light: action spectrum and properties. *Photochem Photobiol* 35: 675-680.

Pospíšil P (2009) Production of reactive oxygen species by photosystem II. *Biochim Biophys Acta Bioenerg* 1787: 1151-1160.

Priya B, Premanandh J, Dhanalakshmi RT, Seethalakshmi T, Uma L, Prabaharan D, Subramanian G (2007) Comparative analysis of cyanobacterial superoxide dismutases to discriminate canonical forms. *BMC Genomics* 8: 435.

Quaite FE, Sutherland BM, Sutherland JC (1992) Quantitation of pyrimidine dimers in DNA from UV-B irradiated alfalfa (*Medicago sativa* L.) seedlings. *Appl Theor Electrophor* 2:171-175.

Rajendran UM, Kathirvel E, Anand N (2007) Desiccation-induced changes in antioxidant enzymes, fatty acids, and amino acids in the cyanobacterium *Tolypothrix scytonemoides*. *World J Microbiol Biotechnol* 23: 251-257.

Rajneesh, Pathak J, Chatterjee A, Singh SP, Sinha RP (2017a) Detection of reactive oxygen species (ROS) in cyanobacteria using the oxidant-sensing probe 2',7'-Dichlorodihydrofluorescein diacetate (DCFH-DA). *Bio-protocol.* DOI:10.21769/BioProtoc.xxxx55555111112000.

Rajneesh, Singh SP, Pathak J, Sinha RP (2017b) Cyanobacterial factories for the production of green energy and value-added products: An integrated approach for economic viability. *Renew Sustainable Energy Rev* 69: 578-596.

Rajneesh, Pathak J, Kannaujiya VK, Singh SP, Sinha, RP (2017c) Analysis of codon usage patterns of photolyase gene in cyanobacteria of various habitats. *3 Biotech* 7: 192.

Rastogi RP, Richa, Kumar A, Tyagi MB, Sinha RP (2010a) Molecular mechanisms of ultraviolet radiation-induced DNA damage and repair. *J Nucleic Acids* Volume 2010, Article ID 592980 DOI:10.4061/2010/592980.

Rastogi RP, Singh SP, Häder DP, Sinha RP (2011) Ultraviolet-B-induced DNA damage and photorepair in the cyanobacterium *Anabaena variabilis* PCC7937. *Env Exp Bot* 74: 280-288.

Rastogi RP, Singh SP, Häder DP, Sinha RP (2010) Detection of reactive oxygen species (ROS) by the oxidant-sensing probe 2′, 7′-dichlorodihydrofluorescein diacetate in the cyanobacterium *Anabaena variabilis* PCC 7937. *Biochem Biophys Res Commun* 397, 603-607.

Richa, Sinha RP (2015a) Biochemical characterization of sunscreening mycosporine-like amino acids from two *Nostoc* species inhabiting diverse habitats. *Protoplasma* 252: 199-208.

Richa, Sinha RP (2015b) Sensitivity of two *Nostoc* species harbouring diverse habitats to ultraviolet-B radiation *Microbiology* 84(3): 398-407.

Richa, Pathak J, Sonker AS, Singh V, Sinha RP (2018) Potential applications of natural bioactive cyanobacterial UV protective compounds. In: La Barre S, Bates S, editors. *Blue technologies: production and uses of marine molecules*: Wiley VCH, Weinheim, Germany. Print Book ISBN: 978-3-527-34138-2.

Richa, Sinha RP, Häder DP (2015) Physiological aspects of UV-excitation of DNA. *Top Curr Chem* 356: 203-248.

Saha SK, Uma L, Subramanian G (2003) Nitrogen stress induced changes in the marine cyanobacterium *Oscillatoria willei* BDU 130511. *FEMS Microbiol Ecol* 45: 263-272.

Sancar A (1996) No end of history for photolyases. *Science* 272(5258): 48-49.

Sass L, Spetea C, Mate Z, Nagy F, Vass I (1997) Repair of UV-B induced damage of Photosystem II *via de novo*synthesis of the D1 and D2 reaction centre subunits in *Synechocystis* sp. PCC 6803. *Photosynth Res* 54: 55-62.

Saxena C, Sancar A, Zhong D (2004) Femtosecond dynamics of DNA photolyase: energy transfer of antenna initiation and electron transfer of cofactor reduction. *J Phys Chem B* 108: 18026-18033.

Schmitz-Hoerner R, Weissenbock G (2003) Contribution of phenolic compounds to the UV-B screening capacity of developing barley primary leaves in relation to DNA damage and repair under elevated UV-B levels. *Phytochemistry* 64: 243-255.

Shick JM, Dunlap WC (2002) Mycosporine-like amino acids and related gadusols: biosynthesis, accumulation, and UV-protective functions in aquatic organisms. *Annu Rev Physiol* 64: 223-262.

Shirkey B, Kovarcik DP, Wright DJ, Wilmoth G, Prickett TF, Helm RF, Gregory EM, Potts M (2000) Active Fe-containing superoxide dismutase and abundant sodF mRNA in *Nostoc commune* (Cyanobacteria) after years of desiccation. *J Bacteriol* 182: 189-197.

Singh G, Babele PK, Sinha RP, Tyagi MB, Kumar A (2013) Enzymatic and non-enzymatic defense mechanisms against ultraviolet-B radiation in two *Anabaena* species. *Process Biochem* 48: 796-802.

Singh SP, Kumari S, Rastogi RP, Singh KL, Sinha RP (2008) Mycosporine-like amino acids (MAAs): chemical structure, biosynthesis and significance as UV-absorbing/screening compounds. *Ind J Exp Biol* 46: 7-17.

Singh SP, Häder DP, Sinha RP (2010a) Cyanobacteria and ultraviolet radiation (UVR) stress: mitigation strategies. *Ageing Res Rev* 9: 79-90.

Singh SP, Klisch M, Sinha RP, Häder DP (2010b) Genome mining of mycosporine-like amino acid (MAA) synthesizing and non-synthesizing cyanobacteria: a bioinformatics study. *Genomics* 95: 120-128.

Singh SP, Rastogi RP, Häder DP, Sinha RP (2014) Temporal dynamics of ROS biogenesis under simulated solar radiation in the cyanobacterium *Anabaena variabilis* PCC 7937. *Protoplasma* 251(5): 1223-1230.

Singh V, Pathak J, Rajneesh, Richa, Sonker AS, Kumar D, Ahmed H, Singh DK, Kannaujiya VK, Sinha RP (2017a) Does cyanobacteria have enough mechanisms to counteract abiotic stresses? In: Sinha RP, Richa, editors. *New Approaches in Biological Research*: Nova Publishers NY, USA. pp. 265-307.

Singh DK, Richa, Kumar D, Chatterjee A, Rajneesh, Pathak J, Sinha RP (2017b) Response of the cyanobacterium *Fischerella* sp. strain HKAR-5 against combined stress UV-B radiation, PAR and pyrogallic acid. *JSM Environ Sci Ecol* 5(3): 1049.

Sinha RP, Klisch M, Gröniger A, Häder DP (1998) Ultraviolet-absorbing/screening substances in cyanobacteria, phytoplankton and macroalgae. *J Photochem Photobiol B: Biol* 47: 83-94.

Sinha RP, Kumari S, Rastogi RP (2008) Impacts of ultraviolet-B radiation on cyanobacteria: photoprotection and repair. *J Sci Res* 52: 125-142.

Sinha RP, Singh SP, Häder DP (2007) Database on mycosporines and mycosporine-like amino acids (MAAs) in fungi, cyanobacteria, macroalgae, phytoplankton and animals. *J Photochem Photobiol B: Biol* 89: 29-35.

Sinha RP, Häder DP (2002) UV-induced DNA damage and repair: a review. *Photochem. Photobiol. Sci.* 1: 225-236.

Smith RC, Mobley CD, Björn LO (2007) Underwater Light. In: Björn LO, editor. *Photobiology. The Science of Life and Light*: Second Edition, Springer, New York, pp. 131-138.

Smith RC, Prézelin BB, Baker KS, Bidigare RR, Boucher NP, Coley T, Karentz D, MacIntyre S, Matlick HA, Menzies D, Ondrusek M, Wan Z, Waters KJ (1992) Ozone depletion: ultraviolet radiation and phytoplankton biology in Antarctic waters. *Science* 255: 952-959.

Soule T, Stout V, Swingley WD, Meeks JC, Garcia-Pichel F (2007) Molecular genetics and genomic analysis of scytonemin biosynthesis in *Nostoc punctiforme* ATCC 29133. *J Bacteriol* 189: 4465- 4472.

Soule T, Garcia-Pichel F, Stout V (2009a) Gene expression patterns associated with the biosynthesis of the sunscreen scytonemin in *Nostoc punctiforme* ATCC 29133 in response to UV-A radiation. *J Bacteriol* 191: 4639- 4646.

Soule T, Palmer K, Gao Q, Potrafka RM, Stout V, Garcia-Pichel F (2009b) A comparative genomics approach to understanding the biosynthesis of the sunscreen scytonemin in cyanobacteria. *BMC Genomics* 10: 336-345.

Stapleton AE, Walbot V (1994) Flavonoids can protect maize DNA from the induction of ultraviolet radiation damage. *Plant Physiology* 105(3): 881-889.

Stolarski R, Bojkov R, Bishop L, Zerefos C, Staehelin J, Zawodny J (1992) Measured trends in stratospheric ozone. *Science* 256: 342-349.

Sutherland BM (1974) Photoreactivating enzyme from humanleukocytes. *Nature* 248(5444): 109-112.

Sutherland BM, PV Bennett (1995) Human white bloodcells contain cyclobutyl pyrimidine dimer photolyase. *Proceedings of the National Academy of Sciences of the United States of America*, 92: 9732-9736.

Todo T, (1999) Functional diversity of the DNA photolyase/blue light receptor family. *Mut Res* 434(2): 89-97.

Torizawa, T, Ueda T, Kuramitsu S, Hitomi K,Todo T,Iwai S,Morikawa K,Shimada I (2004) Investigation of the cyclobutane pyrimidine dimer (CPD) photolyase DNA recognition mechanism by NMR analyses. *J Biol Chem* 279(31): 32950-32956.

Ueda T, Kato A, Kuramitsu S, Terasawa H, Shimada I (2005) Identification and characterization of a second chromophore of DNA photolyase from *Thermus thermophilus* HB27. *J Biol Chem*280: 36237-36243.

Vincent WF, Neale PJ (2000) Mechanisms of UV damage to aquatic organisms. In: de Mora SJ, Demers S, Vernet M, editors. *The Effects of UV Radiation on Marine Ecosystems*: Cambridge University Press, Cambridge, pp. 149-176.

Vincent WF, Roy S (1993) Solar ultraviolet-B radiation and aquatic primary production: damage, protection, and recovery. *Environ Rev* 1: 1-12.

Wai HH, Va L, Oakes M, Nomura M (2000) Complete detection of yeast rDNA repeats and integration of a new rDNA repeat: Use of rDNA deletion strains for functional analysis of rDNA promoter elements *in vivo*. *Nucleic Acid Rev* 28: 3524.

Wang GH, Hu CX, Li DH (2007) The response of antioxidant systems in *Nostocsphaeroides* against UV-B radiation and the protective effects of exogenous antioxidants. *Adv Space Res* 39:1034-1042.

Weatherhead EC, Andersen SB (2006) The search for signs of recovery of the ozone layer. *Nature* 441: 39-45.

Wu H, Gao K, Villafane VE, Watanabe T, Helbling EW (2005) Effects of solar UV radiation on morphology and photosynthesis of filamentous cyanobacterium *Arthrospira platensis*. *Appl Environ Microbiol* 71: 5004-5013.

Xie Z, Wang Y, Liu Y, Liu Y (2009) Ultraviolet-B exposure induces photo-oxidative damage and subsequent repair strategies in a desert cyanobacterium *Microcoleus vaginatus* Gom. *European J Soil Biol* 45: 377-382.

Yasui A, Eker APM, Yasuhira S, Yajima H, Kobayashi T, Takao M, Oikawal A (1994) A new class of DNA photolyases present in various organisms including aplacental mammals *EMBO Journal* 13(24): 6143-6151.

Zeeshan M, Prasad SM (2009) Differential response of growth, photosynthesis, antioxidant enzymes and lipid peroxidation to UV-B radiation in three cyanobacteria. *South African J Bot* 75: 466-474.

Zhang L, Li L, Wu Q (2007) Protective effects of mycosporine-like amino acids of *Synechocystis* sp. PCC 6803 and their partial characterization. *J Photochem Photobiol B* 86: 240-245.

Zhang PY, Yu J, Tang XX (2005) UV-B radiation suppresses the growth and antioxidant systems of two marine microalgae, *Platymonas subcordiformis* (Wille) Hazen and *Nitzschia closterium* (Ehrenb.) W Sm. *J Integr Plant Biol* 47: 683-691.

Zhao W, Ye Z, Zhao J (2007) RbrA, a cyanobacterial rubrerythrin, functions as a FNR-dependent peroxidase in heterocysts in protection of nitrogenase from damage by hydrogen peroxide in *Anabaena* sp. PCC 7120. *Mol Microbiol* 66: 1219-1230.

In: Trends in Life Science Research
Editors: R. P. Sinha and U. P. Shrivastava
ISBN: 978-1-53613-241-0

Chapter 2

MICROBIAL INTERACTIONS IN THE RHIZOSPHERE: THE SIGNIFICANCE OF PLANT BENEFICIAL MICROORGANISMS

Umesh P. Shrivastava*
Department of Botany, Tribhuvan University, Thakur Ram Multiple Campus, Birgunj, Nepal

ABSTRACT

The rhizosphere is a narrow zone of soil, which is influenced by root secretions and it is a hotspot for microorganisms. It has been reported that the concentration of bacteria in the rhizosphere is 10 to 1000 times higher than that in bulk soil. These bacteria residing around rhizosphere are popularly known as rhizobacteria. Rhizobacteria may be beneficial, deleterious or neutral according to their interaction and influence on plants. Plant growth promoting rhizobacteria (PGPR) influence the agricultural crop. This chapter describes the microbial interaction of plants with microorganisms, microorganisms with microorganisms and their significance. Plant microbe interaction is a major area in the field of biotechnology, which is used to search efficient bio-fertilizers for the enhancement of yield of agricultural crops. These rhizobacteria play important roles on crop plants, such as- plant growth stimulation, biocontrol of pathogens, biological N_2-fixation, production of volatile organic compounds, quorum sensing, rhizoremediation, maintenance of soil fertility and nutrient uptake, resistance to various stresses, etc.

Keywords: rhizosphere, rhizobacteria; plant growth promoting rhizobacteria (PGPR), N_2-fixation, biocontrol, bio-fertilizers, plant microbe interaction

* Corresponding Author Email: upshrivastava@gmail.com.

INTRODUCTION

The rhizosphere, which is the narrow zone of soil that is influenced by root secretions, can contain up to 10^{11} microbial cells per gram root (Egamberdieva et al. 2008) and more than 30,000 prokaryotic species (Mendes et al. 2011). Although bacteria were not known to exist until the discovery of microscopic animals by Anton von Leeuwenhoek (1683), their utilization to stimulate plant growth has been exploited since ancient times. Theophrastus (372-287 BC) suggested the mixing of different soil samples for rectifying defects and adding heart to soil (Tisdale and Nelson 1975). It has been reported that the concentration of bacteria in the rhizosphere is 10 to 1000 times higher than that in bulk soil; it is still 100-fold lower than that in the average laboratory medium. Therefore, the lifestyle of rhizobacteria is best characterized as starvation. Virgil reported the establishment of legumes on cultivated land and showed the beneficial effects of legume crops in increasing the fertility of soil (Chew 2002). Hellriegel and Wilfarth (1888) investigated the rhizosphere root colonization in grasses and legumes and showed the ability of soil bacteria to convert atmospheric N_2 into plant usable forms. Based on their experiments on radishes, Kloepper and Schroth (1978) introduced the term 'rhizobacteria' to the soil bacterial community that competitively colonized plant roots and stimulated growth and thereby reducing the incidence of plant diseases. Thereafter, these beneficial rhizobacteria were termed as plant growth-promoting rhizobacteria (PGPR) by Kloepper and Schroth (1981). PGPR can be defined as the essential part of rhizosphere biota that when grown in association with the host plants can stimulate the growth of the host. PGPR seemed as successful rhizobacteria in getting established in soil ecosystem due to their high adaptability in a wide variety of environments, faster growth rate and biochemical versatility to metabolize a wide range of natural and xenobiotic compounds. PGPR is considered as the significant component in the management of agricultural practices with innate genetic potential (Cook 2002). It is believed that PGPR can fulfil at least two of the three criteria such as aggressive colonization, plant growth stimulation and biocontrol (Weller et al. 2002; Vessey 2003). According to Whipps (2001) there are three basic categories of interactions (neutral, negative or positive) generally exists between the rhizobacteria and growing plants. Most rhizobacteria related with plants are commensals in which the microorganisms set up a harmless communication with the host plants displaying no noticeable impact on the development and general physiology of the host (Beattie 2006). In negative connections, the phytopathogenic rhizobacteria produces phytotoxic substances, for example, hydrogen cyanide or ethylene, in this manner, contrarily impact on the development and physiology of the plants. Counter to these deleterious bacteria, there are some PGPRs that can exert a positive plant growth by direct mechanisms such as solubilization of insoluble nutrients, nitrogen fixation, production of growth promoting hormones, etc., or by indirect mechanisms such as stimulation of mycorrhizae development, competitive exclusion of pathogens or removal of phytotoxic substances (Bashan and de-Bashan 2010). However, as per their level of association with the plant root cells, PGPRs can be classified into extracellular plant growth promoting rhizobacteria (ePGPR) and intracellular plant growth promoting rhizobacteria (iPGPR) (Martinez-Viveros et al. 2010). The ePGPRs may exist in the rhizosphere, on the rhizoplane or in the spaces between the cells of root cortex; on the other hand, iPGPRs locates generally inside the specialized nod-ular structures of root cells. The bacterial genera such as *Agrobacterium*, *Arthrobacter, Azotobacter, Azospirillum, Bacillus, Burkholderia, Caulobacter, Chromobacterium, Erwinia, Flavobacterium, Micrococcous, Pseudomonas* and *Serratia* belong to ePGPR (Gray and Smith 2005).

The iPGPR includes the endophytes and *Frankia* species both of which can symbiotically fix atmospheric N_2 with the higher plants (Verma et al. 2010). Endophyte includes a wide range of soil bacterial genera such as *Allorhizobium, Azorhizobium, Bradyrhizobium, Mesorhizobium* and *Rhizobium* of the family Rhizobiaceae that generally invades the root systems in crop plants to form nodules.

In any case, as per their level of relationship with the plant root cells, PGPRs can be arranged into extracellular plant development advancing rhizobacteria (ePGPR) and intracellular plant development advancing rhizobacteria (iPGPR) (Martinez-Viveros et al. 2010). The ePGPRs may exist in the rhizosphere, on the rhizoplane or in the spaces between the cells of root cortex; then again, iPGPRs finds for the most part inside the particular gesture ular structures of root cells. The bacterial genera, for example, *Agrobacterium*, *Arthrobacter*, *Azotobacter*, *Azospirillum*, *Bacillus*, *Burkholderia*, *Caulobacter*, *Chromobacterium*, *Erwinia*, *Flavobacterium*, *Micrococcous*, *Pseudomonas* and *Serratia* has a place with ePGPR (Gray and Smith 2005). The iPGPR incorporates the endophytes and *Frankia* species both of which can harmoniously settle climatic N_2 with the higher plants (Verma et al. 2010). Endophyte incorporates an extensive variety of soil bacterial genera, for example, *Allorhizobium*, *Azorhizobium*, *Bradyrhizobium*, *Mesorhizobium* and *Rhizobium* of the family Rhizobiaceae that for the most part attacks the root frameworks in harvest plants to shape knobs (Wang and Martinez-Romero 2000) and animates development either specifically or in a roundabout way. This gathering of rhizobacteria is generally Gram-negative and pole formed with a lower extent being Gram-positive bars, cocci and pleomorphic. These rhizobacteria are referred as *Bradyrhizobium japonicum* (Guerinot and Chelm 1984), *Rhizobium fredii* (Scholla and Elkan 1984), *Azorhizobium caulinodans* (Dreyfus et al. 1988), *Sinorhizobium fredii* (Chen et al. 1988), *Rhizobium galegae* (Lindstrom 1989), *Sinorhizobium medicae* (Rome et al. 1996), *Rhizobium gallicum* (Amarger et al. 1997), *Rhizobium giardinii* (Amarger et al. 1997), *Allorhizobium undicola* (de Lajudie et al. 1998a), *Mesorhizobium pluriforium* (de Lajudie et al. 1998b), *Rhizobium etli* (Segovia et al. 1993), *Rhizobium ciceri* (Nour et al. 1994), *Sinorhizobium arboris* (Nick et al. 1999) and *Mesorhisobium chacoense* (Velazquez et al. 2001). Some of the important plant species forming symbiotic association with these rhizobial species includes *Acacia* sp., *Arachis hypogaea*, *Cajanus cajan*, *Cercis canadensis*, *Cicer arietinum*, *Glycine max*, *Lens culinaris*, *Lotus corniculatus*, *Medicago sativa*, *Phaseolus vulgaris*, *Pisum sativum* and *Trifolium* sp. (Verma et al. 2010). In addition, several actinomycetes, one of the major components of rhizosphere microbial populations, are also useful because of their significant ecological roles in soil nutrient cycling (Halder et al. 1991; Elliot and Lynch 1995) as well in plant growth-promoting activities (Merzaeva and Shirokikh 2006). Numbers of reports are giving emphasis on the potential of actinomycetes as plant growth-promoting agent (Gomes et al. 2000; Sousa et al. 2008). *Micromonospora* sp., *Streptomyces* sp., *Streptosporangium* sp., and *Thermobifida* sp., are recorded actinomycetes strains known to colonize the plant rhizosphere, showing an immense potentiality as biocontrol agent against a range of root pathogenic fungi (Franco-Correa et al. 2010). Rhizosphere Streptomycetes as potential biocontrol agent of *Fusarium* and *Armillaria* pine rot and as PGPR of Pinus taeda was reported (de Vasconcellos and Cardoso 2009). Evidences are now available on actinobacteria used in the control of *Rhizoctonia solani* and *Pseudomonas solanacearum* in tomato (Sabaratnam and Traquair 2002) and *Colletotrichum musae* in banana (Taechowisan et al. 2003). Soil actinomycetes are also an important source of diverse antimicrobial metabolites (Terkina et al. 2006).

Siderophore production for rhizosphere colonization has also been recorded as one of the important mechanism by certain PGPRs (*Bradyrhizobium japonicum*, *Rhizobium leguminosarum* and *Sinorhizobium meliloti, Pseudomonas sp. strain PN-4D*) (Carson et al. 2000; El-Tarabily and Sivasithamparam 2006; Shrivastava 2013a,b) with plant growth promoting activity. *Pseudomonas* sp. strain PN-4D, *Agrobacterium* sp. strain BN-2A and *Klebsiella* sp. strain BN-4A showed IAA production, phosphate solubilization and ACC deaminase activity with N_2 fixation (Shrivastava and Kumar 2011). Besides, iron-chelating siderophores (Schippers et al. 1988), antibiotics (Weller 1988) and hydrogen cyanides (Stutz et al. 1986) are also likely to be produced by PGPR strains, participating tremendously in the reduction of phytopathogens and deleterious rhizobacteria with a corresponding improvement in plant health. However, regardless of the mechanism of plant growth promotion, PGPR must colonize the rhizosphere or root itself (Glick 1995).

The Rhizosphere and Plant-Microbe Interactions

Plant-microbe interactions may occur at phyllosphere, endosphere and rhizosphere. Phyllosphere is related with the aerial parts of the plants and endosphere with internal transport system. The phyllosphere can be further subdivided into the caulosphere (stems), phylloplane (leaves), anthosphere (flowers), and carposphere (fruits). Rhizosphere, the term, can be defined as any volume of soil specially influenced by the plant roots or in association with the roots and plant-produced material. According to

Bringhurst et al. (2001), rhizosphere includes the region of soil bound by plant roots, often extending a few mm from the root surface. This region of soil is much richer in bacteria than the surrounding bulk soil (Hiltner 1904). Studies based on molecular techniques have estimated more than 4,000 microbial species per gram of soil (Montesinos 2003). Filamentous actinobacteria are also considered as one of the important community in rhizosphere microbiota (Benizri et al. 2001) being able to influence the plant development as well to protect the plant roots against phytopathogens. Plant exudates such as amino acids and sugars provide a rich source of energy and nutrients for the bacteria in rhizosphere, resulting in more microbial populations in the region than outside the region (Haas and Defago 2005). Plant-root interactions in rhizosphere may include root-root, root-insect and root-microbe interactions, resulting in the production of more root exudates that ultimately favours maximum microbial populations (rhizosphere engineering) in this ecologically significant region. Changes in rhizobacterial community structure have been reported with the application of polymerase chain reaction (PCR) and denaturing gradient gel electrophoresis (DGGE) resulting in significant alterations in plant–microbes interactions (Herschkovitz et al. 2005). However, successful root colonization and persistence of PGPRs in plant rhizosphere are required in order to exert their beneficial effect on the plant (Elliot and Lynch 1984). The closeness between the plants and the environment in rhizosphere is thus necessary for better acquisition of water and nutrients by plants as well beneficial interactions of plants with soil-borne microorganisms (Ryan et al. 2009). Cardoso and Freitas (1992) reports that rhizosphere microbial communities are vigorously associated with the biogeochemical cycling of nutrients like C, P, N and S, removal of toxins and production of phytohormones or antibiotics etc. Rhizobacteria may depend on other microbes for nutrient sources as one microbe may convert plant exudates into a form that can be used by another microbe. Thus, rhizosphere has appeared as a multidirectional and dynamic ecological environment of acute plant-microbe interactions (Mayak et al. 2004) ruling essential micro- and macro-nutrients affecting plant growth, although, the process of root colonization is under the influence of various parameters such as bacterial traits, root exudates and several other biotic and abiotic factors (Benizri et al. 2001). In various rhizospheric interactions, the PGPRs are known to colonize the plant root (Andrews and Harris 2000) and accelerate plant growth. Table 1 shows various microorganisms present in the rhizoshere of plants. The colonization of plant rhizosphere by *Azospirillum* sp., *Bacillus subtilis* sp., and *Pseudomonas* sp. has been well studied and reported (Steenhoudt and Vanderleyden 2000; Trivedi et al. 2005). Moreover, immobilized form of PGPR inoculants in comparison to free forms has greater ability of survival and plant root colonization. Glick (1995) and Kaymak (2011) have reported that soil microorganisms, including free-living as well as associative and symbiotic rhizobacteria belonging to the genera like *Acinetobacter*, *Alcaligenes*, *Arthrobacter*, *Azospirillum*, *Azotobacter*, *Bacillus*, *Burkholderia*, *Enterobacter*, *Erwinia*, *Flavobacterium*, *Proteus*,

Pseudomonas, *Rhizobium*, *Serratia*, *Xanthomonas* are the integral parts of rhizosphere biota exhibiting successful rhizosphere colonization. Lugtenberg et al. (2001) described a large number of cell surface molecules as responsible for the effective rhizosphere colonization. Rhizospheric colonization is thus, considered as a crucial step in the application of microorganisms for beneficial purposes such as biofertilization, phytostimulation, biocontrol and phytoremediation, although the colonization of rhizosphere by PGPRs is not a uniform process. In a report by Wenbo et al. (2001), *Kluyvera ascorbata* colonized the upper two-thirds of the surface of canola roots but no bacteria were detected around the root tips. Interactions of microbes with plants and among them in the rhizosphere is shown in a graphic representation (Figure 1). Resource competition, chemical interference, and/or rhizospheric microorganism lead to negative interactions as well as positive interactions between plants is shown in grahic representation in Figure 2.

Table 1. Number of bacterial and archaeal taxa identified in the rhizosphere microbiome

Host	Approach*	Main findings related to rhizosphere microbiome composition	References
Erica andevalensis in a naturally metal-enriched and extremely acidic environment	*16S rRNA* gene clone library	Bacteria: of 101 sequenced clones, the majority was affiliated with the Actinobacteria (38 clones; 12 OTUs) followed by the Acidobacteria (21 clones; 10 OTUs), and Proteobacteria (18 clones; eight OTUs). Archaea: considering 27 clones, the community was composed by Crenarchaeota (21 clones; four OTUs) and Euryarchaeota (six clones; two OTUs)	Mirete et al. (2007)
Maize crop	*nifH* Cluster I clone library	*Azospirillum*, *Bradyrhizobium*, and *Ideonella* were the most abundant genera found in the rhizosphere, comprising c. 5%, 21% and 11% of the clones, respectively. The portion of unidentified bacteria was of 27%	Roesch et al. (2007)
Oat microcosms	*16S rRNA* gene microarray	A total of 1917 taxa were detected, and the community was dominated by Proteobateria and Firmicutes. Less expected rhizosphere-competent phyla were also detected, including Actinobacteria, Verrucomicrobia, and Nitrospira	DeAngelis et al. (2009)
Deschampsia Antarctica and *Colobanthus quitensis* in the Arctic	*16S rRNA* gene pyrosequencing	Firmicutes was the most abundant group found, and Acidobacteria was rarely detected. The predominant genera found were Bifidobacterium (phylum Actinobacteria), Arcobacter (phylum Proteobacteria), and Faecalibacterium (phylum Firmicutes)	Teixeira et al. (2010)

Oak in a forest soil	*16S rRNA* gene pyrosequencing	In one of the rhizosphere samples, 5619 OTUs were identified in the bacterial community. The predominant phyla were Proteobacteria (38%), Acidobacteria (24%), and Actinobacteria (11%). A high proportion of unclassified bacteria (20%) were observed	Uroz et al. (2010)
Sugar beet in agricultural soil	*16S rRNA* gene microarray	A total of 33 346 bacterial and archaeal OTUs were detected, and the community was dominated by Proteobacteria (39%), Firmicutes (20%), and Actinobacteria (9%). The Gamma- and Betaproteobacteria and Firmicutes were identified as the most dynamic taxa associated with disease suppression	Mendes et al. (2011)
Potato in field soil	*16S rRNA* gene microarray	A total of 2432 OTUs were detected in at least one of the samples. The highest number of OTUs belonged to the Proteobacteria (46%), followed by Firmicutes (18%), Actinobacteria (11%), Bacteroidetes (7%), and Acidobacteria (3%).The bacterial families Streptomycetaceae, Micromonosporaceae, and Pseudomonadaceae showed the strongest response at the potato cultivar level	Weinert et al. (2011)
Rhizophora mangle and *Laguncularia racemose* in mangroove	Archaeal *16S rRNA* gene pyrosequencing	About 300 archaeal OTUs were identified. Four classes were found: Halobacteria, Methanobacteria, Methanomicrobia, and Thermoprotei	Pires et al. (2012)
Potato in field soil	Pyrosequencing	A total of 55 121 OTUs were found. Actinobacteria and Alphaproteobacteria were the most abundant groups, followed by Gammaproteobacteria, Betaproteobacteria, Acidobacteria, Gemmatimonadetes, Firmicutes, Verrucomicrobia, Deltaproteobacteria, Cyanobacteria, Bacteriodetes, and the TM7 group	Inceoglu et al. (2011)
Rhizophora mangle in mangroove	*16S rRNA* gene pyrosequencing	Proteobacteria was the most abundant phylum in all samples covering 36-40% of the total sequencing reads	Gomes et al. (2010)
Mannillaria carnea (cactus) in semi-arid environment	*16S rRNA* gene pyrosequencing	Dominant bacterial groups were Acidobacteria, Actinobacteria, Proteobacteria, and Bacteroidetes	Torres-Cortes et al. (2012)
Arabidopsis thaliana in Cologne and Golm soils	*16S rRNA* gene pyrosequencing	About 1000 OTUs were estimated in the rhizosphere and 1000 OTUs in root compartments. The rhizosphere was dominated by Acidobacteria, Proteobacteria, Planctomycetes, and Actinobacteria. Proteobacteria, Actinobacteria, and Bacteroidetes	Bulgarelli et al. (2012)

Source: Mendes et al. 2013.

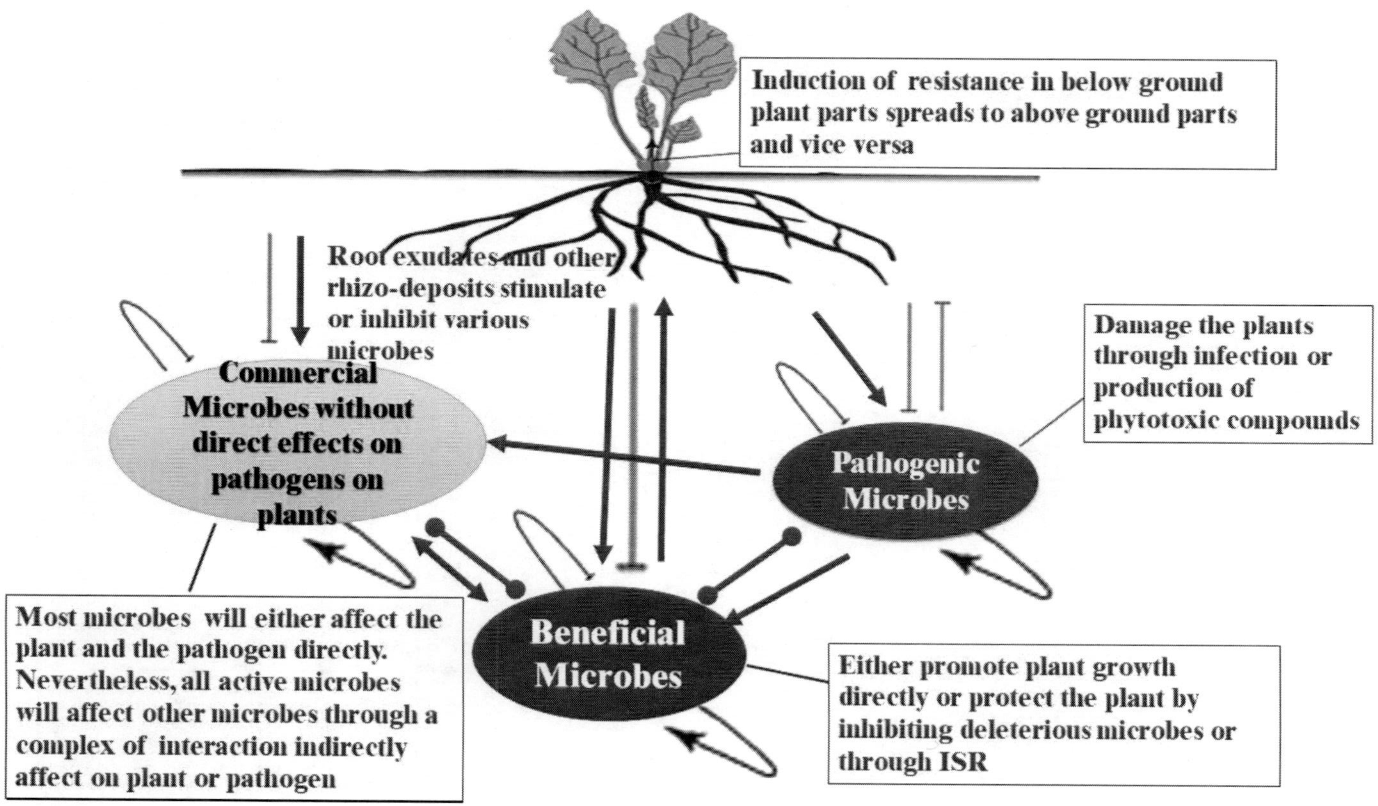

Figure 1. Interactions in the rhizosphere. Plants are able to influence the composition and activation of their rhizosphere microbiome through exudation of compounds that stimulate (green arrows) or inhibit (red blocked arrows). Vice versa, a wide range of soil-borne pathogens are able to affect plant health. Prior to infection, these deleterious microbes are in competition with many other microbes in the rhizosphere for nutrients and space. In this battle for resources, beneficial microbes will limit the success of the pathogen through production of biostatic compounds, consumption of (micro) nutrients or by stimulating the immune system of the plant. Most microbes might neither affect the plant nor the pathogen directly because they occupy different ecological niches (commensal microbes), but are likely to affect every other organism to a certain extent through a complex network of interactions. (Adapted from, Berendsen RL et al. 2013).

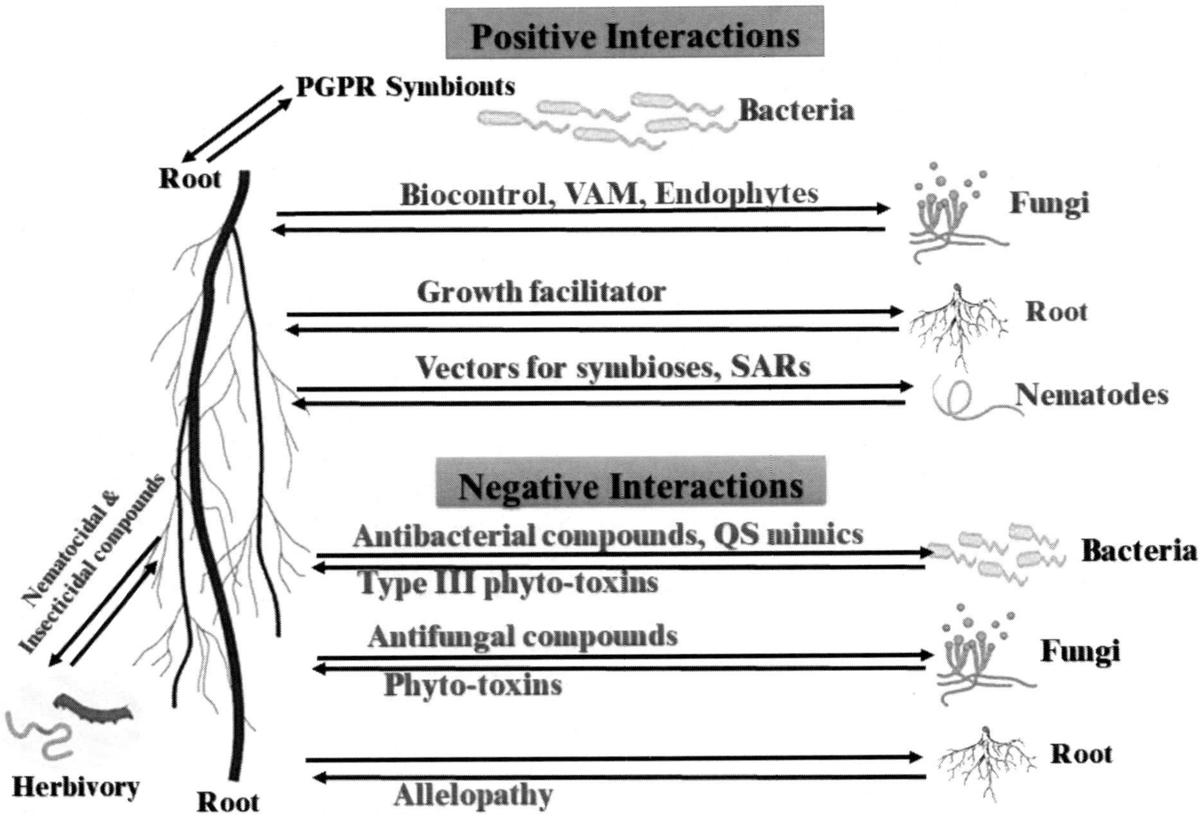

Figure 2. A schematic representation of interactions of chemicals released by plants and microbes in the rhizosphere. Positive and negative interactions are two categories root-mediated rhizospheric interactions. Positive interactions involve root exudate-mediated interactions with plant growth-promoting Rhizobacteria (PGPR). Roots produce chemical signals that attract bacteria and induce chemotaxis. Positive interactions mediated by root exudates also include growth facilitators or growth regulator mimics that support growth of other plants and also perform cross-species signaling with rhizospheric invertebrates. On the contrary, negative interactions mediated by root exudates involve secretion of antimicrobials, phytotoxins, nematicidal, and insecticidal compounds. The arrows in the panels indicate chemical exchange. VAM, vesicular arbuscular mycorrhizas; SARs, systemic acquired resistance. (Adapted from Bais et al. 2006).

PGPRS AND THEIR MECHANISM OF ACTION STIMULATING PLANT GROWTH

The search for efficient PGPRs and their mode of action are major areas of investigation these days and increasing at a rapid rate in order to use the best PGPR strain as commercial biofertilizer. Investigations into the mechanisms of plant growth promotion by PGPR strains indicated that the effective PGPRs increased plant growth basically by changing the whole microbial community structure in rhizosphere (Kloepper and Schroth 1981). It has been observed that mechanisms of plant growth promotion by PGPR includes associative nitrogen fixation, lowering of ethylene levels, production of siderophores and phytohormones, induction of pathogen resistance, solubilization of nutrients, promotion of mycorrhizal functioning, decreasing pollutant toxicity etc (Glick et al. 1999). According to Castro et al. (2009), PGPR strains can promote plant growth and development either directly and indirectly. Direct stimulation of growth includes- biological N_2-fixation, production of phytohormones (auxins, cytokinins and gibberellins), solubilization of minerals like phosphorus and iron, production of siderophores and enzymes and induction of systemic resistance. While indirect stimulation is basically related to biocontrol, including antibiotic production, chelation of available Fe in the rhizosphere, synthesis of extracellular enzymes to hydrolyze the fungal cell wall and competition for niches within the rhizosphere (Zahir et al. 2004; van Loon 2007). PGPR strains, especially, *Pseudomonas fluorescens* and *Bacillus subtilis* are best recorded as the most assuring agents of indirect growth stimulation (Damayanti et al. 2007). Besides, nitrogen transformation, increasing bioavailability of phosphate, iron acquisition, exhibition of specific enzymatic activity and plant protection from harmful pathogens with the production of antibiotics can also successfully improve the quality of crops in agriculture (Spaepen et al. 2007). Thus, based on their mechanism of action, PGPRs can be categorized into three general forms such as biofertilizer, phytostimulator and biopesticide (Table 2). The phenomenon of quorum regulation can affect the expression of each of these traits as PGPRs are reported for their regular interactions with the resident microbial community in rhizosphere (Lugtenberg and Kamilova 2009). Recent investigations on PGPR reveal that it can promote plant growth mainly by following means: (1) Plant growth production by growth hormone like indole acetic acid (IAA) (Mishra et al. 2010); gibberellic acid (Narula et al. 2006), cytokinins (Castro et al. 2008) and growth inhibitor ethylene (Saleem et al. 2007). (2) By producing ACC deaminase enzyme to reduce the level of ethylene in the roots of developing plants (Dey et al. 2004). (3) Asymbiotic N_2-fixation (Ardakani et al. 2010) (4) Exhibition of antagonistic activity against phytopathogenic microorganisms by producing siderophores, b-1,3-glucanase, chitinases, antibiotics, fluorescent pigment and cyanide (Pathma et al. 2011) and (5) solubilization of mineral phosphates and other nutrients (Hayat et al. 2010).

PGPR may use more than one of these mechanisms to enhance plant growth as experimental evidence suggests that the plant growth stimulation is the net result of multiple mechanisms that may be activated simultaneously (Martinez-Viveros et al. 2010). Recently, biochemical and molecular approaches are providing new insight into the genetic basis of these biosynthetic pathways, their regulation and importance in biological control (Joshi and Bhatt 2011). However, to be more effective in the rhizosphere, PGPR must maintain a critical population density for a longer period, although inoculation of plants with PGPR can temporarily enhance the population size.

Table 2. Forms of PGPRs and their mechanism of action stimulating plant growth

PGPR forms	Definition	Mechanism of action	References
Biofertilizer	A substance that contains live microorganisms which, when applied on the seed, plant surface or soil, colonizes the rhizosphere and promote plant growth through increased supply of primary nutrients for the host plant	Biological nitrogen fixation	Vessey (2003)
		Utilization of insoluble phosphorus	Somers et al. (2004)
Phytostimulator	Microorganism, with the ability to produce phytohormones such as indole acetic acid, gibberellic acid, cytokinins and ethylene	Production of phytohormones	Lugtenberg et al. (2002), Somers et al. (2004)
Biopesticide	Microorganisms that promote plant growth by controlling phytopathogenic agents	Production of antibiotics, siderophores, HCN	Vessey (2003)
		Production of hydrolytic enzymes	Somers et al. (2004)
		Acquired and Induced systemic resistance	Chandler et al. (2008)

Potential Role of Plant Beneficial Microorganisms in Agriculture

Plant beneficial microorganisms play important role in agriculture by producing various benefits to agricultural crops. They can be discussed by following subheadings:

Production of Plant Growth Regulators

PGPR can alter root architecture and promote plant development with the production of different phytohormones like IAA, gibberellic acid and cytokinins (Kloepper et al.

2007). Several PGPRs as well as some pathogenic, symbiotic and free living rhizobacterial species are reported to produce IAA and gibberllic acid in the rhizospheric soil and thereby plays a significant role in increasing the root surface area and number of root tips in many plants (Han et al. 2005). Recent investigations on auxin synthesizing rhizobacteria (Spaepen et al. 2007) as phytohormone producer demonstrated that the rhizobacteria could synthesize IAA from tryptophan by different pathways, although the general mechanism of auxin synthesis was basically concentrated on the tryptophan-independent pathways. The phytopathogenic bacteria rather use the indole acetamide pathway to synthesize IAA that has been implicated earlier in the tumor induction in plants. Swain et al. (2007) reported a positive effect of IAA producing strains of *Bacillus subtilis* on *Dioscorea rotundata* L. They applied a suspension of *B. subtilis* on the surface of the plant, which resulted in an increase in the root: stem ratio as well as number of sprouts as compared with the non-inoculated plants. Potentiality of *Azotobacter* spp., to produce high amount of IAA (7.3–32.8 mg/ml) in agriculture was reported by Ahmad et al. (2005). Similarly, significant shoot growths in maize and rice dwarf mutants were promoted by gibberellins-like substances excreted by *Azospirillum* spp. (Boiero et al. 2007). IAA-mediated ethylene production could increase root biomass, root hair number and consequently the root surface area of PGPR inoculated tomato plants (Ribaudo et al. 2006). Involvement of PGPR-formulated cytokinins were also observed in root initiation, cell division, cell enlargement and increase in root surface area of crop plants through enhanced formation of lateral and adventitious roots (Werner et al. 2003). Recently, it has been established that the working pathways of these phytostimulators leading to overall development in crop plants are differently regulated by catabolite repression (Zaied et al. 2009) as physiological regulator of biofilm formation.

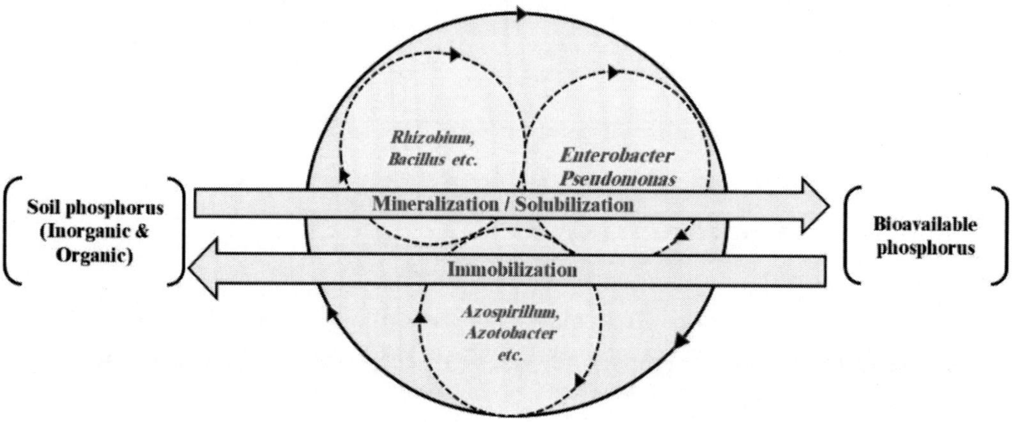

Figure 3. Schematic representation of solubilisation of soil phosphorus by rhizobacteria (Khan et al. 2009).

Phosphorous Solubilization for Plant Growth

Phosphorus is one of the most essential nutrient requirements in plants. Ironically, soils may have large reservoir of total phosphorous (P) but the amounts available to plants are usually a tiny proportion of this total. This low availability of phosphorous to plants is because of the vast majority of soil P is found in insoluble forms, while the plants can only absorb it in two soluble forms, the mono-basic ($H_2PO_4^-$) and the diabasic (HPO_4^{2-}) ions (Glass 1989). Several phosphate solubilizing microorganisms (PSMs) are now recorded to convert the insoluble form of phosphorus to soluble form through acidification, secretion of organic acids or protons (Richardson et al. 2009) and chelation and exchange reactions (Hameeda et al. 2008). Saprophytic bacteria and fungi are reported for the chelation-mediated mechanisms (Whitelaw 2000) to solubilise phosphate in soil. Release of plant root exudates such as organic ligands can also alter the concentration of P in soil solution (Hinsinger 2001). According to Nahas (1996), phosphate solubilization takes place through various microbial processes including organic acid production and proton extrusion. In certain cases, phosphate solubilization is induced by phosphate starvation (Gyaneshwar et al. 1999). A general sketch of phosphorous solubilization in soil is shown in Figure 3. Bacterial genera like *Azospirillum*, *Azotobacter*, *Bacillus*, *Beijerinckia*, *Burkholderia*, *Enterobacter*, *Erwinia*, *Flavobacterium*, *Microbacterium*, *Pseudomonas*, *Rhizobium* and *Serratia* are reported as the most significant phosphate solubilizing bacteria (Sturz and Nowak 2000; Sudhakar et al. 2000; Mehnaz and Lazarovits 2006). Rhizobacteria can solubilize inorganic P sources and enhance growth and yield of crop plants. Besides, examples of some widely reported P-solubilising microbial species intimately associated with a large number of agricultural crops like- potato, tomato, wheat, radish, pulses etc., are *Azotobacter chroococcum* (Kumar and Narula 1999), *Bacillus circulans* and *Cladosporium herbarum* (Singh and Kapoor 1999), *Bradyrhizobium japonicum* (Antoun et al. 1998), *Enterobacter agglomerans* (Kim et al. 1998), *Pseudomonas chlororaphis* and *P. putida* (Cattelan et al. 1999) and *Rhizobium leguminosarum* (Chabot et al. 1998). The ability of PGPRs to solubilize mineral phosphate, therefore, has been of immense interest to agricultural microbiologists since it can enhance the availability of phosphorus for effective plant growth. PGPRs have been recorded to solubilize precipitated phosphates to plants, representing a possible mechanism of plant growth promotion under field conditions (Verma et al. 2001). Synthesis of organic acids by rhizosphere microorganisms could be the possible reason for solubilization of inorganic P sources (Shrivastava 2015).

Production of ACC Deaminase and Regulation of Ethylene Level in Plants

Although ethylene is essential for normal growth and development in plants, at high concentration it can be harmful as it induces defoliation and other cellular processes that may lead to reduced crop performance. Using their 1-amino cyclopro-pane-1-carboxylic acid (ACC) deaminase activity, PGPR can divert ACC from the ethylene biosynthesis pathway in the root system of Arabidopsis thaliana plant (Desbrosses et al. 2009). Thus, rhizobacteria assist in diminishing the accumulation of ethylene levels and reestablish a healthy root system needed to cope with environmental stress. The primary mechanism includes the destruction of ethylene via enzyme ACC deaminase. There are number of publications (Ghosh et al. 2003; Govindasamy et al. 2008; Duan et al. 2009) mentioning rhizosphere bacteria such as *Achromobacter*, *Azospirillum*, *Bacillus*, *Enterobacter*, *Pseudomonas* and *Rhizobium* with ACC deaminase activity. Most of the studies have demonstrated the production of ACC deaminase gene in the plants treated with PGPR under environmental stress. Grichko and Glick (2001) inoculated tomato seeds with *Enterobacter cloacae* and *Pseudomonas putida* expressing ACC deaminase activity and registered an increase in plant resistance. Ghosh et al. (2003) recorded ACC deaminase activity in three *Bacillus* species namely, *Bacillus circulans* DUC1, *Bacillus firmus* DUC2 and *Bacillus globisporus* DUC3 that stimulated root elongation in *Brassica campestris*. Mayak et al. (2004) observed tomato plants inoculated with the bacterium *Achromobacter piechaudii* under water and saline stress conditions and reported a significant increase in fresh and dry weight of inoculated plants. Similar increase in root dry matter and aerial parts in canola (*Brassica napus*) seeds with the inoculation of ACC deaminase gene producing bacterium, *Pseudomonas asplenii* have been reported by Reed and Glick (2005). Shrivastava and Kumar (2013) reported *acdS* gene responsible for ACC deaminase activity in *Klebsiella* sp strain ECI-10A, *Microbacterium* sp strain ECI-12A, *Agrobacterium* sp strain AF-1D, *Pseudomonas* sp strain AF-4B, *Klebsiella* sp strain AF-4C, *Serratia* sp strain AF-5A, *Pseudomonas* sp strain PN-4D, *Agrobacterium* sp strain BN-2A, *Klebsiella* sp strain BN-4A. Genetic modification of PGPR strains expressing ACC deaminase gene are helpful in biological control of various plant diseases. PGPR containing ACC deaminase can boost the plant growth particularly under stressed environmental conditions like salinity, drought, water logging, temperature, pathogenicity and contaminants in response to a multitude of abiotic and biotic stresses (Saleem et al. 2007). In canola, ACC deaminase containing bacteria conferred salt tolerance and promoted plant growth by lowering the synthesis of salt-induced ethylene (Cheng et al. 2007). Recently, a bacterial strain, *Pseudomonas fluorescens* TDK1 containing ACC deaminase is reported to enhance the saline resistance and overall yield in groundnut as compared to those inoculated with *Pseudomonas* strains lacking ACC deaminase activity (Govindasamy et al. 2008). Although efforts have thus, been made to introduce ACC deaminase genes into plants for optimum growth, the genetic modifications for all the plant species are not yet possible due to many handicaps like proprietary rights and international trade agreements on genetically modified (GM) crops and limitations of DNA recombinant technology.

Nitrogen Fixation

Nitrogen (N) is one of the principal plant nutrients, becoming a limiting factor in agricultural ecosystems due to heavy losses by rainfall or mineral leaching. Number of PGPR strains such as *Azoarcus* sp., *Beijerinckia* sp., *Klebsiella pneumoniae*, *Pantoea agglomerans* and *Rhizobium* sp. are reported to fix atmospheric N_2 in soil (Antoun et al. 1998; Riggs et al. 2001) and make it available to plants. Parmar and Dadarwal (1999) recorded fluorescent pseudomonads to promote nodulation in chickpea and latter demonstrated the role of this group of rhizosphere microbiota in N_2 fixation. Recently, Minorsky (2008) reported a PGPR strain, *Pseudomonas fluorescens* B16, exhibiting vigorous colonization in the roots of tomatoes, causing enhancement in plant height, flower number and total fruit weight. A similar investigation on rhizobia to replace the use of nitrogen fertilizer was made by Vessey (2003) and thereby demonstrated a clear picture of improvement in crop yield after the inoculation of rhizobacteria in agricultural soil. PGPR can fix atmospheric N_2 either symbiotically or non-symbiotically. Symbiotic N_2-fixation to legume crops with the inoculation of effective PGPRs are well known (Dobereiner 1997; Barea et al. 2005; Esitken et al. 2006). Various rhizobacterial species like *Azotobacter* sp., *Bacillus* sp., *Beijerinckia* N_2-fixation is limited only to legume crops and various trees and shrubs that form actinorrhizal roots with *Frankia*. On the other hand, nonsymbiotic biological N_2-fixation is basically carried out by free living diazotrophics, belonging to the genera like *Azoarcus* (Reinhold-Hurek et al.1993), *Azospirillum* (Bashan and de-Bashan 2010), *Burkholderia* (Estrada de los Santos et al. 2001), *Gluconacetobacter* (Fuentes-Ramirez et al. 2001) and *Pseudomonas* (Mirza et al. 2006). Besides, combined inoculations of rhizobacterial species to improve the quality of soil are also seemed to be a potent area of research in present day agriculture. For instances, combined inoculations of *Bradyrhizobium* sp., with *Pseudomonas striata* have established enhanced nodule occupancy in soya bean resulting in more biological N_2-fixation (Dubey 1996).

Production of Volatile Organic Compounds

The discovery of rhizobacterial-produced volatile organic compounds (VOCs) constitutes an important mechanism for the elicitation of plant growth by rhizobacteria. Ryu et al. (2003) recorded some PGPR strains namely *Bacillus subtilis* GB03, *B.*

amyloliquefaciens IN937a and *Enterobacter cloacae* JM22 that released a blend of volatile components, particularly, 2, 3-butanediol and acetoin, which promoted growth of *Arabidopsis thaliana*, suggesting that synthesis of bioactive VOCs is a strain-specific phenomenon. Acetoin-forming enzymes have been identified earlier (Forlani et al. 1999) in certain crops like tobacco, carrot, maize and rice although their possible functions in plants were not properly established in that period. It has now been established that the VOCs produced by the rhizobacterial strains can act as signalling molecule to mediate plant–microbe interactions as volatiles produced by PGPR colonizing roots are generated at sufficient concentrations to trigger the plant responses (Ryu et al. 2003). Farmer (2001) identified low-molecular-weight plant volatiles such as terpenes, jasmonates and green leaf components as potent signal molecules for living organisms in different trophic levels. However, to acquire a clear appreciation on the mechanisms of VOCs in signalling plants to register plant defence more investigations into the volatile components in plant-rhizobacteria system should follow.

Rhizosphere Engineering

Rhizosphere microbial populations are enormously affected by the interactions between the plants and the soil environment. Rhizosphere engineering involves the selection of beneficial microbial populations by plant rhizosphere. For instances, some crop species or cultivars select populations of antibiotic-producing strains that play a major role in soils, naturally suppressive to soil-borne fungal pathogens (Ryan et al. 2009). Persistent organic pollutants such as polychlorinated biphenyls (PCBs) are a global problem. Using root-associated microbes in rhizospheric engineering approach, the levels of PCBs can be successfully depleted as these microbes can use plant secondary metabolites such as phenylpropanoids (Narasimhan et al. (2003). Similar technology has been developed by Lugtenberg et al. (2001) during their investigation on the growth of microbes with the ability to metabolize exotic nutrients exuded by plants. One of the earliest success of this technology was based on the favourable partitioning of the exotic nutrient opines, produced by the transgenic plants (Oger et al. 1997) that led to the improved and competitive growth of the metabolizing strains in comparison with the microbes unable to metabolize opines. Rhizosphere engineering ultimately reduces our reliance on agrochemicals by replacing their functions with beneficial microbes, biodegradable biostimulants or transgenic plants (Ryan et al. 2009). It is now possible to create a nutritional bias that may be especially successful in identifying microbial populations due to the general nutrient-limiting conditions in rhizosphere. Molecular microbiological advances are tremendously been exploiting in order to achieve a complete knowledge of the complex chemical and biological interactions that generally occurs in the rhizosphere, ensuring that the strategies to engineer the rhizosphere are safe and ecofriendly to agricultural systems. For example, plants are genetically engineered to modify the rhizosphere pH to release the compounds that could improve nutrient availability, protect plants against biotic and abiotic stresses or encourage the proliferation of beneficial microorganisms. Growth stimulation can be mediated directly through enhanced nutrient acquisition or modulation of phytohormone synthesis. While, indirect stimulation involves the induction of plant antagonism (Ryan et al. 2009). Sundheim et al. (1988) observed that an engineered strain of *Pseudomonas* expressing chitinase gene from *Serratia marcescens* more effectively controlled *Fusarium oxysporum* f. sp. redolens and *Gaeumannomyces graminis* var. *tritici* in vitro. Recently, experiments targeting on the DAPG-producing PGPR strain, *Pseudomonas fluorescens* (phlD$^+$) have demonstrated that plant species can differentially enrich and support different microbial populations (De La Fuente et al. 2006) and genotypes (Landa et al. 2006) in the rhizosphere. Notz et al. (2001) significantly correlated DAPG accumulation by *Pseudomonas fluorescens* CHA0 with the expression of DAPG biosynthesis gene phlA and observed that the expression was significantly greater in the rhizosphere of monocots than dicots. Although the exact mechanism is not totally understood, Di Gregorio et al. (2006) noticed a combined application of Triton X-100 and *Sinorhizobium* sp. Pb002 inoculums for the improvement of lead phytoextraction by *Brassica juncea* in EDTA amended soil.

Plant Beneficial Microorganisms as Biotic Elicitors

Elicitors are chemicals or biofactors of various sources that can trigger physiological and morphological responses and phytoalexin accumulation in plants. It may be abiotic elicitors such as metal ions or inorganic compounds and biotic elicitors, basically derived from fungi, bacteria, viruses, plant cell wall components and chemicals that are released due to antagonistic reaction of plants against phytopathogens or herbivore attack. It has now been observed that the treatment of plants with biotic elicitors can cause an array of defence reactions including the accumulation of a range of plant defensive bioactive molecules such as phytoalexins in the intact plants. Thus, elicitation is being used to induce the expression of genes responsible for the synthesis of antimicrobial metabolites. Rhizosphere microbes are best known to act as biotic elicitors, which can induce the synthesis of secondary products in plants (Sekar and Kandavel 2010). Signal perception is the first committed step towards the biotic elicitor signal transduction pathway in plants. Jasmonic acid and its methyl ester are the signal transducers in a wide range of plant cell cultures that could accumulate rapidly when the suspension cultures of *Rauvolfia canescens* L. and *Eschscholtzia californica* Cham. are treated with a yeast elicitor (Roberts and Shuler 1997). Ajmalicine, serpentine, picrocrocin, crocetin, hyoscyamine and scopolamine, safranal compounds and tanshinone are recorded as the important metabolites produced by PGPR species in eliciting the physiological and morphological responses in crop plants.

Induction of Systemic Disease Resistance

Application of mixtures of different PGPR strains to the seeds or seedlings of certain plants has resulted in increased efficiency of induced systemic resistance (ISR) against several pathogens (Ramamoorthy et al. 2001). Various non-pathogenic, PGPR strains have the ability to induce systemic disease resistance in plants against broad-spectrum phytopathogens (Kloepper et al. 2004; Elbadry et al. 2006). Induction of systemic disease resistance in faba bean (*Vicia faba* L.) against bean yellow mosaic potyvirus (BYMV) via seed bacterization with *Pseudomonas fluorescens* and *Rhizobium leguminosarum* have been investigated by Elbadry et al. (2006). They isolated PGPR strains from the roots of faba bean and examined singly or in combination for the induction of resistance in faba bean against BYMV. The results established a pronounced and significant reduction in percent disease incidence (PDI) as well as in virus concentration (ELISA) in plants treated with *Pseudomonas fluorescens* and *Rhizobium leguminosarum* as compared to the nonbacterized plants. Similarly, induction of systemic resistance by *Pseudomonas putida* strain 89B-27 and *Serratia marcescens* strain 90–166 against Fusarium wilt of cucumber incited by *Fusarium oxysporum* f.sp. *cucumerinum* has been investigated by Liu et al. (1995). Alstroem (1991) observed induced systemic protection of PGPR against the bacterial diseases. He reported that the bean seeds when treated with *Pseudomonas fluorescens* protected the plant against the halo blight disease caused by *Pseudomonas syringae* pv. *phaseolicola*. Kloepper et al. (1993) treated cucumber seeds with rhizobacterial strains like *Pseudomonas putida* 89B-27 and *Serratia marcescens* 90–166 and recorded a significant decrease in incidence of bacterial wilt. Similar investigations on the treatment of cucumber seeds against angular leaf spot disease caused by *Pseudomonas syringae* pv. *lachrymans*, with a large number of PGPR strains such as *Pseudomonas putida* 89B-27, *Flavomonas oryzihabitans* INR-5, *Serratia marcescens* 90–166 and *Bacillus pumilus* INR-7 has been made by Wei et al. (1996). They observed more systemic protection in the plants (indicated by the reduction of total lesion diameter) whose seeds are inoculated with the strains of PGPR as compared to the uninoculated plants. Pieterse et al. (2001) studied rhizobacterial strain, *Pseudomonas fluorescens* to enhance the defensive capacity in plants against broad-spectrum foliar pathogens. Based on their experiments they concluded that *Pseudomonas fluorescens* strain WCS417r could elicit systemic disease resistance in plants through a variety of signal translocation pathways like SA-independent JA-ethylene dependent signalling, ISR-related gene expression, NPR 1-dependent signalling etc. Recently, interactions between *Bacillus* sp. and plants with special reference to induced systemic disease resistance have been elicited by Choudhary and Johri (2009). Several strains of Bacillus like *B. amyloliquefaciens*, *B. subtilis*, *B. pasteurii*, *B. cereus*, *B. pumilus*, *B. mycoides* and *B. sphaericus* (Ryu et al. 2004) are presently recorded to elicit significant reduction in disease incidence on diversity of hosts. Elicitation of resistance by the strains has been demonstrated both in green house and field trials on tomato, bell pepper, muskmelon, watermelon, sugarbeet, tobacco and cucumber. Through the activation of various defence-related enzymes like chitinases, β-1, 3-glucanase, peroxidise (PO), phenylalanine ammonia-lyase (PAL) and polyphenol oxidase (PPO), PGPR strains can induce this type of systemic resistance in plants (Bharathi 2004).

Application of PGPR for Growth Enhancement

Application of PGPR strains in agriculture is a potential issue in increasing international demand for food and improving environmental quality. PGPRs have been continuously used to enhance the plant growth, seed emergence and overall yield of crops in different agro-ecosystems (Minorsky 2008). Inoculation of PGPR species could increase the growth attributes like leaf area, chlorophyll content and consequently, the total biomass of the musa plantlets under nitrogen-free hydroponics (Baset Mia et al. 2010) as compared to the uninoculated control. Dobbelaere et al. (2001) assessed the inoculation effect of *Azospirillum* sp., on the growth of some agriculturally important plants and observed a significant increase in the dry weight of both the root system and aerial parts of the PGPR inoculated plants, resulting in better development and flowering. Esitken et al. (2003) investigated the foliar applications of rhizobacterial microbes in mulberry and apricot and observed better development in total leaf area and chlorophyll production of the inoculated plants. Several PGPR strains such as *Achromobacter xylosoxidans*, *Bacillus subtilis*, *B. licheniformis*, *B. pumilus*, *Brevibacterium halotolerans* and *Pseudomonas putida* are identified as having crucial roles in cell elongation, increasing ACC deaminase activity and plant growth promotion (Sgroy et al. 2009). Total root length, surface area and volume in tomato and cucumber roots increased after inoculation with *Pseudomonas fluorescens* 92rk and P190r (Saravanakumar and Samiyappan 2007). PGPR induces changes in external layers of root cortex due to enhanced divisions of cells in root tips of maize and wheat seedlings (Baset Mia et al. 2010). Seeds of various crops and ornamental plants bacterized with a mixture of PGPR and rhizobia before planting resulted in enhanced growth and disease resistance (Zehnder et al. 2001). Khalid et al. (2004) observed the growth responses of wheat after the inoculation with rhizobacteria and suggested that the growth of wheat, basically depends on a number of factors like plant genotype, nature of PGPR inoculants as well as environmental conditions. There are also reports concerning the root inoculation of apple trees with Bacillus M3 and *Microbacterium* FS01, resulting in significant tree growth and yield (Karlidag et al. 2007). One of the possible mechanisms of enhancing apple trees growth in the study might be due to enhanced production of plant growth regulators and mobilization of available nutrients by PGPRs. Ahanthem and Jha (2007) observed the response of rice crops, inoculated with arbuscular mycorrhizal (AM) fungi and PGPR in soils differing in nitrogen concentrations. They recorded maximum shoot biomass, shoot phosphorus and nitrogen content in the rice plants inoculated with *Azotobacter chroococcum* in combination with *Glomus* spp., than when the plants were inoculated either of them above. Interactions between *Acaulospora* and *Azospirillum* and their synergistic effect on rice growth at different sources and regimes of soil phosphorus have also been made by Ahanthem and Jha (2008). The results thus, indicated the influence of microbial inoculants in reducing the inorganic fertilizer demand by 50%.

Plant Beneficial Microorganisms as Biofertilizer

Biofertilizers are the substances, prepared from living microorganisms which, when applied to the seeds or plant surfaces adjacent to soil can colonize rhizosphere or the interior parts of the plants and thereby promotes root growth. The term, biofertilizer should not be used inter-changeably with green manure, manure, intercrop or organic-supplemented chemical fertilizer. Interestingly some PGPR species have appeared to promote plant growth by acting both as biofertilizer and biopesticide. For instances, strains of *Burkholderia cepacia* have been observed with biocontrol characteristics to *Fusarium* spp., while, can also stimulate growth of maize under iron-poor conditions via siderophore production (Bevivino et al. 1998). *Allorhizobium*, *Azorhizobium*, *Bradyrhizobium*, *Mesorhizobium*, *Rhizobium* and *Sinorhizobium* are reported as the potent PGPR strains for their ability to act as bio-fertilizers (Vessey 2003). The relationship between the PGPR and their host can be categorized into two basic levels of complexity: (1) rhizospheric and (2) endophytic. In rhizospheric relationship, the PGPRs can colonize the rhizosphere, the surface of the root or even the superficial intercellular spaces of plant roots (McCully 2001). It is only due to the changes in different physico-chemical properties of rhizospheric soil such as soil pH, water potential and partial pressure of O_2 and plant exudation as compared to the bulk soil that in turn can affect the ability of PGPR strains to colonize the rhizosphere (Griffiths et al. 1999). In endophytic relationship, PGPR resides within the apoplastic spaces inside the host plants. There is a direct evidence of existence of endophytes in the apoplastic intercellular spaces of parenchyma tissue (Dong et al.1997) and xylem vessel (James et al. 2001). Best examples can be cited from legume-rhizobia symbioses in legumi-nous plants (Vessey 2003). Thus, the means by which the PGPRs enhance nutrient status of host plants and thereby act as biofertilizers can be categorized into five distinct areas such as biological N_2-fixation, increasing the availability of nutrients in rhizosphere, increase in root surface area, enhancing beneficial symbioses of the host and finally the combinations of all the above modes of action. However, the degree of intimacy between the PGPRs and host plant can vary depending on where and how the PGPR colonizes the plant.

Quorum Sensing Signal Interference and Inhibition of Biofilm Formation

Quorum sensing (QS) is a community genetic regulation mechanism that controls microbiological functions of medical, agricultural and industrial importance. Discovery of microbial QS signaling led to identification of numerous enzymatic and non-enzymatic signal interference mechanisms that could quench microbial QS signaling and inhibition of biofilm formation (Ren et al. 2001). QS activation is mediated by a small autoinducer (AI) molecule, responsible for cell–cell communication and the coordinated action in many bacteria, including PGPRs. Commonly reported autoinducer signals are N-acyl homo-serine lactones (AHLs) (von Bodman et al. 2003), although half a dozen of other molecules, including diketopiperazines in several Gram-negative bacteria (Holden et al. 1999), furanosyl borate diester in Vibrio harveyi (Chen et al. 2002) and c-butyrolactone in Streptomyces (Yamada and Nihira 1998) have also been implicated in density-dependent signaling. Investigations on QS signal interference mechanisms thus might significantly broaden the scope of research in modern biotechnology.

Rhizoremediation

The application of PGPRs in rhizoremediation technologies is now being considered as effective, since inoculation of PGPR strains could aid remarkable enhancement in plant growth and development on contaminated agroclimatic conditions. Rhizobacteria can directly assist rhizoremediation by producing IAA, biological nitrogen fixation, solubilizing P and secreting siderophores (Denton 2007). PGPR strains, pseudomonads and *Acinetobacter* enhance uptake of Fe, Zn, Mg, Ca, K and P by crop plants (Esitken et al. 2006). PGPR along with AM fungi are now being utilized in the nutrient poor agricultural soils to increase the solubility of heavy metals and thereby increasing the chances of success in rhizoremediation. Besides, investigations on the application of PGPR strains in decreasing the bioavailability of toxicity resulting in better growth and development in heavy metal contaminated soils through recycling of nutrients, maintaining soil structure, detoxifying chemicals and controlling pests are also well studied (Denton 2007). Studies on certain rhizobacteria in Ni uptake by Alyssum murale indicated that this group of bacteria could release the metal from its nonsoluble phase by decreasing the pH of the environment (Zhuang et al. 2007).

Effects of Plant Beneficial Microorganisms on Root Growth

The treatment of seeds or cuttings in some plants with nonpathogenic bacteria, such as *Alcaligenes*, *Agrobacterium*, *Bacillus*, *Pseudomonas*, *Streptomyces*, etc., induces root formation (Esitken et al. 2003). This phenomenon might be attributed to the production of auxin, inhibition of ethylene synthesis or mineralization of nutrients by efficient PGPRs (Steenhoudt and Vanderleyden 2000). More likely, PGPRs have been reported for their immense potentiality to alter several hormonal pathways that could account for different morphological changes in plants like an increased elongation rate of lateral roots, resulting in more architecture in branched root system of growing plants (Kapulnik et al. 1985). However, considering about interactions between different hormone signalling pathways in plants, it is difficult to determine the exact pathway of elicitation of primary plant rootings by PGPRs. Growth promoting effects of PGPRs on rootings and root growth of *Actinidia deliciosa* stem cuttings was examined by Erturk et al. (2010). *Bacillus* RC23, *Bacillus* RC03, *B. megaterium* RC01, *B. subtilis* OSU142, *B. simplex* RC19, *Comamonas acidovorans* RC41 and *Paenibacillus polymyxa* RC05 were recorded as the successful PGPRs in the experiment. All the bacteria were tested for their IAA activity. Among the rhizobacterial treatments, the highest rooting ratios were obtained at 47.50% for semi-hardwood stem cuttings from *Bacillus* RC03 and *B. simplex* RC19 treatments and 42.50% for hardwood stem cuttings from *Bacillus* RC03. The results suggested the potentiality of PGPR strains to replace the use of synthetic auxins in organic nursery material production. Similarly, Desbrosses et al. (2009) investigated the PGPR-*Arabidopsis* interaction to establish the signalling pathways involved in controlling plant development and observed an ethylene-independent and auxin-independent mechanism, regulating the elongation of root hair in *Arabiopsis*. This is something of great interest since genetic screens for abnormal root hair phenotypes in *Arabidopsis* repeatedly led to the isolation of mutants altered in ethylene or auxin response. They finally concludes that a genetic screen based on inoculation triggered root hair elongation could be a successful tool to unravel mechanisms involved in the control of root hair elongation.

Maintenance of Soil Fertility and Nutrient Uptake

PGPR can change the plant physiology and certain nutritional and physical properties of rhizospheric soil and indirectly influence on the colonization patterns of soil microorganisms in that particular region. Inoculation of rhizobacteria increased uptake of nutrient elements like Ca, K, Fe, Cu, Mn and Zn by plants through stimulation of proton pump ATPase (Mantelin and Touraine 2004). Reports are available on the combinations of *Bacillus* and *Microbacterium* inoculants to improve the uptake of the mineral elements by crop plants (Karlidag et al. 2007). This increase in nutrient uptake by plants might be explained through organic acid production by the plants and PGPRs, decreasing the soil pH in rhizosphere. Ample evidences (Phillips 1980; Forde 2000; Glass et al. 2002) are there on the maintenance of soil fertility by the rhizobacterial isolates to increase the availability of nutrients for plants. Solubilization of unavailable forms of nutrients is one of the essential criteria in facilitating the transport of most of these nutrients (Glick 1995).

Resistance to Water Stress

Drought stress causes limitation to the plant growth and productivity of agricultural crops particularly in arid and semi-arid areas. Inoculation of plants with PGPR can enhance the drought tolerance (Figueiredo et al. 2008) that might be due to the production of IAA, cytokinins, antioxidants and ACC deaminase. Inoculation of seeds of *Phragmites australis* with *Pseudomonas asplenii* improved germination and protect the plants from growth inhibition (Bashan et al. 2008). PGPR are also reported as beneficial to the plants like tomatoes and peppers growing on water deficit soils for conferring resistance to water stress conditions (Aroca and Ruiz-Lozano 2009). More investigations into the mechanisms by which PGPR elicit tolerance to specific stress factors would improve our knowledge on the use of these rhizobacteria in agriculture to provide induced systemic tolerance to water stress.

Antagonistic Activity of Plant Beneficial Microorganisms

Rhizobacteria can suppress the growth of various phytopathogens in variety of ways like competing for nutrients and space, limiting available Fe supply through producing siderophores, producing lytic enzymes and antibiosis (Jing et al. 2007). Among PGPRs, fluorescent pseudomonads are widely reported for their broad-spectrum antagonistic activity against number of phytopathogens. Han et al. (2005) have reported *Delftia tsuruhatensis* strain, HR4, which suppressed the growth of various plant pathogens like *Pyricularia oryzae*, *Rhizoctonia solani* and *Xanthomonas oryzae*. Deliveries of microbial antagonists with urban and agricultural wastes are believed to be the most effective means in suppressing root pathogens of avocado and citrus (Sultana et al. 2006). Recently, different PGPR strains of *Rhizobium meliloti* have been reported to produce siderophores (Arora et al. 2001) in iron stress conditions and thereby added an advantage to exclude the pathogen, *Macrophomina phaseolina*, causing charcoal rot of groundnut. Application of *Pseudomonas aeroginosa* in combination with common medicinal plant *Launaea nudicaulis* also holds good promises for effective control of root infecting fungi of mungbean (Mansoor et al. 2007).

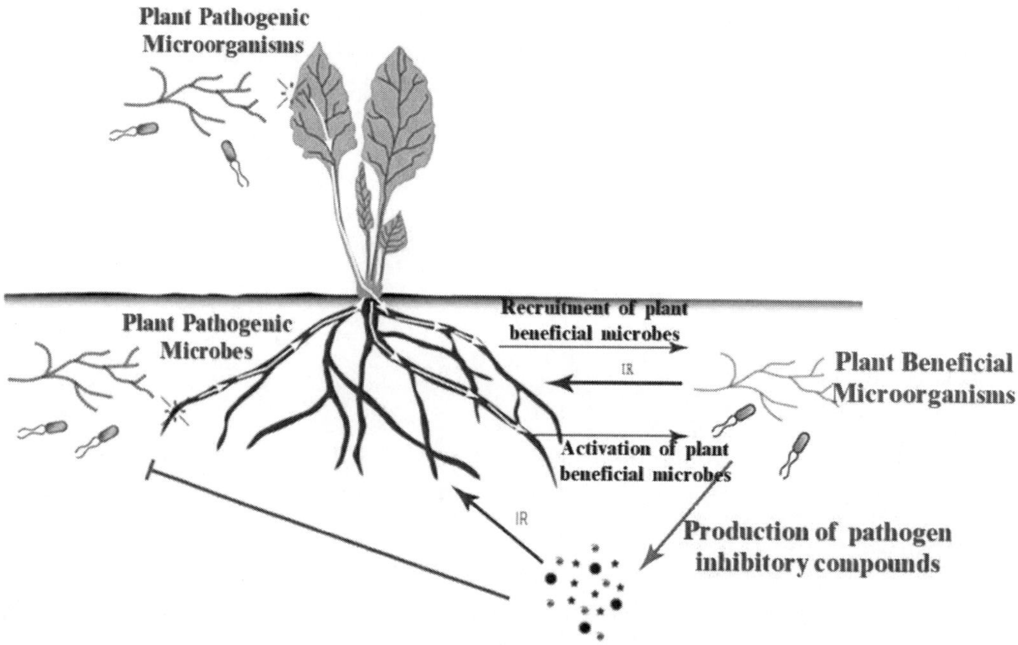

Figure 4. Model showing microbiome to the deliverance enlisting and activation of beneficial effects by the plant after microbial attack. Infected plants perceive pathogen invasion in roots or shoot and subsequently increase the secretion of microbe-stimulatory compounds in non-infected roots. These stimulants can recruit and activate plant-beneficial microorganisms. Beneficial microorganisms can induce resistance (IR) directly or produce pathogen-inhibitory compounds. Some pathogen-inhibitory compounds are known to induce resistance themselves (Adapted from Bais et al. 2006).

Plant Beneficial Microorganisms as Biocontrol Agent

Competition for nutrients, niche exclusion, induced systemic resistance and production of anti-fungal metabolites (AFMs) is the probable means responsible for biocontrol activity of PGPRs (Bloemberg and Lugtenberg 2001). Most of the PGPRs are recorded to produce AFMs, of which phenazines, pyrrolnitrin, 2, 4-diacetylphloroglucinol (DAPG), pyoluteorin, viscosinamide and tensin are the frequently detected classes. Among PGPRs, *Pseudomonas* is the best-characterized biocontrol agent at molecular level. *P. fluorescens* strain, WCS374 has been recorded to suppress Fusarium wilt in radish leading to an average increase of 40% in yield (Bakker et al. 2007). The individual genes such as phzO and phzH responsible for the presence of functional group on phenazine compound have been detected (Chin-A-Woeng et al. 2001). More recently, informations have been generated on the biosynthesis of pyoluteorin in *Pseudomonas fluorescens* Pf-5 and 2, 4-di-acetylphloroglucinol in *P. fluorescens* Q2-87 (Kidarsa et al. 2011). Biocontrol activity of *Streptomyces* spp. are reported by Kumar et al. (2009) indicating the tremendous potentiality of PGPRs as an alternative in controlling plant diseases in agriculture than that of conventional fungicides. *Azospirillum*, *Azotobacter*,

Bacillus, *Enterobacter*, *Paenibacillus*, *Pseudomonas* and *Streptomyces* are recorded as the potent genera of rhizobacteria acting against the pathogens like tomato mottle virus, tobacco necrosis virus, *Rhizoctonia bataticola*, *Myzus persicae*, *Acyrthosiphon kondoi*, *Fusarium avenaceum* etc. Besides, experiments on the dual effect of PGPR and AM fungi on *Fusarium oxysporum* f. sp., *melongenae* causing brinjal wilt has been made by Kalita et al. (2009). PGPR strains such as *Azotobacter* sp., *Azospirillum* sp., and *Pseudomonas fluorescens* and AM fungi like *Glomus fasciculatum*, *G. mossae* and *Gigaspora margarita* are recorded as the most promising microbes to suppress the wilt disease of brinjal, in vitro. The microbial inoculants when used as composite inoculum exhibited maximum efficiency in the suppression of diseases with the characteristic increase in chlorophyll content, total number of leaves, shoot height and thereby facilitating overall crop yield than when inoculated singly. However, application of these PGPR strains did not affect populations of beneficial indigenous rhizosphere bacteria including the fluorescent pseudomonads and the siderophore producing bacterial strains. Plants recruit plant-beneficial microbes to their roots in response to the attack is illustrated in Figure 4.

Selection and Characterization of Plant Beneficial Microorganisms

PGPR strains have diverse applications in agriculture, horticulture and forestry. The process of applying rhizobacteria in soil and plant parts to eradicate bacterial and fungal pathogens was pioneered in Soviet Union by 1958 (Suslow et al. 1979) even though the selection of efficacious PGPR strains at that period was highly complicated. Specific PGPR strains are initially selected from several hundreds of root-colonizing bacteria isolated from excised roots of field grown plants. Potential PGPRs are then selected for their ability to inhibit the growth of various phytopathogens or miscellaneous rhizosphere bacteria and fungi in vitro. Pure cultures of antagonistic rhizobacterial strains are screened in greenhouse trials. Seed or seed pieces of test plants are then treated with bacterial suspension (10^8 cfu/ml) and planted in replicated pot tests. During the experiment, those PGPRs that consistently caused statistically significant increases in root or shoot development or both are selected for further testing in agricultural field. Recently, selections of efficacious PGPR strains have been made by mass screening technique (Compant et al. 2005). Here, primary screenings of new isolates are done based on physiological, nutritional and biochemical characteristics as in Bergey's Manual of Determinative Bacteriology (Holt et al. 1994). While DNA and RNA homology tests are also considered as most reliable tools for the characterization of potent PGPR strains (Bashan et al. 1993). Now a days, application of protein profile analysis technique (Maiti et al. 2009) proved to be useful for patenting procedures. Restriction fragment length polymorphism (RFLP) analysis (Osborn et al. 2000) by using probe-target sequence and restriction endonuclease digestion pattern are useful for specific strain identification. However, gas chromatographic analysis of cellular fatty acid is very useful and proved to be accurate enough for identifying effective bacterial strains (Sasser 1990).

COMMERCIALIZATION OF PLANT BENEFICIAL MICROORGANISMS

The success and commercialization of PGPR strains depend on the linkages between the scientific organizations and industries. According to Nandakumar et al. (2001), different stages in the process of commercialization include isolation of antagonist strains, screening, pot tests and field efficacy, mass production and formulation development, fermentation methods, formulation viability, toxicology, industrial linkages and quality control. Thus, isolation of an effective strain is a prime criterion for better agricultural development, which is usually done from pathogen suppressive soils either by dilution plate technique or by baiting the soil with fungal structures like sclerotia (Nakkeeran et al. 2005). The selection of best antagonistic strain is carried out by screening for antimicrobial action against different soil borne pathogens apart from the target pathogen. The plant, pathogen and antagonists are then coexposed to controlled environmental conditions. Promising antagonists are further tested for their efficacy in field trials along with standard recommended fungicides (Pengnoo et al. 2000). Mass production is achieved through liquid (Manjula and Podile 2001), semisolid and solid fermentation techniques (Lewis 1991). Moreover, commercial suc-cess of PGPR strains requires economical and viable market demand, consistent and broad-spectrum action, safety and stability, longer shelf life, low capital costs and easy availability of career materials. Thus, the first requirement for entrepreneurship requires a patent application of the identified strain. Quality control in this step is crucial to retain the confidence of farmers on the efficacy of the antagonistic strain. Research inventions from China, Russia and several other western countries have now proved the potential use of PGPRs towards plant disease management. The first commercial product of *Bacillus subtilis* was developed during 1985 in US. 60-75% of cotton, peanut, soya bean, corn, vegetables and small grain crops raised in US are now treated with commercial product of *B. subtilis*, which become effective against soil borne pathogens such as *Fusarium* and *Rhizoctonia* (Nakkeeran et al. 2005). In China, PGPRs have been successfully applied over two decades about an area of 20 million hectares of different crop plants for commercial development. Owing to the potentiality of *Bacillus* spp., more than 20 different commercial products of *Bacillus* origin are sold in China to mitigate soil borne diseases (Backman et al. 1997). Besides, *Bacillus* spp., certain other PGPR strains belonging to the genera such as *Agrobacterium*, *Azospirillum*, *Bulkholderia*, *Pseudomonas* and

Streptomyces are also used for the production of several commercial products, which are generally being applied against several target pathogens like *Botrytis cinerea*, *Penicillium* spp., *Mucor pyroformis*, *Geotrichum candidum*, *Erwinia amylovora*, russet inducing bacteria, *Fusarium* spp., *Rhizoctonia* spp., *Pythium* spp., *Fusarium* sp., *Phytophthora* spp., and *P. tolassii* (Nakkeeran et al. 2005). Some of the important PGPR strains along with their commercial products as formulated by Chet and Chernin (2002) and Glick et al. (1999). Since PGPRs have its own potentiality in controlling plant diseases and pest management, these commercial products such as Diegall, Galltrol-A, ZeaNit, Epic, Quantum 4000, Victus, Mycostop etc., have, therefore, been registered for the practical use of farming community. Besides, the potentiality of PGPR inoculants in beneficial improvement of agricultural plants in developing countries can never be ignored. In India, more than 40 stakeholders from different provinces have registered themselves for the mass production of PGPRs with Central Insecticide Board (CSI), Faridabad, Haryana through collaboration with Tamil Nadu Agricultural University, Coimbatore, India.

FUTURE PROSPECTS AND CHALLENGES

PGPR inoculants are able to fulfil various beneficial interactions in plants leading to promising solutions for sustainable and environment-friendly agriculture. The applications of rhizosphere soil of agricultural crops with suitable bacterial populations have established appreciable hopes in both the laboratory and greenhouse experiment. Further, improved understanding on the way by which PGPRs promote plant growth can lead to expanded exploitation of these 'biofertilizers' to reduce the potential negative environmental effects associated with various agricultural productions. In recent decade, an effort of applying genetically engineered PGPRs to remediate complex contaminated soil (Denton 2007) and thereby increasing the productivity of crop plants in agriculture is another attractive idea of research. The rhizobacterial community can be specifically engineered to point various pollutants at co-contaminated sites to provide customized rhizoremediation system (Wu et al. 2006). Recent progress of molecular biology and biotechnology in the understanding of rhizobacterial interactions with the nodules of crop plants will encourage a suitable area of research in PGPR mechanisms relating to rhizosphere colonization. Reports are now available from genetically engineered *Arabidopsis thaliana* plants to remove lead and cadmium contaminates after inoculating with rhizobacterial population. However, combined applications of transgenic plants with PGPRs have proved another promising future (Ali and Hj 2010) in advancing rhizoremediation technologies. Efforts have, therefore, been concentrated on the production of transgenic plants, which can increase remediation efficiency by expressing a particular PGPR protein (Zhuang et al. 2007). Farwell et al. (2007) compared the growth of a transgenic canola, *Brassica napus* inoculated with PGPR strain,

Pseudomonas putida UW4, to normal canola and found a significant growth in the former. Besides, the symbiotic relationship of *Pseudomonas putida* and sunflower seeds with synthetic phytochelatins has been investigated by Wu et al. (2006). They observed that the engineered strain could protect the sunflower plants against the toxic effects of cadmium at 300 lM concentration. Thus, biotechnology can be applied to improve the efficacy of PGPR strains through transgenics for agricultural improvement. However, modern technology based on the transformations of 1-aminocyclopropane-1-carboxylic acid deaminase gene, which directly stimulates plant growth by cleaving the immediate precursor of plant ethylene into Pseudomonas fluorescens CHAO not only increased the plant growth but also accelerated biocontrol properties of PGPR species (Holguin and Glick 2001). Genomic tinkering of naturally occurring PGPR strains with effective genes (Nakkeeran et al. 2005) could lead to accentuated expression of genomic products and thereby alleviating the attack of both pests and diseases on field crops that would further facilitate for better introduction of a single bacterium with multiple modes of action to benefit the growers. Thus, future success of industries producing microbial inoculants, especially PGPRs, will depend on innovative business management, product marketing, extension education and extensive research. Further optimization is required for better fermentation and formulation processes of effective PGPR strains to introduce in agriculture.

CONCLUSION

It becomes clear that by development and formulations of PGPRs in biological promotion of different characteristics of plant growth. Most of the PGPR have significant potential to increase plant height, root length and dry matter production in various agricultural crops like rice, maize, potato, tomato, wheat, etc. The development of stable formulations of antagonistic PGPRs in sustainable agricultural systems thus, can establish as another promising approach replacing the use of chemical fertilizers. Besides, PGPRs are able to protect natural environments as well as biological resources by playing a significant role in integrated pest management system (IPM). In addition to this, certain PGPR strains have also the ability to activate octadecanoid, shikimate and terpenoid pathways (Gouws 2009) which in turn assists in the alterations of VOCs production in host plants. In accordance with their mode of action, PGPRs can be classified as phytostimulators, biofertilizers and biopesticides with certain bacteria having overlapping applications. However, screening systems for choosing the best rhizobacterial strain for studies on the biology of presented PGPR with the efficient PGPR and other microbial species in the plant rhizosphere will require more appropriate knowledge. The contribution of ACC deaminase quality, siderophore, phosphate, phytohormones like IAA, cytokinin, gibberellins and so forth., noduation, disease suppression and their coordinated expression appeared to be dependable in improving the plant development, yield and supplement take-up of different product plants in various agro ecosystems exist. Thus, it is becoming increasingly apparent that most of the PGPR strains can promote plant growth by several mechanisms, though most studies currently focus on individual mechanisms and have not yet been able to sort out the relative contributions of different processes that are also responsible for successful plant growth promotion. However, carefully controlled field trials of crop plants inoculated along with rhizobacteria are necessary for maximum commercial exploitation of PGPR strains.

REFERENCES

Ahanthem S, Jha DK (2007) Response of rice crop inoculated with arbuscular mycorrhizal fungi and plant growth promoting rhizobacteria to different soil nitrogen concentrations. *Mycorrhiza News* 18(4): 15-20.

Ahanthem S, Jha DK (2008) Interactions between *Acaulospora* and *Azospirillum* and their synergistic effect on rice growth at different sources and regimes of soil phosphorus. *Mycorrhiza News* 20(2):6-12.

Ahmad F, Ahmad I, Khan MS (2005) Indole acetic acid production by the indigenous isolates of *Azotobacter* and fluorescent *Pseudomonas* in the presence and absence of tryptophan. *Turk. J. Biol*. 29: 29-34.

Ali K, Hj SZ (2010) Phytoremediation of heavy metals with several efficiency enhancer methods. *Afr. J. Biotechnol*. 9(25): 3689-3698.

Alstroem S (1991) Induction of disease resistance in common bean susceptible to halo blight bacterial pathogen after seed bacterization with rhizosphere *Pseudomonads*. *J. Gen. Appl. Microbiol*. 37(6): 495-501.

Amarger N, Macheret V, Laguerre G (1997) *Rhizobium gallicum* sp. nov. and *Rhizobium giardinii* sp. nov., from *Phaseolus vulgaris* nodules. *Int. J. Syst. Bacteriol*. 47(4): 996-1006.

Andrews JH, Harris RF (2000) The ecology and biogeography of microorganisms on plant surfaces. *Annu. Rev. Phytopathol*. 38:145-180. doi:10.1146/annurev.phyto.38.1.145

Antoun H, Beauchamp CJ, Goussard N, Chabot R, Lalande R (1998) Potential of *Rhizobium* and *Bradyrhizobium* species as plant growth promoting rhizobacteria on non-legumes: Effect on radishes (*Raphanus sativus* L.). *Plant Soil* 204: 57-67. doi: 10.1023/A:1004326910584

Ardakani SS, Heydari A, Tayebi L, Mohammedi M (2010) Promotion of cotton seedlings growth characteristics by development and use of new bioformulations. *Int. J. Bot*. 6(2): 95-100.

Aroca R, Ruiz-Lozano JM (2009) Induction of plant tolerance to semi-arid environments by beneficial soil microorganisms-a review. In: Lichtouse E (ed) *Climate change, intercropping, pest control and beneficial microorganisms, sustainable agriculture reviews*. Springer, The Netherlands 2: 121-135.

Arora NK, Kang SC, Maheshwari DK (2001) Isolation of sidero-phore-producing strains of *Rhizobium meliloti* and their biocontrol potential against *Macrophomina phaseolina* that causes charcoal rot of groundnut. *Curr. Sci.* 81(6): 673-677.

Backman PA, Wilson M, Murphy JF (1997) Bacteria for biological control of plant diseases. In: Rechcigl NA, Rechcigl JE (eds) *Environmentally safe approaches to crop disease control*. Lewis Publishers, Boca Raton, FL, pp 95-109.

Bais HP, Weir TL, Perry LG, Gilroy S, Vivanco JM (2006) The role of root exudates in rhizosphere interactions with plants and other organisms. *Annu. Rev. Plant Biol.* 57: 233-266.

Bakker PAHM, Pieterse CMJ, van Loon LC (2007) Induced systemic resistance by fluorescent *Pseudomonas* spp. *Phytopathology* 97(2):239-243.

Barea JM, Pozo MJ, Azcon R, Aguilar CA (2005) Microbial cooperation in the rhizosphere. *J. Exp. Bot.* 56(417): 1761-1778.

Baset Mia MA, Shamsuddin ZH, Wahab Z, Marziah M (2010) Effect of plant growth promoting rhizobacterial (PGPR) inoculation on growth and nitrogen incorporation of tissue-cultured musa plantlets under nitrogen-free hydroponics condition. *Aust. J. Crop. Sci.* 4(2): 85-90.

Bashan Y, de-Bashan LE (2010) How the plant growth-promoting bacterium *Azospirillum* promotes plant growth-a critical assessment. *Adv. Agron.* 108: 77-136.

Bashan Y, Holguin G, Lifshitz R (1993) Isolation and characterization of plant growth-promoting rhizobacteria. In: Glick BR, Thomp-son JE (eds) *Methods in plant molecular biology and biotechnology*. CRC Press, Boca Raton, FL, pp 331-345.

Bashan Y, Puente ME, de-Bashan LE, Hernandez JP (2008) Environmental uses of plant growth-promoting bacteria. In: Barka EA, Clement C (eds) *Plant-microbe interactions*. Trivandrum, Kerala, India, pp 69-93.

Beattie GA (2006) Plant-associated bacteria: survey, molecular phylogeny, genomics and recent advances. In: Gnanamanickam SS (eds) *Plant-associated bacteria*. Springer, The Netherlands, pp 1-56. doi: 10.1007/978-1-4020-4538-7_1

Benizri E, Baudoin E, Guckert A (2001) Root colonization by inoculated plant growth-promoting rhizobacteria. *Biocontrol Sci Technol* 11(5):557-574.

Bevivino A, Sarrocco S, Dalmastri C, Tabacchioni S, Cantale C, Chiarini L (1998) Characterization of a free-living maize rhizosphere population of *Burkholderia cepacia*: effect of seed treatment on disease suppression and growth promotion of maize. *FEMS Microbiol. Ecol.* 27(3): 225-237. doi:10.1111/j.15746941. 1998.tb00539.x

Bharathi S (2004) *Development of botanical formulations for the management of major fungal diseases of tomato and onion*. PhD Thesis, Tamil Nadu Agricultural University, Coimbatore, India, pp. 152.

Bloemberg GV, Lugtenberg BJJ (2001) Molecular basis of plant growth promotion and biocontrol by rhizobacteria. *Curr. Opin. Plant Biol.* 4: 343-350.

Boiero L, Perrig D, Masciarelli O, Penna C, Cassan F, Luna V (2007) Phytohormone production by three strains of *Bradyrhizobium japonicum* and possible physiological and technological implications. *Appl. Microbiol. Biotechnol.* 74: 874-880. doi:10.1007/s00253-006-0731-9

Bringhurst RM, Cardon ZG, Gage DJ (2001) Galactosides in the rhizosphere: utilization by *Sinorhizobium meliloti* and development of a biosensor. *Proc. Natl. Acad. Sci. USA* 98: 4540-4545.

Bulgarelli D, Rott M, Schlaeppi K (2012) Revealing structure and assembly cues for *Arabidopsis* root-inhabiting bacterial microbiota. *Nature* 488: 91-95.

Cardoso EJBN, Freitas SS (1992) A rizosfera. In: Cardoso EJBN, Tsai SM, Neves PCP (eds) Microbiologia do solo. Sociedade Brasileira de Ciencia do Solo, [The rhizosphere. In: Cardoso EJBN, Tsai SM, Neves PCP (eds) *Soil microbiology*. Brazilian Society of Soil Science] Campinas, pp. 41-57.

Carson KC, Meyer JM, Dilworth MJ (2000) Hydroxamate siderophores of root nodule bacteria. *Soil Biol. Biochem.* 32: 11-21.

Castro RO, Cantero EV, Bucio JL (2008) Plant growth promotion by *Bacillus megaterium* involves cytokinin signalling. *Plant Signal Behav.* 3(4): 263-265.

Castro RO, Cornejo HAC, Rodriguez LM, Bucio JL (2009) The role of microbial signals in plant growth and development. *Plant Signal Behav.* 4(8): 701-712.

Cattelan AJ, Hartel PG, Fuhrmann JJ (1999) Screening for plant growth-promoting rhizobacteria to promote early soybean growth. *Soil Sci. Soc. Am. J.* 63: 1670-1680.

Chabot R, Beauchamp CJ, Kloepper JW, Antoun H (1998) Effect of phosphorus on root colonization and growth promotion of maize by bioluminescent mutants of phosphate-solubilizing *Rhizobium legu-minosarum* biovar *phaseoli*. *Soil Biol. Biochem.* 30: 1615-1618.

Chandler D, Davidson G, Grant WP, Greaves J, Tatchell GM (2008) Microbial biopesticides for integrated crop management: an assessment of environmental and regulatory sustainability. *Trends Food Sci. Tech.* 19: 275-283.

Chen WX, Yan GH, Li JL (1988) Numerical taxonomic study of fast growing soybean rhizobia and a proposal that *Rhizobium fredii* be assigned to *Sinorhizobium* gen. nov. *Int. J. Syst. Bacteriol.* 38(4): 392-397. doi: 10.1099/00207713-38-4-392

Chen X, Schauder S, Potier N, Van Dorsselaer A, Pelczer I, Bassler BL, Hughson FM (2002) Structural identification of a bacterial quorum-sensing signal containing boron. *Nature* 415: 545-549.

Cheng Z, Park E, Glick BR (2007) 1-Aminocyclopropane-1-carboxylate deaminase from *Pseudomonas putida* UW4 facilitates the growth of canola in the presence of salt. *Can. J. Microbiol.* 53(7): 912-918.

Chet I, Chernin L (2002) Biocontrol, microbial agents in soil. In: Bitton G (ed) *Encyclopedia of environmental microbiology*. Willey, New York, USA, pp 450-465.

Chew K (2002) *Georgics*. Hackett Publishing Company, Indianapolis, USA, pp. 152.

Chin-A-Woeng TF, Thomas-Oates JE, Lugtenberg BJ, Bloemberg GV (2001) Introduction of the phzH gene of *Pseudomonas chlororaphis* PCL1391 extends the range of biocontrol ability of phenazine-1-carboxylic acid producing *Pseudomonas* spp. strains. *Mol. Plant Microbe. Interact.* 14(8): 1006-1015.

Choudhary DK, Johri BN (2009) Interactions of *Bacillus* spp. and plants with special reference to induced systemic resistance (ISR). *Microbiol. Res.* 164: 493-513.

Compant S, Duffy B, Jerzy N, Clement C, Barka EA (2005) Use of plant growth-promoting bacteria for biocontrol of plant diseases: principles, mechanisms of Action, and future prospects. *Appl. Environ. Microbiol.* 71(9): 4951-4959.

Cook RJ (2002) Advances in plant health management in the twentieth century. *Annu. Rev. Phytopathol.* 38: 95-116.

Damayanti TA, Pardede H, Mubarik NR (2007) Utilization of root-colonizing bacteria to protect hot-pepper against Tobacco Mosaic Tobamovirus. *Hayati. J. Biosci.* 14(3): 105-109.

De La Fuente L, Landa BB, Weller DM (2006) Host crop affects rhizosphere colonization and competitiveness of 2, 4-diacetylph-loroglucinol-producing *Pseudomonas fluorescens*. *Phytopathology* 96: 751-762. doi:10.1094/PHYTO-96-0751

de Lajudie P, Laurent-Fulele E, Willems A, Torek U, Coopman R, Collins MD, Kersters K, Dreyfus B, Gillis M (1998a) Allorhizobium undicola gen. nov., sp. nov., nitrogen-fixing bacteria that efficiently nodulate *Neptunia natans* in Senegal. *Int. J. Syst. Bacteriol.* 48: 1277-1290.

de Lajudie P, Willems A, Nick G, Moreira F, Molouba F, Hoste B, Torck U, Neyra M, Collins MD, Lindström K, Dreyfus B (1998b) Characterization of tropical tree rhizobia and description of *Mesorhizobium plurifarium* sp. nov. *Int. J. Syst. Bacteriol.* 48: 369-382. doi:10.1099/00207713-48-2-369.

de Vasconcellos RLF, da Silva MCP, Ribeiro CM, Cardoso EJBN (2010) Isolation and screening for plant growth-promoting (PGP) actinobacteria from *Araucaria angustifolia* rhizosphere soil. *Sci. Agric.* 67: 743-746.

DeAngelis KM, Ji PS, Firestone MK, Lindow SE (2005) Two novel bacterial biosensors for detection of nitrate availability in the rhizosphere. *Appl. Environ. Microbiol.* 71: 8537-8547.

Denton B (2007) Advances in phytoremediation of heavy metals using plant growth promoting bacteria and fungi. MMG 445 *Basic Biotechnol.* 3: 1-5.

Desbrosses G, Contesto C, Varoquaux F, Galland M, Touraine B (2009) PGPR-*Arabidopsis* interactions is a useful system to study signalling pathways involved in plant developmental control. *Plant Signal Behav*. 4: 321-323.

Dey R, Pal KK, Bhatt DM, Chauhan SM (2004) Growth promotion and yield enhancement of peanut (*Arachis hypogaea* L.) by application of plant growth-promoting rhizobacteria. *Microbiol. Res*. 159: 371-394.

Di Gregorio S, Barbafieri M, Lampis S, Sanangelantoni AM, Tassi E, Vallini G (2006) Combined application of Triton X-100 and *Sinorhizobium* sp. Pb002 inoculum for the improvement of lead phytoextraction by *Brassica juncea* in EDTA amended soil. *Chemosphere* 63: 293-299.

Dobbelaere S, Croonenborghs A, Thys A, Ptacek D, Vanderleyden J, Dutto P, Labandera-Gonzalez C, Caballero-Mellado J, Aguirre JF, Kapulnik Y, Brener S (2001) Responses of agronomically important crops to inoculation with *Azospirillum*. *Aust. J. Plant Physiol*. 28: 871-879.

Dobereiner J (1997) Biological nitrogen fixation in the tropics: social and economic contributions. *Soil Biol. Biochem*. 29(5): 771-774.

Dong Z, McCully ME, Canny MJ (1997) Does Acetobacter diazo-trophicus live and move in the xylem of sugarcane stems? Anatomical and physiological data. *Ann. Bot*. 80: 147-158.

Dreyfus B, Garcia JL, Gillis M (1988) Characterization of *Azorhizobium caulinodans* gen. nov., sp. nov., a stem-nodulating nitrogen-fixing bacterium isolated from *Sesbania rostrata*. *Int. J. Syst. Bacteriol*. 38: 89-98.

Duan J, Muller KM, Charles TC, Vesely S, Glick BR (2009) 1-Aminocyclopropane-1-carboxylate (ACC) deaminase genes in rhizobia from Southern Saskatchewan. *Microb. Ecol*. 57: 423-436. doi:10.1007/s00248-008-9407-6.

Dubey SK (1996) Combined effect of *Bradyrhizobium japonicum* and phosphate-solubilizing *Pseudomonas striata* on nodulation, yield attributes and yield of rainfed soybean (*Glycine max*) under different sources of phosphorus in Vertisols. *Ind. J. Microbiol*. 33: 61-65.

Egamberdieva D, Kamilova F, Validov S, Gafurova L, Kucharova Z, Lugtenberg B (2008) High incidence of plant growth-stimulating bacteria associated with the rhizosphere of wheat grown on salinated soil in Uzbekistan. *Environ. Microbiol*. 10: 1-9.

Elbadry M, Taha RM, Eldougdoug KA, Gamal-Eldin H (2006) Induction of systemic resistance in faba bean (*Vicia faba* L.) to bean yellow mosaic potyvirus (BYMV) via seed bacterization with plant growth promoting rhizobacteria. *J. Plant Dis. Protect*. 113(6): 247-251.

Elliot LF, Lynch JM (1984) *Pseudomonads* as a factor in the growth of winter wheat (*Triticum aestivum* L.) *Soil Biol. Biochem*. 16: 69-71. doi:10.1016/0038-0717(84)90128-7.

Elliot LF, Lynch JM (1995) The international workshop on establishment of microbial inocula in soils: cooperative research project on biological resource management of the organization for economic cooperation and development (OECD). *Am. J. Alt. Agric*. 10: 50-73.

El-Tarabily KA, Sivasithamparam K (2006) Non-streptomycete actinomycetes as biocontrol agent of soil-borne fungal plant pathogens and as plant growth promoters. *Soil Biol. Biochem*. 38: 1505-1520.

Erturk, Y, Ercisli S, Haznedar A, Cakmakci R (2010) Effects of plant growth promoting rhizobacteria (PGPR) on rooting and root growth of kiwifruit (*Actinidia deliciosa*) stem cuttings. *Biol. Res*. 43: 91-98.

Esitken A, Karlidag H, Ercisli S, Turan M, Sahin F (2003) The effect of spraying a growth promoting bacterium on the yield, growth and nutrient element composition of leaves of apricot (*Prunus armeniaca* L. cv. Hacihaliloglu). Aust. *J. Agric. Res*. 54(4): 377-380.

Esitken A, Pirlak L, Turan M, Sahin F (2006) Effects of floral and foliar application of plant growth promoting rhizobacteria (PGPR) on yield, growth and nutrition of sweet cherry. *Sci. Hort*. 110: 324-327.

Estrada de los Santos P, Bustillos-Cristales MR, Caballero-Mellado J (2001) *Burkholderia*, a genus rich in plant-associated nitrogen fixers with wide environmental and geographic distribution. *Appl. Environ. Microb*. 67: 2790-2798.

Farmer EE (2001) Surface to air signals. *Nature* 411: 854-856.

Farwell AJ, Vesely S, Nero V, Rodriguez H, McCormack K, Shah S, Dixon DG, Glick BR (2007) Tolerance of transgenic canola plants (*Brassica napus*) amended with plant growth-promoting bacteria to flooding stress at a metal-contaminated field site. *Environ. Pollut*. 147: 540-545.

Figueiredo MVB, Burity HA, Martinez CR, Chanway CP (2008) Alleviation of drought stress in the common bean (*Phaseolus vulgaris* L.) by co-inoculation with *Paenibacillus polymyxa* and *Rhizobium tropici*. *Appl. Soil Ecol*. 40: 182-188.

Forde BG (2000) Nitrate transporters in plants: structure, function and regulation. *Biochem. Biophys. Acta* 1465: 219-235.

Franco-Correa M, Quintana A, Duque C, Suarez C, Rodríguez MX, Barea JM (2010) Evaluation of actinomycete strains for key traits related with plant growth promotion and mycorrhiza helping activities. *Appl. Soil Ecol*. 45: 209-217.

Fuentes-Ramirez LE, Bustillos-Cristales R, Tapia-Hernandez A, Jiménez-Salgado T, Wang ET, Martínez-Romero E, Caballero-Mellado J (2001) Novel nitrogen-fixing acetic acid bacteria, *Gluconacetobacter johannae* sp. nov. and *Gluconacetobacter azotocaptans* sp. nov., associated with coffee plants. *Int. J. Syst. Evol. Micribiol*. 51: 1305-1314.

Ghosh S, Penterman JN, Little RD, Chavez R, Glick BR (2003) Three newly isolated plant growth-promoting bacilli facilitate the seedling growth of canola, *Brassica campestris. Plant Physiol. Biochem.* 41: 277-281. doi:10.1016/S0981-9428(03)00019-6.

Glass ADM (1989) *Plant nutrition: an introduction to current concepts*. Jones and Bartlett Publishers, Boston, p 234.

Glass ADM, Britto DT, Kaiser BN, Kinghorn JR, Kronzucker HJ, Kumar A, Okamoto M, Rawat S, Siddiqi MY, Unkles SE, Vidmar JJ (2002) The regulation of nitrate and ammonium transport systems in plants. *J. Exp. Bot.* 53: 855-864.

Glick BR (1995) The enhancement of plant growth by free living bacteria. *Can. J. Microbiol.* 41: 109-117.

Glick BR, Patten CL, Holguin G, Penrose DM (1999) *Biochemical and genetic mechanisms used by plant growth-promoting bacteria.* Imperial College Press, London

Gomes NCM, Cleary DFR, Pinto FN, Egas C, Almeida A, Cunha A, Mendonça-Hagler LC, Smalla K (2010) Taking root: enduring effect of rhizosphere bacterial colonization in mangroves. *PLoS ONE* 5: e14065. doi: 10.1371/journal.pone.0014065

Gomes RC, Semedo LTAS, Soares RMA, Alviano CS, Linhares LF, Coelho RRR (2000) Chitinolytic activity of actinomycetes from a cerrado soil and their potential in biocontrol. *Lett. Appl. Microbiol.* 30: 146-150.

Gouws LM (2009) *The molecular analysis of the effects of lumichrome as a plant growth promoting substance.* Dissertation, University of Stellenbosch, South Africa.

Govindasamy V, Senthilkumar M, Gaikwad K, Annapurna K (2008) Isolation and characterization of ACC deaminase gene from two plant growth-promoting rhizobacteria. *Curr. Microbiol.* 57(4): 312–317.

Gray EJ, Smith DL (2005) Intracellular and extracellular PGPR: commonalities and distinctions in the plant-bacterium signaling processes. *Soil Biol. Biochem.* 37: 395-412.

Grichko VP, Glick BR (2001) Amelioration of flooding stress by ACC deaminase-containing plant growth-promoting bacteria. *Plant Physiol. Bioch.* 39: 11-17.

Griffiths BS, Ritz K, Ebblewhite N, Dobson G (1999) Soil microbial community structure: Effects of substrate loading rates. *Soil Biol. Biochem.* 31: 145-153. doi:10.1016/S0038-0717(98)00117-5.

Guerinot ML, Chelm BK (1984) Isolation and expression of the *Bradyrhizobium japonicum* adenylate cyclase gene (cya) in *Escherichia coli*. *J. Bacteriol.* 159: 1068-1071.

Gyaneshwar P, Parekh LJ, Archana G, Poole PS, Collins MD, Hutson RA, Kumar GN (1999) Involvement of a phosphate starvation inducible glucose dehydrogenase in soil phosphate solubilization by *Enterobacter asburiae*. *FEMS Microbiol. Lett.* 171: 223-229.

Haas D, Defago G (2005) Biological control of soil-borne pathogens by fluorescent *Pseudomonads*. *Nat. Rev. Microbiol.* 3: 307-319.

Halder AK, Mishra AK, Chakarbartty PK (1991) Solubilization of inorganic phosphates by *Bradyrhizobium*. *Ind. J. Exp. Biol.* 29: 28-31.

Hameeda B, Harini G, Rupela OP, Wani SP, Reddy G (2008) Growth promotion of maize by phosphate-solubilizing bacteria isolated from composts and macrofauna. *Microbiol. Res.* 163: 234-242.

Hayat R, Ali S, Amara U, Khalid R, Ahmed I (2010) Soil beneficial bacteria and their role in plant growth promotion: a review. *Ann. Microbiol.* 60(4): 579-598. doi:10.1007/s13213-010-0117-1.

Hellriegel H, Wilfarth H (1888) *Untersuchungen uber die Stickstoff-nahrung der Gramineen und Leguminosen.* Beilageheft zu der Zeitschrift des Vereins fur Rubenzucker-Industrie Deutschen Reichs, [Studies on the nitrogenous food of gramineae and legumes. Supplement to the journal of the association for Rubenzucker industry German Reichs], p 234.

Herschkovitz Y, Lerner A, Davidov Y, Rothballer M, Hartmann A, Okon Y, Jurkevitch E (2005) Inoculation with the plant-growth-promoting rhizobacterium *Azospirillum brasilense* causes little disturbance in the rhizosphere and rhizoplane of maize (*Zea mays*). *Microb. Ecol.* 50(2): 277-288.

Hiltner L (1904) Uber neuere Erfahrungen und Probleme auf dem Gebiet der Bodenbakteriologie und unter besonderer Ber-ucksichtigung der Grundungung und Brache. *Arbeiten der Deutschen Landwirtschaftlichen Gesellschaft* [On recent experiences and problems in the field of soil bacteriology and with particular reference to foundations and fallow land. *Works of the German Agricultural Society*] 98: 59-78.

Hinsinger P (2001) Bioavailability of soil inorganic P in the rhizosphere as affected by root induced chemical changes: a review. *Plant Soil* 237: 173-195.

Holden MTG, Ram Chhabra S, De Nys R, Stead P, Bainton NJ, Hill PJ, Manefield M, Kumar N, Labatte M, England D, Rice S (1999) Quorum-sensing cross-talk: isolation and chemical characterization of cyclic dipeptides from *Pseudomonas aeruginosa* and other gram-negative bacteria. *Mol. Microbiol.* 33: 1254-1266.

Holguin G, Glick BR (2001) Expression of the ACC Deaminase Gene from *Enterobacter cloacae* UW4 in *Azospirillum brasilense*. *Microb. Ecol.* 41(3): 281-288.

Holt JG, Krieg NR, Sneath PHA, Staley JT, Williams ST (1994) *Bergey's manual of determinative bacteriology*, 9th edn. Williams and Wikins co, Baltimore, USA, p 566.

Inceoglu O, Abu Al-Soud W, Salles JF, Semenov AV, van Elsas JD (2011) Comparative analysis of bacterial communities in a potato field as determined by pyrosequencing. *PLoS ONE* 6(8): e23321. doi:10.1371/journal.pone. 0023321.

Jing YD, He ZL, Yang XE (2007) Role of soil rhizobacteria in phytoremediation of heavy metal contaminated soils. *J. Zhejiang Univ. Sci.* 8(3): 192-207.

Joshi P, Bhatt AB (2011) Diversity and function of plant growth promoting rhizobacteria associated with wheat rhizosphere in North Himalayan region. *Int. J. Environ. Sci.* 1(6): 1135-1143.

Kalita RB, Bhattacharyya PN, Jha DK (2009) Effects of plant growth promoting rhizobacteria and arbuscular mycorrhizal fungi on *Fusarium oxysporum* causing brinjal wilt. *JAPS* 4: 29-35.

Kapulnik Y, Okon Y, Henis Y (1985) Changes in root morphology of wheat caused by *Azospirillum* inoculation. *Can. J. Microbiol.* 31: 881-887. doi:10.1139/m85-165.

Karlidag H, Esitken A, Turan M, Sahin F (2007) Effects of root inoculation of plant growth promoting rhizobacteria (PGPR) on yield, growth and nutrient element contents of leaves of apple. *Sci. Hort.* 114(1): 16-20.

Kaymak HC (2011) Potential of PGPR in Agricultural Innovations. In: Maheshwari DK (ed) *Plant growth and health promoting bacteria. Microbiol monographs*. Springer, Berlin 18:45-79.

Khalid A, Arshad M, Zahir ZA (2004) Screening plant growth-promoting rhizobacteria for improving growth and yield of wheat. *J. Appl. Microbiol.* 96(3): 473-480.

Kidarsa TA, Goebel NC, Zabriskie TM, Loper JE (2011) Phloroglucinol mediates cross-talk between the pyoluteorin and 2, 4 diacetylphloroglucinol biosynthetic pathways in *Pseudomonas fluorescens* Pf-5. *Mol. Microbiol.* 81(2): 395-414.

Kim KY, Jordan D, McDonald GA (1998) Effect of phosphate-solubilizing bacteria and vesicular-arbuscular mycorrhizae on tomato growth and soil microbial activity. *Biol. Fertil. Soils.* 26: 79-87. doi:10.1007/s003740050347.

Kloepper JW, Gutierrez-Estrada A, Mclnroy JA (2007) Photoperiod regulates elicitation of growth promotion but not induced resistance by plant growth-promoting rhizobacteria. *Can. J. Microbiol.* 53(2): 159-167.

Kloepper JW, Ryu CM, Zhang S (2004) Induced systemic resistance and promotion of plant growth by *Bacillus* spp. *Phytopathology* 94(11): 1259-1266.

Kloepper JW, Schroth MN (1978) Plant growth-promoting rhizobacteria on radishes. In: *Proceedings of the 4th international conference on plant pathogenic bacteria.* Gilbert-Clarey, Tours, pp 879-882.

Kloepper JW, Schroth MN (1981) Relationship of *in vitro* antibiosis of plant growth promoting rhizobacteria to plant growth and the displacement of root microflora. *Phytopathology* 71: 1020-1024.

Kloepper JW, Tuzun S, Liu L, Wei G (1993) Plant growth-promoting rhizobacteria as inducers of systemic disease resistance. In: Lumsden RD, Waughn JL (Eds) Pest management: biologically based technologies. *American Chemical Society Books,* Washington, DC, pp 156-165.

Kumar A, Salini P, Shrivastava JN (2009) Production of peptide antifungal antibiotic and biocontrol activity of *Bacillus subtilis*. *Ind. J. Exp. Biol.* 47: 57-62.

Kumar V, Narula N (1999) Solubilization of inorganic phosphates and growth emergence of wheat as affected by *Azotobacter chroococcum* mutants. *Biol. Fertil. Soils* 28: 301-305. doi: 10.1007/s003740050497.

Landa BB, Mavrodi OV, Schroeder KL, Allende-Molar R, Weller DM (2006) Enrichment and genotypic diversity of phlD-containing fluorescent *Pseudomonas* spp., in two soils after a century of wheat and flax monoculture. *FEMS Microbiol. Ecol.* 55: 351-368.

Lewis JA (1991) Formulation and delivery system of biocontrol agents with emphasis on fungi Beltsville symposia in agricultural research. In: Keister DL, Cregan PB (eds) *The rhizosphere and plant growth* 14: 279-287.

Lindstrom K (1989) *Rhizobium galegae*, a new species of legume root nodule bacteria. *Int. J. Syst. Bacteriol.* 39: 365-367.

Liu L, Kloepper JW, Tuzun S (1995) Induction of systemic resistance in cucumber against *Fusarium* wilt by plant growth promoting rhizobacteria. *Phytopathology* 85: 695-698.

Lugtenberg B, Kamilova F (2009) Plant growth-promoting rhizobacteria. *Annu Rev Microbiol* 63:541-556.

Lugtenberg BJ, Chin-A-Woeng TF, Bloemberg GV (2002) Microbe-plant interactions: principles and mechanisms. *Antonie Van Leeuwenhoek*. 81: 373-383.

Lugtenberg BJ, Dekkers L, Bloemberg GV (2001) Molecular determinants of rhizosphere colonization by *Pseudomonas*. *Annu. Rev. Phytopathol.* 39: 461-490.

Maiti B, Shekar M, Khusiramani R, Karunasagar I, Karunasagar I (2009) Evaluation of RAPD-PCR and protein profile analysis to differen-tiate *Vibrio harveyi* strains prevalent along the southwest coast of India. *J. Genet.* 88(3): 273-279.

Manjula K, Podile AR (2001) Chitin-supplemented formulations improve biocontrol and plant growth promoting efficiency of *Bacillus subtilis* AF1. *Can. J. Microbiol.* 47(7): 618-625.

Mansoor F, Sultana V, Haque SE (2007) Enhancement of biocontrol potential of *Pseudomonas aeroginosa* and *Paecilomyces lilacinus* against root rot of mungbean by a medicinal plant *Launaea nudicaulis* L. *Pak. J. Bot.* 39(6): 2113-2119.

Mantelin S, Touraine B (2004) Plant growth-promoting bacteria and nitrate availability: impacts on root development and nitrate uptake. *J. Exp. Bot* 55:27-34. doi:10.1093/jxb/erh010.

Martinez-Viveros O, Jorquera MA, Crowley DE, Gajardo G, Mora ML (2010) Mechanisms and practical considerations involved in plant growth promotion by rhizobacteria. *J. Soil Sci. Plant Nutr.* 10:293-319.

Mayak S, Tirosh T, Glick BR (2004) Plant growth-promoting bacteria confer resistance in tomato plants to salt stress. *Plant Physiol. Biochem.* 42(6):565-572. doi:10.1016/j.plaphy.2004.05.009.

McCully ME (2001) Niches for bacterial endophytes in crop plants: a plant biologist's view. *Aust. J. Plant Physiol.* 28: 983-990.

Mehnaz S, Lazarovits G (2006) Inoculation effects of *Pseudomonas putida*, *Gluconacetobacter azotocaptans*, and *Azospirillum lipoferum* on corn plant growth under greenhouse conditions. *Microb. Ecol.* 51(3): 326-335.

Mendes R, Kruijt M, de Bruijn I, Dekkers E, van der Voort M, Schneider JH, Piceno YM, DeSantis TZ, Andersen GL, Bakker PAHM, Raaijmakers JM (2011) Deciphering the rhizosphere microbiome for disease-suppressive bacteria. *Science* 332: 1097-1100.

Merzaeva OV, Shirokikh IG (2006) Colonization of plant rhizosphere by actinomycetes of different genera. *Microbiology* 75: 226-230. doi:10.1134/S0026261706020184.

Minorsky PV (2008) On the inside. *Plant Physiol.* 146: 323-324.

Mirete S, de Figueras CG, Gonzalez-Pastor JE (2007) Novel nickel resistance genes from the rhizosphere metagenome of plants adapted to acid mine drainage. *Appl. Environ. Microbiol.* 73: 6001-6011.

Mirza MS, Mehnaz S, Normand P, Prigent-Combaret C, Moënne-Loccoz Y, Bally R, Malik KA (2006) Molecular characterization and PCR detection of a nitrogen fixing *Pseudomonas* strain promoting rice growth. *Biol. Fertil. Soils.* 43: 163-170. doi: 10.1007/s00374-006-0074-9

Mishra M, Kumar U, Mishra PK, Prakash V (2010) Efficiency of plant growth promoting rhizobacteria for the enhancement of *Cicer arietinum* L. growth and germination under salinity. *Adv. Biol. Res.* 4(2): 92-96.

Montesinos E (2003) Plant associated microorganisms: a view from the scope of microbiology. *Int. Microbiol.* 6: 221-223. doi: 10.1007/s10123-003-0141-0.

Nahas E (1996) Factors determining rock phosphate solubilization by microorganisms isolated from soil. *World J. Microbiol. Biotechnol.* 12:567-572. doi:10.1007/BF00327716.

Nakkeeran S, Fernando WGD, Siddiqui ZA (2005) Plant growth promoting rhizobacteria formulations and its scope in commercialization for the management of pests and diseases. In: Siddiqui ZA (ed) *PGPR: biocontrol and biofertilization.* Springer, Dordrecht, pp 257-296.

Nandakumar R, Babu S, Viswanathan R, Sheela J, Raguchander T, Samiyappan R (2001) A new bio-formulation containing plant growth promoting rhizobacterial mixture for the management of sheath blight and enhanced grain yield in rice. *Biocontrol.* 46(4): 493-510. doi:10.1023/A:1014131131808.

Narasimhan K, Basheer C, Bajic VB, Swarup S (2003) Enhancement of plant-microbe interactions using a rhizosphere metabolomics-driven approach and its application in the removal of polychlorinated biphenyls. *Plant Physiol.* 132: 146-153.

Narula N, Deubel A, Gans W, Behl RK, Merbach W (2006) Paranodules and colonization of wheat roots by phytohormone producing bacteria in soil. *Plant soil Environ.* 52(3): 119-129.

Nick G, de Lajudie P, Eardly BD, Suomalainen S, Paulin L, Zhang X, Gillis M, Lindström K (1999) *Sinorhizobium arboris* sp. nov. and *Sinorhizobium kostiense* sp. nov., isolated from leguminous trees in Sudan and Kenya. *Int. J. Syst. Bacteriol.* 49: 1359-1368.

Notz R, Maurhofer M, Schnider-Keel U, Duffy B, Haas D, Defago G (2001) Biotic factors affecting expression of the 2, 4-diacetylphloroglucinol biosynthesis gene phlA in *Pseudomonas fluorescens* biocontrol strain CHA0 in the rhizosphere. *Phytopathology* 91: 873-881. doi:10.1094/PHYTO.2001.91.9.873.

Nour SM, Fernandez MP, Normand P, Cleyet-Marel JC (1994) *Rhizobium ciceri* sp. nov., consisting of strains that nodulate chickpeas (*Cicer arietinum* L.). *Int. J. Syst. Bacteriol.* 44: 511–522.

Oger P, Petit A, Dessaux Y (1997) Genetically engineered plants producing opines alter their biological environment. *Nat. Biotechnol.* 15: 369-372.

Osborn AM, Moore ER, Timmis KN (2000) An evaluation of terminal-restriction fragment length polymorphism (T-RFLP) analysis for the study of microbial community structure and dynamics. *Environ. Microbiol.* 2(1): 39-50.

Parmar N, Dadarwal KR (1999) Stimulation of nitrogen fixation and induction of flavonoid like compounds by rhizobacteria. *J. Appl. Microbiol.* 86: 36-44.

Pathma J, Kennedy RK, Sakthivel N (2011) Mechanisms of fluorescent *Pseudomonads* that mediate biological control of phytopathogens and plant growth promotion of crop plants. In: Maheshwari DK (ed) *Bacteria in Agrobiology: Plant Growth Responses.* Springer, Berlin, pp 77-105. doi: 10.1007/978-3-642-20332-9-4.

Pengnoo A, Kusongwiriyawong C, Nilratana L, Kanjanamaneesathian M (2000) Greenhouse and field trials of the bacterial antagonists in pellet formulations to suppress sheath blight of rice caused by *Rhizoctonia solani*. *Biocontrol* 45(2): 245-256. doi:10.1023/A:1009948404423.

Phillips DA (1980) Efficiency of symbiotic nitrogen fixation in legumes. *Annu. Rev. Plant Physiol.* 31: 29-49.

Pieterse CMJ, Johan A, Pelt V, Saskia CM, Van-Wees SC, Ton J, Léon-Kloosterziel KM, Keurentjes JJ, Verhagen BWM, Knoester M, Van der Sluis I, Bakker PAH, Van Loon LC (2001) Rhizobacteria-mediated induced systemic resistance: triggering, signaling and expression. *Eur. J. Plant Pathol.* 107(1): 51-61. doi: 10.1023/A:1008747926678.

Pires ACC, Cleary DFR, Almeida A, Cunha Â, Dealtry S, Mendonça-Hagler LCS, Smalla K, Gomes NC (2012) Denaturing gradient gel electrophoresis and barcoded pyrosequencing reveal unprecedented archaeal diversity in mangrove sediment and rhizosphere samples. *Appl. Environ. Microbiol.* 78: 5520-5528.

Ramamoorthy V, Viswanathan R, Raguchander T, Prakasam V, Samiyappan R (2001) Induction of systemic resistance by plant growth promoting rhizobacteria in crop plants against pests and diseases. *Crop Prot.* 20: 1-11.

Reed ML, Glick BR (2005) Growth of canola (*Brassica napus*) in the presence of plant growth-promoting bacteria and either copper or polycyclic aromatic hydrocarbons. *Can. J. Microbiol.* 51:1061-1069.

Reinhold-Hurek B, Hurek T, Gillis M, Hoste B, Vancanneyt M, Kersters K, De Ley J (1993) *Azoarcus* gen. nov., nitrogen-fixing Proteobacteria associated with roots of Kallar grass (*Leptochloa fusca* (L.) Kunth), and description of two species, *Azoarcus indigens* sp. nov. and *Azoarcus communis* sp. nov. *Int. J. Syst. Bacteriol.* 43: 574-584.

Ren D, Sims JJ, Wood TK (2001) Inhibition of biofilm formation and swarming of *Escherichia coli* by (5Z)-4-bromo-5-(bromomethylene)-3-butyl-2(5H)-furanone. *Environ. Microbiol.* 3(11): 731-736.

Ribaudo C, Krumpholz E, Cassan F, Bottini R, Cantore M, Cura A (2006) *Azospirillum* sp. promotes root hair development in tomato plants through a mechanism that involves ethylene. *J. Plant Growth Regul.* 24: 175-185. doi:10.1007/s00344-005-0128-5.

Richardson AE, Barea JM, McNeill AM, Prigent-Combaret C (2009) Acquisition of phosphorus and nitrogen in the rhizosphere and plant growth promotion by microorganisms. *Plant Soil* 321: 305-339. doi:10.1007/s11104-009-9895-2.

Riggs PJ, Chelius MK, Iniguez AL, Kaeppler SM, Triplett EW (2001) Enhanced maize productivity by inoculation with diazotrophic bacteria. *Aust. J. Plant Physiol.* 28: 829-836. doi:10.1071/PP01045.

Roberts SC, Shuler ML (1997) Large-scale plant cell culture. *Curr. Opin. Biotechnol.* 8: 154-159.

Roesch LFW, Fulthorpe RR, Riva A Casella G, Hadwin AK, Kent AD, Daroub SH, Camargo FAO, Farmerie WG, Triplett EW (2007) Pyrosequencing enumerates and contrasts soil microbial diversity. *ISME J.* 1: 283-290.

Rome S, Fernandez MP, Brunel B, Normand P, Cleyet-Marel JC (1996) *Sinorhizobium medicae* sp. nov., isolated from annual *Medicago* spp. *Int. J. Syst. Bacteriol.* 46: 972-980.

Ryan PR, Dessaux Y, Thomashow LS, Weller DM (2009) Rhizosphere engineering and management for sustainable agriculture. *Plant Soil* 321: 363-383. doi:10.1007/s11104-009-0001-6.

Ryu CM, Farag MA et al. (2003) Bacterial volatiles promote growth in *Arabidopsis*. *Proc. Natl. Acad. Sci. USA* 100: 4927-4932. doi:10.1073/pnas.0730845100.

Ryu CM, Farag MA, Hu CH, Reddy MS, Wei HX, Kloepper JW, Paré PW (2004) Bacterial volatiles induce systemic resistance in *Arabidopsis*. *Plant Physiol.* 134: 1017-1026.

Sabaratnam S, Traquair JA (2002) Formulation of a *Streptomyces* biocontrol agent for the suppression of *Rhizoctonia* damping-off in tomato transplants. *Biol. Control* 23: 245-253.

Saleem M, Arshad M, Hussain S, Bhatti AS (2007) Perspective of plant growth promoting rhizobacteria (PGPR) containing ACC deaminase in stress agriculture. *J. Ind. Microbiol. Biotechnol.* 34(10): 635-648.

Saravanakumar D, Samiyappan R (2007) ACC deaminase from *Pseudomonas fluorescens* mediated saline resistance in groundnut (*Arachis hypogea*) plants. *J. Appl. Microbiol.* 102(5): 1283-1292.

Sasser M (1990) Identification of bacteria through fatty acid analysis. In: Klement Z, Rudolph K, Sands D (eds) *Methods in phytobacteriology*. Akademiai Kiato, Budapest, pp 199-204.

Schippers B et al. (1988) Biological control of pathogens with rhizobacteria. *Philos. Trans. R. Soc. B-Biol Sci.* 318: 283-293. doi: 10.1098/rstb.1988.0010.

Scholla MH, Elkan GH (1984) *Rhizobium fredii* sp. nov., a fast-growing species that effectively nodulates soybeans. *Int J Syst Bacteriol* 34:484-486.

Segovia L, Young JPW, Matinez-Romero E (1993) Reclassification of American *Rhizobium leguminosarum* Biovar *Phaseoli* type I strains as *Rhizobium etli* sp. nov. *Int. J. Syst. Bacteriol.* 43: 374-377.

Sekar S, Kandavel D (2010) Interaction of plant growth promoting rhizobacteria (PGPR) and endophytes with medicinal plants-new avenues for phytochemicals. *J. Phytol.* 2: 91-100.

Sgroy V, Cassan F, Masciarelli O, Del Papa MF, Lagares A, Luna V (2009) Isolation and characterization of endophytic plant growth-promoting (PGPB) or stress homeostasis-regulating (PSHB) bacteria associated to the halophyte *Prosopis strombulifera*. *Appl. Microbiol. Biotechnol.* 85: 371-381. doi:10.1007/s00253-009-2116-3

Shrivastava UP (2013a) Characterization of diazotrophic rhizobacteria under various conditions. *Int. J. Appl. Sci. Biotechnol.* 1(3): 110-117. DOI: http://dx.doi.org/10.3126/ijasbt.v1i3.8607

Shrivastava UP (2013b) Isolation and characterization of plant growth promoting rhizobacteria (PGPR) from rice rhizosphere of Varanasi and east Champaran district of India. *Academic Voices* 3: 81-90 DOI: http://dx.doi.org/10.3126/av.v3i1.9542.

Shrivastava UP (2015) Isolation and characterization of phosphate solubilizing diazotrophic bacteria from rhizosphere of rice plant of Indo-Nepal border. *Academic Voices* 5: 79-87. DOI: http://dx.doi.org/10.3126/av.v5i0.15856.

Shrivastava UP, Kumar A (2011) Biochemical characterization of siderophore producing plant growth promoting rhizobacteria (PGPR) isolated from rice rhizosphere. *Nep. J. Integrated Sci.* 1: 31-37.

Shrivastava UP, Kumar A (2013) Characterization and optimization of 1-Aminocyclopropane-1-Carboxylate Deaminase (ACCD) activity in different rhizospheric PFPR along with *Microbacterium* sp. strain ECI-12A. *Int. J. Appl. Sci. Biotechnol.* 1(1): 11-15. DOI: http://dx.doi.org/10.3126/ijasbt.v1i1.7921.

Singh S, Kapoor KK (1999) Inoculation with phosphate-solubilizing microorganisms and a vesicular-arbuscular mycorrhizal fungus improves dry matter yield and nutrient uptake by wheat grown in a sandy soil. *Biol. Fertil. Soils.* 28: 139-144. doi:10.1007/s003740050475.

Somers E, Vanderleyden J, Srinivasan M. (2004) Rhizosphere bacterial signalling: a love parade beneath our feet. *Crit. Rev. Microbiol.* 30: 205-235.

Sousa CS, Soares ACF, Garrido MS (2008) Characterization of *Streptomycetes* with potential to promote plant growth and biocontrol. *Sci. Agric.* 65: 50-55. http://dx.doi.org/10.1590/S0103-9016 2008000100007.

Spaepen S, Vanderleyden J, Remans R (2007) Indole-3-acetic acid in microbial and microorganism-plant signaling. In: Unden F (ed) *FEMS microbiol rev.* Blackwell Publishing Ltd., New York pp 1-24. doi:10.1111/j.1574-6976.2007.00072.x.

Steenhoudt O, Vanderleyden J (2000) *Azospirillum*, a free-living nitrogen-fixing bacterium closely associated with grasses: genetic, biochemical and ecological aspects. *FEMS Microbiol. Rev.* 24: 487-506.

Sturz AV, Nowak J (2000) Endophytic communities of rhizobacteria and the strategies required to create yield enhancing associations with crops. *J. Appl. Soil Ecol.* 15: 183-190.

Stutz EW, Defago G, Kern H (1986) Naturally occurring fluorescent pseudomonads involved in suppression of black root rot of tobacco. *Phytopathology* 76: 181-185. doi:10.1094/Phyto-76-181.

Sudhakar P, Chattopadhyay GN, Gangwar SK, Ghosh JK (2000) Effect of foliar application of *Azotobacter*, *Azospirillum* and *Beijerinckia* on leaf yield and quality of mulberry (*Morus alba*). *J. Agric Sci.* 134: 227-234.

Sultana V, Ara J, Parveen G, Haque SE, Ahmad VU (2006) Role of Crustacean chitin, fungicides and fungal antagonists on the efficacy of *Pseudomonas aeruginosa* in protecting Chilli from root rot. *Pak. J. Bot.* 38(4): 1323-1331.

Sundheim L, Poplawsky AR, Ellingboe AH (1988) Molecular cloning of two chitinase genes from *Serratia marcescens* and their expression in *Pseudomonas* species. *Physiol. Mol. Plant Pathol.* 33: 483-491.

Suslow TV, Kloepper JW, Schroth MN, Burr TJ (1979) Beneficial bacteria enhance plant growth. *Calif. Agric*. 33: 15-17.

Swain MR, Naskar SK, Ray RC (2007) Indole-3-acetic acid production and effect on sprouting of yam (*Dioscorea rotundata* L.) minisetts by *Bacillus subtilis* isolated from culturable cowdung microflora. *Pol. J. Microbiol*. 56: 103-110.

Teixeira LCRS, Peixoto RS, Cury JC, Sul WJ, Pellizari VH, Tiedje J, Rosado AS (2010) Bacterial diversity in rhizosphere soil from Antarctic vascular plants of Admiralty Bay, maritime Antarctica. *ISME J*. 4: 989-1001.

Terkina IA, Parfenova VV, Ahn TS (2006) Antagonistic activity of actinomycetes of Lake Baikal. *Appl Biochem Microbiol* 42:173-176.

Tisdale SL, Nelson WL (1975) *Soil fertility and fertilizers*, 3rd edn. Macmillan Publishing, New York, p 694.

Torres-Cortes G, Millan V, Fernandez-Gonzalez AJ, Aguirre-Garrido JF, Ramírez-Saad HC, Fernández-López M, Toro N, Martínez-Abarca F (2012) Bacterial community in the rhizosphere of the cactus species *Mammillaria carnea* during dry and rainy seasons assessed by deep sequencing. *Plant Soil* 357: 275-288.

Trivedi P, Pandey A, Palni LMS (2005) Carrier-based preparations of plant growth-promoting bacterial inoculants suitable for use in cooler regions. *World J. Microbiol. Biotechnol*. 21: 941-945. doi: 10.1007/s11274-004-6820-y.

Uroz S, Buee M, Murat C, Frey-Klett P, Martin F (2010) Pyrosequencing reveals a contrasted bacterial diversity between oak rhizosphere and surrounding soil. *Environ. Microbiol. Rep*. 2: 281-288.

van Loon LC (2007) Plant responses to plant growth-promoting rhizobacteria. *Eur. J. Plant Pathol*. 119: 243-254. doi:10.1007/s10658-007-9165-1.

Velazquez E, Igual JM, Willems A, Fernández MP, Muñoz E, Mateos PF, Abril A, Toro N, Normand P, Cervantes E, Gillis M (2001) *Mesorhizobium chacoense* sp. nov., a novel species that nodulates *Prosopis alba* in the Chaco Arido region (Argentina). *Int. J. Syst. Evol. Microbiol*. 51: 1011-1021.

Verma JP, Yadav J, Tiwari KN, Lavakush SV (2010) Impact of plant growth promoting rhizobacteria on crop production. *Int. J. Agric. Res*. 5: 954-983.

Verma SC, Ladha JK, Tripathi AK (2001) Evaluation of plant growth promoting and colonization ability of endophytic diazotrophs from deep water rice. *J. Biotechnol*. 91: 127-141.

Vessey JK (2003) Plant growth promoting rhizobacteria as biofertil-izers. *Plant Soil* 255: 571–586.

von Bodman SB, Bauer WD, Coplin DL (2003) Quorum-sensing in plant-pathogenic bacteria. *Annu. Rev. Phytopathol*. 41: 455-482.

Wang ET, Martinez-Romero E (2000) *Sesbania herbacea-Rhizobium huautlense* nodulation in flooded soils and comparative characterization of *S. herbacea*-nodulating rhizobia in different environments. *Microb. Ecol*. 40: 25-32.

Wei L, Kloepper JW, Tuzun S (1996) Induced systemic resistance to cucumber diseases and increased plant growth by plant growth-promoting rhizobacteria under field conditions. *Phytopathology* 86: 221-224.

Weinert N, Piceno Y, Ding GC, Meincke R, Heuer H, Berg G, Schloter M, Andersen G, Smalla K (2011) PhyloChip hybridization uncovered an enormous bacterial diversity in the rhizosphere of different potato cultivars: many common and few cultivar-dependent taxa. *FEMS Microbiol Ecol.* 75: 497-506.

Weller DM (1988) Biological control of soil borne plant pathogens in the rhizosphere with bacteria. *Annu. Rev. Phytopathol.* 26: 379-407. doi:10.1146/annurev.py.26.090188.002115

Weller DM, Raaijmakers JM, Gardener BB, Thomashow LS (2002) Microbial populations responsible for specific soil suppressive-ness to plant pathogens. *Annu. Rev. Phytopathol.* 40: 309-348.

Wenbo M, Kelly Z, Glick BR (2001) Biological activity and colonization pattern of the bioluminescence-labeled plant growth-promoting bacterium Kluyvcra ascorbata SUD165/26. *FEMS Microbiol. Ecol.* 35(2): 137-44.

Werner T, Motyka V, Laucou V, Smets R, Onckelen HV, Schmulling T (2003) Cytokinin-deficient transgenic *Arabidopsis* plants show multiple developmental alterations indicating opposite functions of cytokinins in the regulation of shoot and root meristem activity. *Plant Cell* 15: 2532-2550.

Whipps JM (2001) Microbial interactions and biocontrol in the rhizosphere. *J. Exp. Bot.* 52: 487-511.

Whitelaw MA (2000) Growth promotion of plants inoculated with phosphate-solubilizing fungi. *Adv. Agron.* 69: 99-151. doi:10.1016/ S0065-2113(08)60948-7.

Wu CH, Wood TK, Mulchandani A, Chen W (2006) Engineering plant-microbe symbiosis for rhizoremediation of heavy metals. *Appl. Environ. Microbiol.* 72(2): 1129-1134. doi:10.1128/AEM. 72.2.1129-1134.

Yamada Y, Nihira T (1998) Microbial hormones and microbial chemical ecology. In: Barton DHR, Nakanishi K (eds) *Comprehensive natural products chemistry.* Elsevier Sciences, Amsterdam 8: 377-413.

Zahir ZA, Muhammad A, Frankenberger WT (2004) Plant growth promoting rhizobacteria: applications and perspectives in agriculture. *Adv. Agron.* 81: 97-168. doi:10.1016/S0065-2113(03)81003-9.

Zaied KA, El-Diasty ZM, El-Rhman MMA, El-Sanossy ASO (2009) Effect of horizontal DNA transfer between *Azotobacter* strains on protein patterns of *Azotobacter transconjugants* and bio-chemical traits in bioinoculated Okra (*Abelmoschus Esculentus*, L.). *Aust. J. Basic Appl. Sci.* 3(2): 748-760.

Zehnder GW, Murphy JF, Sikora EJ, Kloepper JW (2001) Application of rhizobacteria for induced resistance. *Eur. J. Plant Pathol*. 107: 39-50.

Zhuang X, Chen J, Shim H, Bai Z (2007) New advances in plant growth-promoting rhizobacteria for bioremediation. *Environ. Int*. 33(3): 406-413.

In: Trends in Life Science Research
Editors: R. P. Sinha and U. P. Shrivastava

ISBN: 978-1-53613-241-0

Chapter 3

HERBAL CURES FOR PSORIASIS, AN AUTOIMMUNE DISORDER

Shashi Pandey Rai[1,*], Deepika Tripathi[1], Neha Pandey[1], Anjana Kumari[1], Krishna Kumar Rai[1], Apoorva[1] and Sanjay Kumar Rai[2]

[1]Laboratory of Morphogenesis, Centre of Advanced Study in Botany, Banaras Hindu University, Varanasi, India
[2]Rajendra Agricultural University, Bihar, Pusa, Samastipur, India

ABSTRACT

Psoriasis is a persistent, long-lasting (chronic), immune-mediated inflammatory skin problem that changes the life cycle of skin cells There are several factors that contribute to the development and/or aggravation of the disease. The causes of psoriasis pathogenesis are generally a complex interaction among genetic, epigenetic, immunological, and environmental components that involve T cells (T lymphocytes) of the immune system which migrates to the dermis and trigger the release of several cytokines like interleukins (IL), tumor necrosis factor (TNF), and interferons (IF), thereby causing inflammation and rapid production of keratinocytes. Psoriasis also develops due to various stresses like oxidative stress, caused as a result of reactive oxygen species (ROS); psychological stress; and skin injuries. Lipoxygenases (LOXs), a class of enzymes producing inflammatory mediators, are also known to govern psoriasis pathogenesis. Excessive use of certain drugs such as lithium salt, beta blockers and the antimalarial drug chloroquininine can aggravate the disease as well. The bioactive compounds present in medicinal plants have remarkable antioxidant properties required for the protection of the cells against the damaging effects of ROS and LOXs. Thus, they prevent the body against the adverse effects caused by oxidative stress and/or

[*] Corresponding Author Email: shashi.bhubotany@gmail.com.

inflammatory responses. These plants are therefore capable of preventing and/or curing pathological conditions of skin and inflammatory diseases. In this chapter, the plant extracts/phytocompounds that have efficient antioxidant, anti-inflammatory, and, immuno-modulatory properties (the significant attributes essential for combating psoriasis), and can be further exploited as potent drug candidates for curing psoriasis were described to prospect them for drug discovery.

Keywords: psoriasis, lipoxygenases, reactive oxygen species, medicinal plants

INTRODUCTION

Psoriasis is an inflammatory, hyper-proliferative autoimmune human skin disorder that affects 1% to 3% of the total world's population, which accounts for approximately 80 to 135 million people worldwide. It is a chronic inflammatory skin condition, in which although the symptoms may come and go but the disease persists for years. Clinical manifestations of the disease involve periodic flare-ups of thick patches of red and inflamed skin covered by silvery and flaky scales on the surface of the skin. It often occurs in the individuals between the ages of 15 to 40 years and can develop with lifetime duration of more than 30 years. The symptoms of psoriasis can have severity range from barely noticeable to outbreaks of lesions covering almost whole of the body. There are several types of psoriasis, namely, plaque psoriasis, guttate psoriasis, erythrodermic psoriasis, inverse psoriasis, pustular psoriasis, nail psoriasis, scalp psoriasis, and psoriatic arthritis. Of all the types, plaque psoriasis is the most common type that accounts for nearly 80-90% of all the cases. Psoriasis is not contagious but can be inherited from parents.

There are several causes and factors that contribute to the development of the disease. Psoriasis is often initiated or aggravated by certain specific triggers. Production of free radicals or reactive oxygen species (ROS) like superoxide ion, hydrogen peroxide, and hydroxyl radical in the body as a result of the oxidation process is the major cause of psoriasis pathogenesis. Lipoxygenases (LOXs), the key enzymes responsible for the biosynthesis of inflammatory mediators like leukotrienes and lipoxins, also play an important role in the pathophysiology of the disease. Moreover, psoriasis is believed to result from certain genetic factors like interleukin (IL)-1β, IL-6, IL-17, IL-22, IL-23, tumor necrosis factor (TNF)-α, interferon (IFN)-γ, transforming growth factor (TGF)-β, toll-like receptor (TLR)-2, STAT-3, 15-LOX-2, CCHCR1, StAR, and VDR that influence the immune response in the skin. Other triggers of psoriasis are streptococcal infections of the upper respiratory tract, alcohol and tobacco consumption, environmental contaminants, radiation, and psychological stress that may contribute in the exacerbation of the disease. Consumption of various drugs including chloroquine, chlorpropamide, lithium, and practaol aggravate the ailment as well.

Numerous conventional treatments for psoriasis have been used since long. Some of them include topical treatment using cortisone creams, tar, calcipotriol, and anthralin; systemic treatment using drugs like methotrexate, acitretin, cyclosporin, psoralens; and UV treatment. Some recent advances in the medical sector for the disease have provided a platform for the design of new biological agents and immune-response modifiers that are devised to specifically target key mechanisms underlying psoriasis pathogenesis. Biologics are a heterogeneous group of monoclonal antibodies, fusion proteins, and recombinant cytokines that can modify and regulate the significant and specific mechanisms of the disease. Although, a large scale and successful use of all these treatments is prevalent for the prevention of psoriasis, but there are innumerable risks, side effects, and limitations accompanied with their usage. Side effects generally include irritation of the skin, temporary skin discoloration, sensitivity to sunlight, risk of liver-kidney damage. Furthermore, use of these treatments can also lead to increased risk of significant morbidity including skin cancers, lymphoma, and liver diseases.

Medicinal plants, since ages, have been used as antioxidant, anti-inflammatory, and immunomodulatory agents. Plant based drugs are highly beneficial in treating various human ailments as they possess the components of therapeutic value. Furthermore, due to their easy availability, cheaper cost, and non-toxic nature as compared to the modern medicine, there is an increasing demand of these plants in the pharmaceutical industries. Most of the medicinal plants are rich in antioxidant and anti-inflammatory compounds that protect the cells against the damaging effects of ROS and LOXs, thereby preventing oxidative stress and/or adverse inflammatory responses. Thus, they can be highly beneficial in combating inflammatory condition and other symptoms related to psoriasis.

Psoriasis: An Autoimmune Mediated Skin Disorder

Psoriasis is an autoimmune skin disorder characterized by periodic flare-ups of patches of thick, red, inflamed skin covered by silvery and flaky scales on the extensor surfaces that involve excessive itching and pain (Nestle et al. 2009). Histologically, it is characterized by cutaneous inflammation, increased epidermal proliferation, hyperkeratosis, angiogenesis, abnormal keratinization, shortened maturation time, and parakeratosis (Lee et al. 2000). The cardinal features of psoriatic skin are erythrema, indurations and scaling. Proriasis is a chronic skin condition, thereby meaning that the disease remains even though symptoms may come and go. It is typically a lifelong condition that affects 1% to 3% of the world's population (Griffiths and Barker 2007). The people suffering from this disease are affected both physically and psychologically, which in turn reduce their quality of life. Psoriasis can get worse over time, but it is not possible to predict who will go on to develop extensive psoriasis or those in whom the disease may appear to vanish (Nestle et al. 2009). It is seen equally in both genders and in all ages of people. The disease is not contagious but it may be genetic. Various organizations such as 'The International Federation of Psoriasis Associations' (IFPA), 'The Psoriasis International Network', 'The Psoriasis Association', and 'Group for Research and Assessment of Psoriasis and Psoriatic Arthritis (GRAPPA)' are involved in global campaigning for improved medical care, greater public awareness, and increased research for the improvement of the lives of people suffering from psoriasis all over the world.

History: Incidence and Prevalence of the Disease

The term 'Psoriasis' was coined by Ferdinand von Hebra, an Austrian physician and dermatologist, in the year 1841, from the Greek word 'psora' meaning 'to itch'. Psoriasis affects the population worldwide. Lomholt (1964) conducted a comprehensive field study in the Faroe Islands, Denmark, and found 2.8%of the inhabitants to be affected. This rate of prevalence was higher than that in central Europe, where approximately 1.5% population was found affected (Farber and Nall 1998). In the United States, approximately 7.5 million people have psoriasis. Researches indicate that Caucasians develop psoriasis more frequently than other races. Ethnic factors have also been found to influence the prevalence of the disease particularly within the United States (Christophers 2001). The prevalence among blacks (0.45%-0.7%) was recorded to be far lower than that in the remainder of the US population (1.4%-4.6%) (Gelfand et al.2005). Evidences of a genetic predisposition to psoriasis are present, although the inheritance pattern is still unclear (Henseler and Christophers 1985). The illness develops in about half of the siblings of persons with psoriasis when both parents are affected, but prevalence falls to 16% when only one parent is affected and to 8% when neither of the parent is affected (Farber et al. 1974).Around 71% of patients with childhood psoriasis have a positive family history (Altmuller et al. 2001).Psoriasis can begin at any age but generally it is observed to start either around the age of 20 or between 50 and 60.According to some reports, the severity of psoriasis is strongly linked to an increased incidence of some other diseases mainly affecting the lungs, heart, kidneys, liver, and pancreas (Gelfand et al. 2009). Furthermore, psoriasis has been found to be significantly associated with a wide range of disorders, including chronic pulmonary disease (COPD), diabetes, myocardial infarction, peripheral vascular disease, peptic ulcer disease, renal disease, and other rheumatologic diseases (Al-Mutairiet al. 2010).

Different Types of Psoriasis and Its Clinical Appearance

Psoriasis is characterized by fast multiplying skin cells that divide up to 10 times faster than normal skin cells. When these skin cells reach to the surface raised, red plaques of dead cells covered with white scales form. Psoriasis begins as a small scaling papule. When multiple papules coalesce, they form scaling plaques. These plaques tend to occur in the scalp, elbows, and knees. Generally, there are nine types of psoriasis, as mentioned in Figure 1. The main type of psoriasis is plaque psoriasis which accounts for approximately 80-90% of all the cases (Henseler and Christophers 1985). Other forms are guttate psoriasis, erythrodermic psoriasis, inverse psoriasis, pustular psoriasis, nail psoriasis, scalp psoriasis, and psoriatic arthritis (Gladman 2009). Psoriasis is most likely to affect the scalp, elbows, knees, groin, and lower back (Griffiths and Barker 2007).

Other frequently affected areas include palms of the hands, soles of the feet, legs, back, face, fingernails, and toenails. However, the involvement of different parts of the body could be according to the different types of psoriasis, as shown in Figure 2.

All types of psoriasis usually share certain features like the malfunctioning of the immune system and the tendency to run in families (Liu et al. 2007). An immune suppressed patient can suffer from more than one type of psoriasis at the same time. However, each type of psoriasis has its unique characteristics and psoriatic individual may experience physical discomfort and disability. Itching and inflammation can interfere with basic functions, such as self-care, walking, and sleep. Sometimes psychological distress can lead to depression and social isolation. The symptoms, clinical appearances, and manifestations of the different types of psoriasis are described in Table 1.

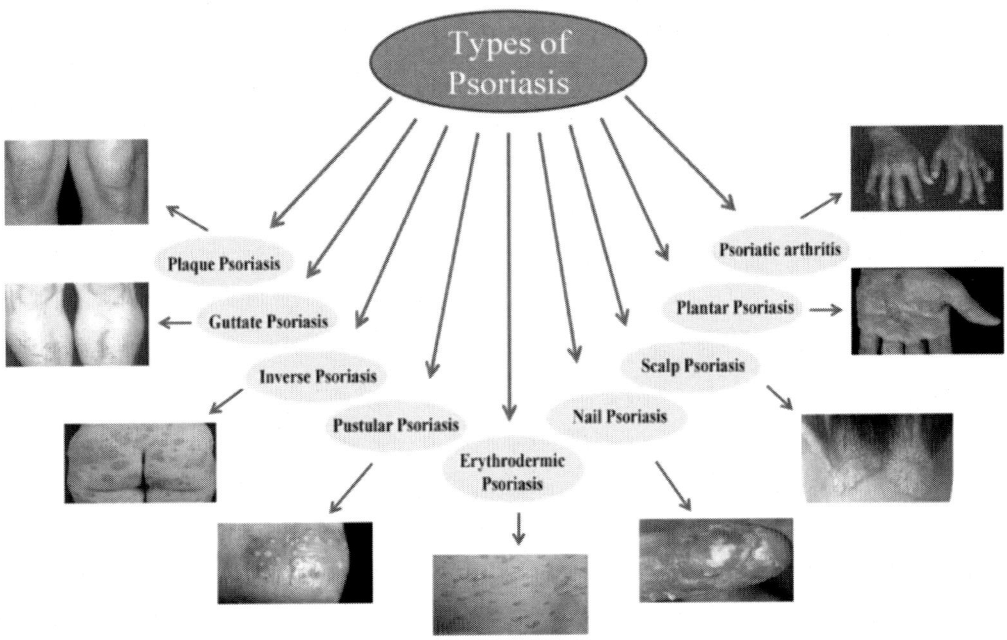

Figure 1. Different types of psoriatic lesions on the skin.

Table 1. Symptoms, clinical appearances, areas affected, and manifestations of the different types of psoriasis

Types of Psoriasis	ICD-10 Code	Symptoms	Clinical appearance	Areas affected	Severity/Manifestations
Erythrodermic psoriasis	L40.85	Inflammated skin accompanied by severe itching, swelling, and pain	Widespread inflammation and exfoliation of skin	Whole of the body	Can be fatal as it disrupts body's ability to regulate temp and perform barrier functions, may be precipitated by severe sunburn or use of oral steroids
Guttate psoriasis	L40.4	Small drop-like red lesions	Numerous small teardrop-shaped spots	Trunk, limbs, scalp	Associated with streptococcal throat infection, generally occurs below the age of 30
Inverse psoriasis (Flexural psoriasis)	L40.83-4	Large, dry, smooth and vividly red plaques occur in the folds of the skin	Smooth inflamed patches of skin	Skin folds, around genitals, between thigh and groin, armpits, under breasts	Aggravated by friction and sweat, vulnerable to fungal infections
Nail psoriasis	L40.86	Produces variety of changes in the appearance of fingernails and toenails	Discoloring under the nail, pitting of the nails, thickening of the skin under the nail, loosening (onycholysis) and crumbling (onychodystrophy) of the nail plate	Finger and toe nails, nail plate, nail matrix (tissue from which the nail grows), nail bed (tissue under the nail), and skin at the base of the nail	Damage to the nail bed, loss of the nail
Palmo-Plantar Psoriasis	L40.3	Well-circumscribed, red, scaly, plaques	Skin red, dry and thickened, often with painful cracks (fissures).	Palms and soles	Palmo-plantar psoriasis affected almost 24% 0f peoples.
Plaque psoriasis (Psoriasis vulgaris)	L40.0	Inflamed red skin covered with slivery white scaly skin that sometimes itch or burn	Circular to oval shaped red plaques	Mostly on elbows, knees, trunk, scalp but may be found on any part of the skin	Affects 80-90% of people, most common form, chronic stationary psoriasis, persistent for years
Psoriatic arthritis	L40.5	Deterioration and pain in joints of the body, inflammation of the fingers and toes known as dactylitis	Sausage-shaped swelling of fingers and toes	Joints and connective tissue of hands and feet, legs and spine, fingers and toes, hips, knees and spine (spondylitis), tendons	10-15% people who have psoriasis could also be affected by this arthritis, can cause joint damage
Pustular psoriasis	L40.1-3, L40.82	White blisters surrounded of red skin	Raised bumps filled with non-infectious pus	Hands and feet (palmoplantarpustulosis); widespread patches on any part of body	May be triggered by medications, infections, emotional stress, or exposure to certain chemicals
Scalp psoriasis	L40.8	Redness and scales on skin of the scalp, involvement of forehead and face	Sharply bordered erythematous/ squamous lesions present covered by silver or white scales	Scalp, forehead, face	1.5-2% prevalence, lesions can spread over the anterior hairline onto the forehead, may affect postauricular region

* The International Statistical Classification of Diseases and Related Health Problems 10th Revision (ICD-10) is a coding of diseases and signs, symptoms, abnormal findings, complaints, social circumstances and external causes of injury or diseases, as classified by the World Health Organization (WHO).

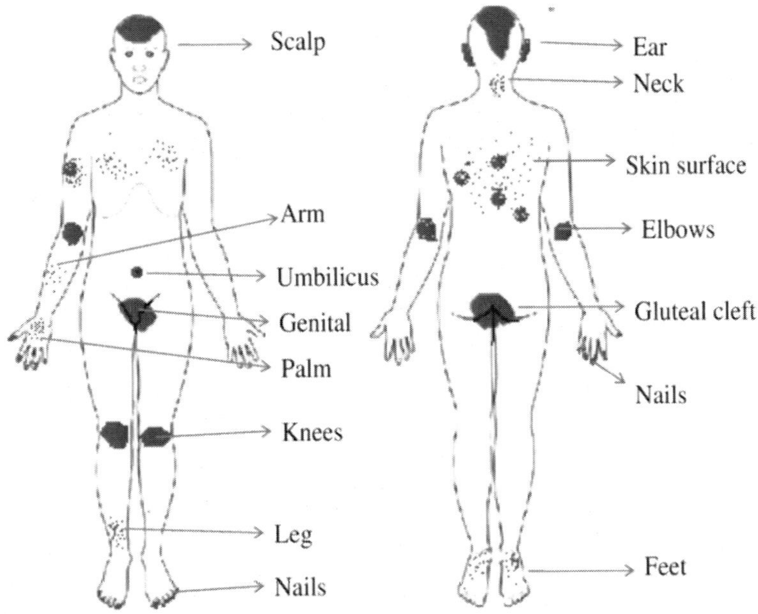

Figure 2. General body parts commonly affected in psoriasis.

MAJOR CAUSES OF PSORIASIS

Psoriasis is primarily a disorder of excessive growth and reproduction of skin cells, which is due to a fault of the epidermis and its keratinocytes (Pastore et al. 2006). The disorder is, however, immune-mediated wherein T cells (T lymphocytes) of the immune system become active, migrate to the dermis and trigger the release of cytokines like TNF-α (tumor necrosis factor-alpha), in particular, thereby causing inflammation and the rapid production of skin cells (Simonart and Heenen 2007). Infections, especially streptococcal infections of the upper respiratory tract, have been found to be associated with the disease (Owen et al. 2000). In addition, exacerbation of psoriasis has been observed in patients infected with the human immunodeficiency virus (HIV). Other significant triggers of the disease include oxidative stress (caused due to reactive oxygen species (ROS)) and psychological stress, skin injuries, and changes in season and climate (Choi and Koo 2003). Use of certain medicines like lithium salt, beta blockers, ACE (Angiotensin converting enzyme) inhibitors, and the antimalarial drugs like chloroquininine can aggravate the disease (Tsankov et al. 2000). Excessive alcohol consumption, smoking, and obesity can be some other factors contributing in the exacerbation of psoriasis (Hayes and Koo 2010). The disease has a large hereditary component, and many genes are associated with it. All the factors responsible for causing or triggering psoriasis are given in Figure 3.

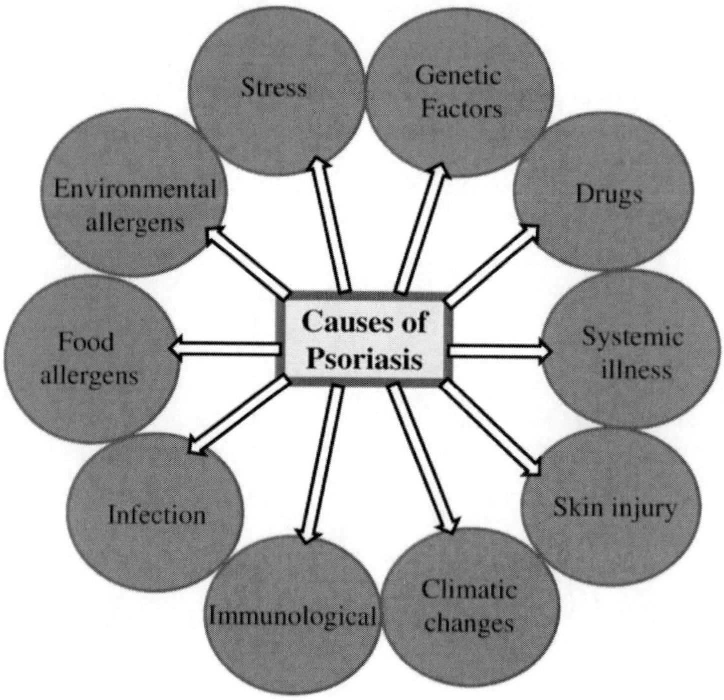

Figure 3.Causes of psoriasis.

Generation of Reactive Oxygen Species (ROS)

A large number of environmental factors like contaminants, radiations, and tobacco smoke etc. tremendously increase the production of free radicals in the body (Bluhm et al. 1971). Free radicals are also generated in the body as a result of the normal metabolic processes like in the mitochondrial respiratory chain, through liver mixed function oxidases, and xanthine oxidase activity. These free radicals, also known as ROS, adversely affect the cells by severely damaging and modifying cellular bio-molecules, such as DNA, thereby leading to serious consequences for the cell (Cooke et al. 2003). They, thus, inflict a state of oxidative stress on the whole body. However, skin being the outermost covering of the body, is the major target of the oxidative stress originating from the environment. Cellular damage due to ROS production has implications in the etiology and pathophysiology of many human diseases including autoimmune pathologies, gastrointestinal inflammation, and neurodegenerative disorders like Alzheimer's disease, Parkinson's disease, and multiple sclerosis (Sarma et al. 2010). Psoriasis is also one such disease that thrives in the body due to improper balance of oxidants and antioxidants (Zhou et al. 2009). Antioxidants are agents that can minimize cell damage thereby preventing diseases (Kinghorn et al. 2004). Although endogenous antioxidants naturally present in the body are capable of attenuating the harmful effects of ROS, but the enormous and prolonged presence of free radicals can override ROS defence mechanisms and mediate numerous cellular responses that contribute to the development of diseases like psoriasis.

ROLE OF LIPOXYGENASES (LOXS)

Lipoxygenases are non-heme iron-containing enzymes that catalyze the stereo-specific incorporation of molecular oxygen into polyunsaturated fatty acids (PUFA) (Brash 1999). They are classed into several subcategories including 5-, 12- and 15-lipoxygenases. They catalyze generation of leukotrienes by oxygenation of arachidonic acid in granulocytes, macrophages, and mast cells (Morris et al. 2001). Leukotrienes, the potent biological mediators of inflammation, are known to play an important role in the pathophysiology of several inflammatory diseases like rheumatoid arthritis, asthmatic responses, Alzheimer's dementia, as well as psoriasis (Werz 2002). Inhibitors of leukotriene synthesis may inhibit the inflammatory processes involved in psoriasis. 15-LOX-2 has been implicated in interferon-gamma-induced inflammatory processes in normal human epidermal keratinocytes and in psoriatic skin (Setsu et al. 2006).

GENETIC VARIATIONS AND PSORIASIS

Genetic variation in the major histocompatibility complex increases risk of developing psoriasis. However, only ~10% of individuals with this risk factor develop PS, indicating that other genetic effects and environmental triggers are important. Pathogenesis of psoriasis is induced by a cascade of inflammatory mediators in the dermis that arise from dendritic cells, macrophages, and T cells (Palfreeman et al. 2013). A predominant infiltration of the dermis by CD4+ T-helper (Th) cells produce pro-inflammatory cytokines such as interferon (IFN)-γ, tumor necrosis factor-alpha (TNF-α), interleukin (IL)-17A, IL-22 IL-1β, IL-6, and IL-23 (Zaba et al. 2009). Apart from these, some other inflammatory mediators like STAT-3 (Signal transducer and activator of transcription), TLR-2 (Toll-like receptor), growth factors like TGF-β (Transforming growth factor), and genes like CCHCR1 (Coiled-coil alpha-helical rod protein 1), StAR (Steroidogenic acute regulatory protein), and VDR (Vitamin D receptor) significantly accelerate the inflammatory processes in psoriasis (Honma et al. 2012). Furthermore, nine loci on different chromosomes associated with psoriasis have been identified through classic genome-wide linkage analysis (Veal et al. 2001). These are known as psoriasis susceptibility loci 1 through 9 (PSORS1 through PSORS9). Out of the nine, the major determinant is PSORS1 which is located in the major-histocompatibility-complex (MHC) region on chromosome 6 and accounts for nearly 35-50% of its heritability (Burden et al. 1998). Many of the genes in these loci are involved in causing inflammation (Watson et al. 1972).

Epigenetic Risk Factors for Psoriasis

Epigenetics of human diseases have potential role in common non-neoplastic human diseases, especially the autoimmune diseases. Epigenetic regulations are essential for the function of immune system. A failure to maintain epigenetic homeostasis in the immune response due to stress and environmental influences leads to aberrant gene expression, contributing to immune dysfunction and in some cases the development of autoimmunity in genetically predisposed individuals. Epigenetic changes *via* DNA methylation and/or histone modification, are genetic regulatory changes that are not associated with alterations in the DNA sequences. This epigenetics can explain the link between psoriasis risk alleles, causes, risk factors and disease development, as alleles possess various regulatory mechanisms for epigenetic modification (Zeng et al. 2017). There are multiple genes involved in psoriasis pathogenesis that demonstrate abnormal methylation patterns involved in differentiation and proliferation of epidermal cell cycle. Various other symptoms such as apoptosis, inflammation, and IFN-γ and TNF-α signaling are also associated and controlled by epigenetic regulatory mechanisms (Ovejero-Benito et al. 2017).

Pathogenesis of Psoriasis

Psoriasis vulgaris is a common skin autoimmune, multifactorial disorder mainly identified by formation of inflamed skin, raised plaques that constantly shed scales (dead skin cells) derived from fast uncontrolled growth of skin epithelial cells. The disease is defined by multistep cellular changes in the skin such as hyperplasia and ectasia, and infiltration of T lymphocytes, neutrophils, and other types of leucocyte of affected epidermal keratinocytes of skin. Psoriatic skin exhibits a lot of pathological changes induced by several external and internal factors in cutaneous cell types. Although the main mechanism underlying the pathogenesis of psoriasis remains obscure, experimental evidences suggest T-lymphocyte immunopathogenesis as the primary cause behind it (Sabat et al. 2007). Investigations show that some of the compounds specifically targeting T-cell functions are helpful in alleviating psoriasis (Ellis and Krueger 2001). Innate immune cells of the body produce key cytokines like TNF-α, IFN-α, IFN-γ, IL-1β, and

IL-6, among others, that activate myeloid dendritic cells (Stavropoulos et al. 2008). These cells on activation present antigens and secrete mediators such as IL-12 and IL-23, leading to the differentiation of T helper cells type 17 (Th17) and type 1 (Th1) (Fitch et al. 2007). T cells, in turn, secrete mediators like IL-17A, IL-17F, and IL-22 (Wolk et al. 2009) that activate keratinocytes and induce the production of pro-inflammatory cytokines TNF-α, IL-1β, and IL-6 (Murphy et al. 2000); chemokines (CXCL8-CXCL11 and CCL20), and antimicrobial peptides etc. (Figure 4).

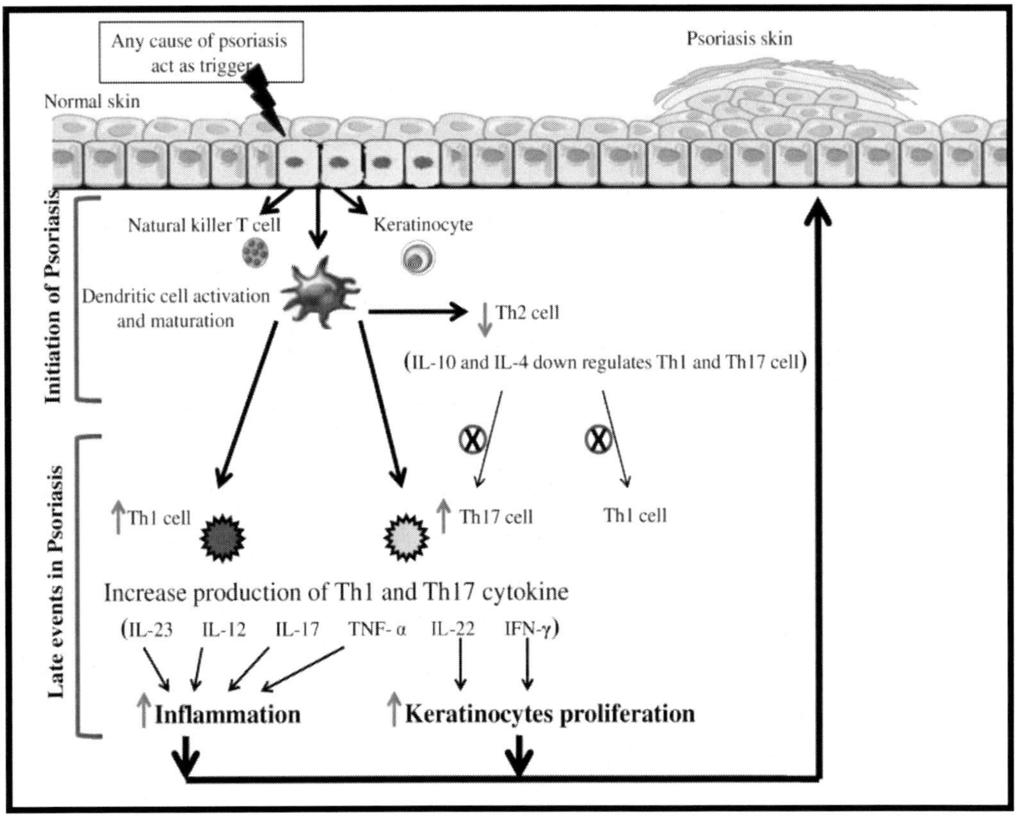

Figure 4. Layout of the immunopathogenic mechanism involving key cells and mediators responsible for initiation and psoriasis disease development.

Treatments Used for Psoriasis and Their Limitations

Treatment options mainly depend on the extent of the disease, its severity as well as the emotional response to the disease. They include topical agents, phototherapy and systemic agents. All of these different treatments may be used alone or in combination with one another depending upon disease severity.

Table 2. Types of treatments used for curing psoriasis and their limitations

Types of treatments	Treating agents	Course of action/Mode of administration	Limitations/Side effects
Oral treatment	Methotrexate; Acitretin; Cyclosporin; Psoralens (Methoxsalen (Oxsoralen-Ultra) and trioxsalen (Trisoralen))	*Methotrexate:* taken once a week, works largely via modulating the immune system *Acitretin*: (derivative of vitamin A) helps program the skin cells back to normal, has effects on the immune system *Cyclosporin*: works via its effects on the immune system. Is taken orally two or three times a day *Psoralens*: taken by mouth one to two hours before PUVA therapy or sunlight exposure, therapy is given 2-3 times per week, with maintenance treatments every 2-4 weeks until remission	*Methotrexate*: causes liver and kidney damage *Acitretin*: dryness and irritation of the skin, lips, eyes, nose, and mucous membrane surfaces; thinning hair; elevation of cholesterol and triglyceride levels; liver toxicity; bone change *Cyclosporine*: increased risk of skin cancer, infection, or lymphoma; may cause damage to the kidneys; high blood pressure *Psoralens*: make skin more sensitive to light, increased risk of sunburn, skin cancer, cataracts; PUVA therapy causes nausea, itching, and burning
Phototherapy	Narrowband UVB, Broadband UVB and PUVA	A measured dose of the appropriate wavelength of light is delivered by fluorescent tubes, should only be administered by practitioners skilled in its use *Narrowband UVB*: incorporates the sun's most beneficial rays for treating psoriasis and leads to faster and more prolonged benefits compared to broadband phototherapy and natural sunlight *PUVA*: UVA treatment combined with the administration of psoralens	Long term damage to the skin, possible increased risk of skin cancers, and increased skin ageing
Topical treatment	Cortisone creams; Tar; Calcipotriol; Anthralin (Dithranol)	*Cortisone*: choice of steroid varies depending on the severity of the lesions and their location (e.g., scalp, face or skin folds) *Tar*: Used for over one hundred years as creams and shampoos *Calcipotriol*: Medication related to vitamin D, application usually at night to minimize odour during the day *Anthralin*: used in a thick ointment sometimes under dressings, more commonly applied at higher concentrations for 10-15 minutes	*Cortisone*: prolonged use leads to thinning of the skin *Tar*: makes skin more sensitive to the sun, stain clothing, smelly *Calcipotriol*: can cause irritation, particularly on sensitive skin areas like face or groin, excess use leads to increased level of calcium in the blood *Anthralin*: Irritation of the skin, temporary skin discolouration, permanent staining of fabric

Conventional Treatments

A large number of methods have been used for treatment of psoriasis, that mainly include topical treatment (application of medicated ointments or creams on the surface of body) using cortisone (steroid) creams, tar, calcipotriol, and anthralin; systemic treatment (medications taken internally by oral pills or through injections) using drugs like methotrexate, acitretin, cyclosporin, and psoralens (Heydendael et al. 2003); UV treatment (exposure of skin to ultraviolet radiation) through narrow band UVB, broad band UVB, and PUVA (combination treatment of psoralens (P) with UVA) (Stern 2007). Besides these, some other natural therapies are also common for curing the disease (Table 2).

Moreover, nutritional and alternative supplementary diet have also been reported to be beneficial against psoriasis (Table 3).Although people often overlook the role of nutrition in the treatment of psoriasis which may enhance the success of treatment. Nutritional supplementation may provide a viable treatment alternative in patients affected with auto immune induced psoriasis. However, long-term usage of chemical supplements can be hazardous due to the risk of serious cumulative organ toxicity (Tristani-Firouzi and Krueger 1998) and can have limitations due to the lack of efficacy over time (Ashcroft et al. 2000). These treatments are often accompanied by various side effects including irritation of the skin, temporary skin discoloration, sensitivity to sunlight, risk of sunburn and skin cancer, increased risk of liver-kidney damage and cataract, infection, and lymphoma. High doses of the drugs can be potentially toxic.

It's unfortunately there are very few reports related to possible benefits of dietary modifications in curing/relieve from psoriasis. Nutritional and dietary supplements could be very simple cost effective ways to control the symptoms and pathogenesis of psoriasis.

Table 3. Nutrient and alternative supplement therapies for psoriasis

Dietary supplements	Dietary changes
• Essential Fatty Acids Omega 3 fatty acids, particularly fish oil supplements. The recommended dosage is 1000 mgs twice daily • Milk thistle stops the breakdown of substances that contribute to psoriasis, tea or tincture made with milk thistle to help alleviate the symptoms of psoriasis • Bioactive Whey Protein Isolate • XP-828L is a novel dietary supplement made of a protein extract derived from bovine whey that has recently been shown to be beneficial in psoriasis • Vitamins B12 (100 to 1000 mcg/day); Vitamin E (400 to 800 IU/day); Folate (400 mcg/day); Vitamin D • Selenium (200 mcg/day) and Zinc (30 mg/day) is effective in treating psoriasis	• Eliminating alcohol consumption • Avoiding simple sugars, foods high in acid, fat and foods that are known allergens • Intake of low-calorie or gluten-free diets, • Antioxidants are beneficial for psoriasis sufferers because of their anti-inflammatory properties and their defense against oxidative stress.

Modern Biological Therapies for Psoriasis

The ongoing researches in the field of psoriasis are widespread. Recent advances in psoriasis research have provided a platform for the rational design of new biological agents and immune-response modifiers that specifically target key mechanisms of psoriasis pathogenesis (Lowes et al. 2007). Biologics are basically a heterogeneous group of monoclonal antibodies, fusion proteins, and recombinant cytokines, which are designed to modify and regulate the significant and specific mechanisms underlying psoriasis immunopathogenesis (Papoutsaki et al. 2009). Each of the pathogenetic mechanisms in psoriasis involving cytokines such as TNF-α, IL-22, IFNγ, IL-6, IL-12, and IL-23 represent optimal target for highly specific therapeutical intervention (Rácz and Prens 2009). Based on their mode of action, the biological therapies are classified into three categories, which are T-cell modulating agents (e.g., alefacept and efalizumab); the inhibitors of TNF-α blockers, e.g., adalimumab, certolizumab, etanercept, golimumab and infliximab); and the inhibitors of IL-12 and IL-23 (e.g., ustekinumab and briakinumab) (Weger 2010). Three of these agents, namely, alefacept, efalizumab, and etanercept are currently approved by the Food and Drug Administration (FDA) for the psoriasis treatment (Lebwohl et al. 2003), while several others (e.g., infliximab) are in the final phase of clinical development. Although biologics are generally safe and their clinical trials have been proven efficient for the treatment of psoriasis, however considering the possibility that certain uncommon and unfortunate events might occur due to their long-term exposure, their usage can be risky (Weger, 2010).

Psoriasis is increasing globally at an alarming rate and many approaches have been studied in the areas of conventional as well as complementary and alternative medicine in order to find a cure for this dreadful disease. Many treatments available such as phototherapy, administration of systemic agents, applications of steroidal topical agents, and use of new biologics including antibodies as drugs, have no doubt provided relief to the patients, but the use of all these agents exhibit many side effects like hepatotoxicity, nephrotoxicity, and bone marrow suppression (Jain and Singh 2013). They can also predispose patients to infections like tuberculosis (Raychaudhuri et al. 2009) and increased risk of developing malignancies like lymphoma, leukemia, and melanomas (Pallavicini et al. 2010). Moreover, treatment of severe psoriasis with many of the effective agents may involve an increased risk of significant morbidity including skin cancers and liver diseases (Dommasch and Gelfand 2009).Since psoriasis has a chronic recurrent nature, it is a challenge to completely cure or at least suppress the disease for a longer period (Pauls et al. 2001).

Potential Role of Medicinal Plants: Natural Therapeutics

Medicinal plants play a pivotal role in the ancient as well as modern health care system and there is a growing interest all over the world for discovering their unexplored reservoir (Kumar 2004). Plant based drugs are highly beneficial in treating various human ailments because they contain the components of therapeutic value (WHO Reports 1993). There is an increasing demand of these plants in the pharmaceutical industries due to their easy availability, cheaper cost, and non-toxic nature as compared to the modern medicine (Agbor and Ngogang 2005). Most of the medicinal plants are rich in antioxidant compounds that protect the cells against the damaging effects of ROS (Krishnaiah et al. 2011). Antioxidants reduce the oxidative stress in cells and are therefore useful to prevent pathological conditions of human body including cancer, cardiovascular diseases, asthma, arthritis, and other inflammatory diseases (Devasagayam et al. 2004). Many herbal extracts from different plant parts like stems, roots, bark, leaves, fruits, and seeds of several important medicinal species have been found rich in antioxidant properties (Tiwari 2001). The antioxidant activity of the plants is mainly attributed to the flavones, isoflavones, flavonoids, anthocyanin, coumarin, lignans, catechins, and isocatechins present in them (Aqil et al. 2006). ROS scavenging through chemoprevention and quenching of lipid peroxidation chain reactions by natural antioxidant compounds present in medicinal plants helps in preventing, delaying, or ameliorating many of the disorders (Gonzalez et al. 1998).

Lipoxygenases are very important agents in the inflammatory response as they catalyze oxygenation of arachidonic acid in granulocytes, macrophages and mast cells, thereby biosynthesizing inflammatory mediators that give rise to innumerable inflammatory disorders (Yamamoto 1992). The products of LOX catalysis, such as hydroperoxyeicosatetraenoic acids (HPETE), hydroxyeicosatetraenoic acids (HETE), leukotrienes, and lipoxins are involved in the development of many diseases like rheumatoid arthritis, cardiovascular diseases, pulmonary diseases, as well as psoriasis (Haeggström and Funk 2011). LOXs have been targeted *in-vitro, in-vivo,* and *in-silico*, by many researchers till date, in order to study their inhibition mechanisms and to discover their potent inhibitors. LOX inhibition property of the medicinal plants can be attributed to the antioxidants present in them (Schneider and Bucar 2005). Antioxidants inhibit the LOX activity by inhibiting lipid hydroperoxide formation due to scavenging of lipidoxy- or lipidperoxy-radicals formed during enzymatic peroxidation. This limits the availability of lipid hydroperoxide substrate which is required for the catalytic cycle of LOX (Rackova et al. 2007). Another way of LOX inhibition by antioxidants can be competitive inhibition through chelation of its non-heme bound iron (Lin et al. 2001) or by reduction of its ferric form (Gutierrez-Lugo et al. 2004). A large number of biological agents, natural bioactive compounds, and immune-response modifiers can be derived from medicinal plant extracts that are rich in antioxidants, which hence could be used as LOX inhibitors to treat many inflammatory disorders (Rackova et al. 2007).

Current researches are therefore focused on multidisciplinary application of methodologies for exploring the potential of medicinal plants and studying the inhibitory effects of the bioactive compounds present in them (Dev 1999). There are many medicinal plants that are known to possess antioxidant and anti-inflammatory properties as well as dermo-protective properties (Sharma et al. 2003). Some of the plants with their common names and families are given in Table 4. These plants also possess various pharmacological attributes essentially required for the treatment of psoriasis.

The five significant pharmacological properties that are prerequisites for a plant to be used as a potent drug against psoriasis are anti-inflammatory, immunomodulatory, antioxidant, antimicrobial, and wound-healing (Tabassum and Hamdani 2014). Each of the enlisted plant possesses either one or more than one of the five essential medicinal properties required for curing psoriasis. The plants are categorized on the basis of their pharmacological properties in Table 4.

Table 4. Some medicinal plants commonly used against skin disorders

S. No.	Plant's scientific name	Common name	Family	Medicinal activity
1	*Achyranthes aspera*	Prickly chaff flower	Amaranthaceae	Anti-inflammatory, Anti-oxidant, Antimicrobial
2	*Allium cepa L.*	Onion	Liliaceae	Anti-inflammatory, Anti-oxidant
3	*Allium sativum*	Garlic	Liliaceae	Antimicrobial
4	*Aloe vera L.*	Aloe	Liliaceae	Healing capacity
5	*Amaranthus hypochondriacus L.*	Chaulai	Amaranthaceae	Immunomodulatory
6	*Andrographis paniculataBurm.f.*	Kalmegh	Acanthaceae	Anti-inflammatory, Antimicrobial
7	*Arnica Montana*	Arnica	*Compositae*	*Healing capacity, anti-inflammatory*
8	*Artemisia annua L.*	Sweet wormwood	Asteraceae	Antimicrobial
9	*Azadirachta indica A. Juss.*	Neem	Meliaceae	Immunomodulatory
10	*Bacchinia variegate*	Kachanar	Fabaceae	Anti-inflammatory
11	*Berberis vulgaris*	Berberry	Berberidaceae	Anti-oxidant
12	*Beta vulgaris*	Beet root	Brassicaceae	Anti-oxidant, Anti-inflammatory
13	*Calendula officinalis L.*	Marigold	Asteraceae	Anti-oxidant
14	*Celastrus paniculatusWilld.*	Malkangani	Celastraceae	Anti-inflammatory, Immunomodulatory, Anti-oxidant
15	*Crocus sativus L.*	Saffron	Iridaceae	Anti-inflammatory
16	*Curcuma longa L.*	Turmeric	Zingiberaceae	Healing capacity, Anti-inflammatory, Anti-oxidant, Antimicrobial
17	*Mahonia aquifolium*	Oregon graps	Berberidaceae	Anti-oxidant
18	*Nigella sativa L.*	Kalonji	Apiaceae	Antimicrobial
19	*Ocimum sanctum*	Tulsi	Lamiaceae	Anti-inflammatory, Immunomodulatory
20	*Origanum vulgare*	Oregano	Lamiaceae	Anti-oxidant, Antimicrobial
21	*Piper nigrum L.*	Black pepper	Piperaceae	Antimicrobial
22	*Psoralea corylifoliaL.*	Bakuchi	Fabaceae	Antimicrobial, Anti-inflammatory
23	*Tinospora cordifolia (Thunb.) Miers*	Giloe	Menispermaceae	Healing capacity, Anti-oxidant
24	*Triticum aestivum L.*	Wheat grass	Poaceae	Anti-oxidant, Immunomodulatory
25	*Withania somnifera*	Ashwagandha	Solanaceae	Anti-oxidant
26	*Zingiber officinale Roscoe*	Ginger	Zingiberaceae	Anti-oxidant, Antimicrobial, Anti-inflammatory

CONCLUSION

The medications which are commonly applied to treat psoriasis have a number of side effects, which has shifted the researchers' attention to safer natural alternatives. PUVA therapy which is widely accepted worldwide for psoriasis treatment has active moiety psoralen derived from natural source, *Psoralea corylifolia.* Keeping in mind, severe side effects associated with synthetic drugs available for the treatment of psoriasis, herbal drugs seem to be a viable approach to explore for newer, safer and effective antipsoriatic drugs. Plants such as *Curcuma longa*, *Aloe vera*, and *Celastrus paniculatus* have been used in the traditional systems to offer relief in skin disorders like psoriasis, and proved scientifically for their traditional claims. Inclusion of such plant drugs in the arsenal of modern therapeutics has revived the faith of researchers in natural resources. Further chemical, genetic and environmental knowledge could be a useful tool for obtaining several bioactive compounds from these medicinal plants with high clinical and pharmaceutical value. Several modern techniques such as chromatography, DNA fingerprinting, GC-MS analysis, Metabolite profiling could be an effective tool for herbal product characterization, authentication, and standardization of all the supply chain steps for curing psoriasis.

Finally, although emphasis has been placed on mapping common variants related to the psoriasis, recent studies have demonstrated that rare variants also play an important role in complex trait etiology and their identification should have a greater impact on risk assessment, disease prevention, and treatment due to their large genetic effects. The development of second-generation sequencing (NexGen) technologies has made it possible to identify the rare causative variants/mutations responsible for skin disease. These efforts will lead to identification of new psoriasis susceptibility genes and associated signal pathways, which eventually pave the way for new therapeutic bioactive targets of the disease. There is a great need for coordinated efforts to perform the necessary clinical trials related to the efficacy and safety of herbal bioactive compounds, both alone and in combination with conventional drug therapies.

REFERENCES

Agbor AG, Ngogang YJ (2005) Toxicity of herbal preparations. *Cam J Ethnobot* 1: 23-28.

AL-Mutairi N, AL-Farag S, AL-Mutairi A, AL-Shiltawy M (2010) Comorbidities associated with psoriasis: an experience from the Middle East. *J Dermatol* 37, 2: 146-155.

Altmüller J, Palmer LJ, Fischer G, Scherb H, Wjst M (2001) Genomewide scans of complex human diseases: true linkage is hard to find. *Am J Hum Genet* 69, 5: 936-950.

Aqil F, Ahmad I, Mehmood Z (2006) Antioxidant and free radical scavenging properties of twelve traditionally used Indian medicinal plants. *Turkish J Biol* 30, 3: 177-183.

Ashcroft DM, Li Wan Po A, Griffiths CEM (2000) Therapeutic strategies for psoriasis. *J Clin Pharm Ther* 25, 1: 1-10.

Bluhm AL, Weinstein J, Sousa JA (1971) Free radicals in tobacco smoke. *Nature* 229, 5285: 500-500.

Brash AR (1999) Lipoxygenases: Occurrence, functions, catalysis, and acquisition of substrate. *J Bio Chem* 274: 23679-82.

Burden AD, Javed S, Bailey M, Hodgins M, Connor M, Tillman D (1998) Genetics of psoriasis: paternal inheritance and a locus on chromosome 6p. *J Invest Dermatol* 110: 958-960.

Choi J, Koo JY (2003) Quality of life issues in psoriasis. *J Am Acad Dermatol* 49: S57-S61.

Christophers E (2001) Psoriasis-epidemiology and clinical spectrum. *Clin Exp Dermatol* 26: 314-320.

Cooke MS, Evans MD, Dizdaroglu M and LunecJ (2003) Oxidative DNA damage: mechanisms, mutation, and disease. *The FASEB J* 17, 10: 1195-1214.

Dev S (1999) Ancient-modern concordance in Ayurvedic plants: some examples. *Environ Health Perspect* 107, 10: 783-9.

Devasagayam TPA, Tilak JC, Boloor KK, Sane KS, Ghaskadbi SS, Lele RD (2004) Review: Free radical and antioxidants in human health: current status and future prospects. *J Assoc Physicians India* 53: 794-804.

Dommasch E, Gelfand JM (2009) Is there truly a risk of lymphoma from biologic therapies? *Dermatologic ther* 22, 5: 418-430.

Ellis CN, Krueger GG (2001) Treatment of chronic plaque psoriasis by selective targeting of memory effector T lymphocytes. *N Engl J Med* 345: 248-255.

Farber EM, Nall ML (1998) Epidemiology: natural history and genetics. In: Roenigk HH, Maibach HI, eds. *Psoriasis*. New York: Marcel Dekker: 107-58.

Farber EM, Nall ML, Watson W (1974) Natural history of psoriasis in 61 twin pairs. *Arch Dermatol* 109: 207-211.

Fitch E, Harper E, Skorcheva I, Kurtz SE, Blauvelt A (2007) Pathophysiology of psoriasis: Recent advances on IL-23 and Th17 cytokines. *Curr Rheumatol Rep* 9, 6: 461-467.

Gelfand JM, Dommasch ED, Shin DB, Azfar RS, Kurd SK, Wang X, Troxel AB (2009) The risk of stroke in patients with psoriasis. *J Invest Dermatol* 129: 2411-2418.

Gelfand JM, Stern RS, Nijsten T,Feldman SR, Thomas J, Kist J, Rolstad T, Margolis DJ (2005) The prevalence of psoriasis in African Americans: results from a population-based study. *J Am Acad Dermatol* 52: 23-26.

Gladman DD (2009) Psoriatic arthritis. *DermatolTher* 22, 1: 40-55.

Gonzalez MJ, Riordan NH, Riordan HD (1998) Antioxidants as Chemopreventive Agents for Breast Cancer. *BioMedicina* 1, 4: 120-126.

Griffiths CE, Barker JN (2007) Pathogenesis and clinical features of psoriasis. *Lancet* 370, 9583: 263-71.

Gutierrez-Lugo MT, Deschamps JD, Holman TR, Suarez E, Timmermann BN (2004) Lipoxygenase inhibition by anadanthoflavone, a new flavonoid from the aerial parts of *Anadenantheracolumbrina*. *Planta Med* 70, 3: 263-265.

Haeggström JZ, Funk CD (2011) Lipoxygenase and leukotriene pathways: biochemistry, biology, and roles in disease. *Chem Rev* 111: 5866-5898.

Hayes J, Koo J (2010) Psoriasis: depression, anxiety, smoking, and drinking habits. *Dermatologic Ther* 23, 2: 174-180.

Henseler T, Christophers E (1985) Psoriasis of early and late onset: characterization of two types of psoriasis vulgaris. *J Am Acad Dermatol* 13: 450-456.

Heydendael VMR, Spuls PI, Opmeer BC, de Borgie CA, Reitsma JB, Goldschmidt WF, Bossuyt PM, Bos JD, de Rie MA (2003) Methotrexate versus cyclosporine in moderate-to-severe chronic plaque psoriasis. *N Engl J Med* 349: 658-665.

Honma M, Minami-Hori M, Takahashi H, Iizuka H (2012) Podoplanin expression in wound and hyperproliferative psoriatic epidermis: regulation by TGF-β and STAT-3 activating cytokines, IFN-γ, IL-6, and IL-22. *J Dermatol Sci* 65, 2: 134-140.

Jain, A, Singh, JA (2013) Harms of Tumor Necrosis Factor Inhibitors in Rheumatic diseases: A focused Systematic Review of the Literature. *Immunother* 5, 3: 265-299.

Kinghorn, AD, Su BN, Jang DS, Chang LC, Lee D, Gu JQ, Carcache-Blanco EJ, Pawlus AD, Lee SK, Park EJ, Cuendet M, Gills JJ, Bhat K, Park HS, Mata-Greenwood E, Song LL, Jang M, Pezzuto JM (2004) Natural Inhibitors of Carcinogenesis. *Planta Med* 7, 8: 691-705.

Krishnaiah D, Sarbatly R, Nithyanandam R (2011) A review of the antioxidant potential of medicinal plant species. *Food Bioprod Process* 89, 3: 217-233.

Kumar P (2004) Valuation of medicinal plants for pharmaceutical uses. *Curr Sci* 86, 7: 930-937.

Lebwohl M, Tyring SK, Hamilton TK, Toth D, Glazer S, Tawfik NH, Walicke P, Dummer W, Wang X, Garovoy MR, Pariser D (2003) A novel targeted T-cell modulator, efalizumab, for plaque psoriasis. *N Engl J Med* 349: 2004-2013.

Lee YA, Ruschendorf F, Windemuth C, Schmitt-Egenolf M, Stadelmann A, Nürnberg G, Ständer M, Wienker TF, Reis A, Traupe H (2000) Genome wide scan in German families reveals evidence for a novel psoriasis-susceptibility locus on chromosome 19p13. *Am J Hum Genet* 67: 1020-1024.

Lin JK, Tsai SH, Lin-Shiau SY (2001) Anti-inflammatory and antitumor effects of flavonoids and flavanoids. *Drugs Future* 26: 145-157.

Liu Y, Krueger JG, Bowcock AM (2007) Psoriasis: genetic associations and immune system changes. *Genes Immun* 8, 1: 1-12.

Lomholt G (1964) Prevalence of skin diseases in a population; a census study from the Faroe Islands. *Dan Med Bull* 11: 1-7.

Lowes MA, Bowcock AM, Krueger JG (2007) Pathogenesis and therapy of psoriasis. *Nature* 445, 7130: 866-873.

Murphy JE, Robert C, Kupper TS (2000) Interleukin-1 and cutaneous inflammation: a crucial link between innate and acquired immunity. *J Invest Dermatol* 114: 602-608.

Nestle FO, Kaplan DH, Barker J (2009) Psoriasis. *N Engl J Med* 361, 5: 496-509.

Ovejero-Benito MC, Cabaleiro T, Sanz-García A, Llamas-Velasco M, Rodríguez MS, Prieto-Pérez R, Talegón M, Román M, Ochoa D, Reolid A, Daudén E, Abad-Santos F (2017) Epigenetic biomarkers associated with anti-TNF drugs response in moderate-to- severe psoriasis. *Br J Dermatol* Accepted Author Manuscript. doi:10.1111/bjd.15504.

Owen CM, Chalmers RJ, O'Sullivan T, Griffiths CE (2000) Antistreptococcal interventions for guttate and chronic plaque psoriasis. *Cochrane Database Syst Rev* 2: CD001976.

Palfreeman AC, McNamee KE, McCann FE (2013) New developments in the management of psoriasis and psoriatic arthritis: a focus on apremilast. *Drug Des Devel Ther* 7: 201-210.

Pallavicini FB, Caporali R, Sarzi-Puttini P, Atzeni F, Bazzani C, Gorla R, Marchesoni A, Favalli EG, Montecucco C (2010) Tumour necrosis factor antagonist therapy and cancer development: Analysis of the LORHEN registry. *Autoimmun Rev* 9: 175-80.

Papoutsaki M, Costanzo A, Chimenti S (2009) Emerging treatments in the management of psoriasis: biological targeting with ustekinumab. *Clin Cosmet Investig Dermatol* 2: 95-103.

Pastore S, Mascia F, Mariani V, Girolomoni G (2006) Keratinocytes in skin inflammation. *Expert Rev Dermatol* 1, 2: 279-291.

Pauls K, Schön M, Kubitza RC, Homey B, Wiesenborn A, Lehmann P, Ruzicka T, Parker CM, Schön MP (2001) Role of integrin E (CD103)β7 for tissue-specific epidermal localization of CD8+ T lymphocytes. *J Invest Dermatol* 117: 569-575.

Rackova L, Oblozinsky M, Kostalova D, Kettmann V, Bezakova L (2007) Free radical scavenging activity and lipoxygenase inhibition of *Mahoniaaquifolium* extract and isoquinoline alkaloids. *J Inflammation* 4: 15-21.

Rácz E, Prens EP (2009) Molecular pathophysiology of psoriasis and molecular targets of antipsoriatic therapy. *Expert Rev Mol Med* 11: e38.

Raychaudhuri SP, Nguyen CT, Raychaudhuri SK, Gershwin ME (2009) Incidence and nature of infectious disease in patients treated with anti-TNF agents. *Autoimmun Rev* 9: 67-81.

Sabat R, Philipp S, Höflich C, Kreutzer S, Wallace E, Asadullah K, Volk H, Sterry W, Wolk K (2007) Immunopathogenesis of psoriasis. *Exp Dermatol* 16, 10: 779-798.

Sarma AD, Mallick AR, Ghosh AK (2010) Free Radicals and Their Role in Different Clinical Conditions: An Overview. *Int J PharmaSci Res* 1, 3: 185-192.

Schneider I, Bucar F (2005) Lipoxygenase inhibitors from natural plant sources. Part 1: Medicinal plants with inhibitory activity on arachidonate 5-lipoxygenase and 5-lipoxygenase [sol] cyclooxygenase. *Phytother Res* 19, 2: 81-102.

Setsu N, Matsuura H, Hirakawa S, Arata J, Iwatsuki K (2006) Interferon-gamma-induced 15-lipoxygenase-2 expression in normal human epidermal keratinocytes and a pathogenic link to psoriasis vulgaris. *Eur J Dermatol* 16, 2: 141-145.

Sharma L, Agarwal G, Kumar A (2003) Medicinal plants for skin and hair care. *Indian J Tradit Know* 2, 1: 62-68.

Simonart T, Heenen M (2007) T cell/keratinocyte interactions in psoriasis: where is the trigger? *Br J Dermatol* 156, 1: 171-172.

Stavropoulos PG, Goules AV, Avgerinou G, Katsambas AD (2008) Immunopathogenesis of psoriatic arthritis: Recent advances. *Cent Eur J Med* 3, 3: 245-253.

Stern RS (2007) Psoralen and Ultraviolet A Light Therapy for Psoriasis. *New Engl J Med* 357, 7: 682-690.

Tabassum N, Hamdani M (2014) Plants used to treat skin diseases. *Pharmacogn Rev* 8, 15: 52-60.

Tiwari A (2001) Imbalance in antioxidant defence and human diseases: Multiple approach of natural antioxidant therapy, *Curr Sci* 81, 9: 1179-87.

Tristani-Firouzi P, Krueger GG (1998) Efficacy and safety of treatment modalities for psoriasis. *Cutis* 61: 11-21.

Tsankov N, Angelova I, Kazandjieva J (2000) Drug-induced psoriasis: recognition and management. *Am J Clin Dermatol* 1: 159-165.

Veal CD, Clough RL, Barber RC, Mason S, Tillman D, Ferry B, Jones AB, Ameen M, Balendran N, Powis SH, Burden AD, Barker JN, Trembath RC (2001) Identification of a novel psoriasis susceptibility locus at 1p and evidence of epistasis between PSORS1 and candidate loci. *J Med Genet* 38: 7-13.

Watson W, Cann HM, Farber EM, Nall, ML (1972) The genetics of psoriasis. *Arch Dermatol* 105: 197-207.

Weger W (2010) Current status and new developments in the treatment of psoriasis and psoriatic arthritis with biological agents. *Br J Pharmacol* 160, 4: 810-820.

Werz O (2002) 5-lipoxygenase: cellular biology and molecular pharmacology. *Curr Drug Targets Inflamm Allergy* 1: 23-44.

WHO (1993) Research guidelines for evaluating the safety and efficacy of herbal medicines. Manila, *World Health Organization regional office for the Western Pacific.*

Wolk K, Witte E, Warszawska K, Schulze-Tanzil G, Witte K, Philipp S, Kunz S, Döcke WD, Asadullah K, Volk HD, Sterry W, Sabat R (2009) The Th17 cytokine IL-22 induces IL-20 production in keratinocytes: A novel immunological cascade with potential relevance in psoriasis. *Eur J Immunol* 39, 12: 3570-81.

Yamamoto S (1992) Mammalian lipoxygenases: molecular structures and functions. *Biochim Biophys Acta* 1128: 117-131.

Zaba LC, Fuentes-Duculan J, Eungdamrong NJ, Abello MV, Novitskaya I, Pierson KC, Gonzalez J, Krueger JG, Lowes MA (2009) Psoriasis is characterized by accumulation of immunostimulatory and Th1/Th17 cell-polarizing myeloid dendritic cells. *J Invest Dermatol* 129, 1: 79-88.

Zeng J, Luo S, Huang Y, Lu Q (2017) Critical role of environmental factors in the pathogenesis of psoriasis. *J Dermatol* doi:10.1111/1346-8138.13806.

Zhou Q, Mrowietz U, Rostami-Yazdi M (2009) Oxidative stress in the pathogenesis of psoriasis. *Free Radic Biol Med* 47, 7: 891-905.

In: Trends in Life Science Research
Editors: R. P. Sinha and U. P. Shrivastava

ISBN: 978-1-53613-241-0

Chapter 4

THE BASICS OF STEM CELLS AND THEIR ROLE IN VISION

Mukesh Damala*[1,2]*, Abhinav Reddy Kethiri*[2,3]*,
***Fatemeh Tavakkoli*[2,4]*, Enoch Raju*[2] *and Vivek Singh*[2,*]**

[1]University of Hyderabad, Hyderabad, India
[2]Prof. Brien Holden Eye Research Center, Center for Ocular Regeneration, Tej Kohli Cornea Institute, L.V. Prasad Eye Institute, Hyderabad, India
[3]Manipal Academy of Higher Education, Manipal, India
[4]Centre for Genetic Disorders, Banaras Hindu University, Varanasi, India

ABSTRACT

The advances in the field of stem cell research have high medical significance. The recent findings and stem cell-based applications have paved the way for many therapeutic treatments of much clinical impact. In spite of the controversies over ethics and principles in the field of stem cell research, stem cells were being used in clinical applications addressing many unanswered medical conditions and disorders. This chapter discusses the properties of stem cells, their origin and events in their evolution over the time to the recent types of stem cells and their potency. The current statistics of active clinical trials and their applications in the field of medicine were addressed with ocular surface as the major focus. Additionally, the principles and importance of ethics related to the use of human stem cells were briefly discussed. This chapter provides an overview and a basic understanding of the concept of stem cells, their translational role in ocular disorders, helping to explore more about the rapidly advancing field of stem cell biology.

Keywords: stem cell, potency, limbus, blastocyst, embryo, retina

* Corresponding Author: Vivek Singh, Ph.D., SSR- Stem cell biology laboratory and Center for Ocular Regeneration, Kallam Anji Reddy Campus, L V Prasad Eye Institute, Hyderabad, Telangana, India – 500034. Phone: +91-40-30612286; Email: singhvivekbhu@gmail.com; viveksingh@lvpei.org.

Introduction

Over the last couple of decades, the biology of Stem cells (SC) is gaining tremendous prominence in regenerative medicine and health, due to the vast advances in stem cell technologies and therapies that offer a way to treat many incurable diseases.

Origin of "Stem Cell"

The origin of the term "stem cell" dates back to 1868, where the word "stem cell" was referred to several cell types in various studies. German biologist, Ernst Haeckel, was the first person to have used the term "Stammzelle" (German for stem cell) to describe the ancestor unicellular organism from which he presumed all multi-cellular organisms evolved (Ramalho-Santos et al. 2007). Later, he proposed that the fertilized egg which gives rise to all the cells of an organism can also be called as a stem cell. Extending Haeckel's statement, Theodor Boveri, in 1892, proposed that the cells along the germline lineage between the fertilized egg and committed germ cells are called stem cells. In the same year, Häcker had used the term "stem cell" that refers to a large cell which had under gone asymmetric cell division where one of the daughter cells gave rise to the mesoderm and the other gave rise to the germ cells during development. But the term has got its popularity, from the book 'The Cell in Development and Inheritance', by Wilson (1896), who used the term "stem cell", referring to the unspecialized mother cell of the germline. He was credited as having coined the term stem cell, by many, from the impact of his book, it created all over the world (Ramalho-Santos et al. 2007). Table 1 depicts the chronology of various events of stem cell research.

Defining Stem Cells

A cell that has the ability to proliferate and differentiate in to other cell types required for an organism is called the stem cell (Morrison 1997). Simply, stem cells are those that can form themselves and other cells as well. Hence, they are referred to as "stem" since many types of cells can originate from them.

The two properties that differentiate the stem cells from normal cells of the body are self-renewal and potency. Self-renewal describes the cell's proliferation ability to divide for a number of generations without undergoing any differentiation. The property of potency describes the differentiation and specialization property of trans-differentiating into other cells. Stem cells are unspecialized in nature and do not commit to a specific cell lineage. However, they could change in to different cell types of a specific organ or the organ system and this process of converting in to other cell type is known as differentiation.

It can be mediated by the niche surrounding the cell or the external stimuli from the environment. This process involves an asymmetric division of the stem cell giving rise to two daughter cells of which, one would act as a copy of mother cell and the other differentiates in to specific cell lineage (Figure 1). From there, the progenitor cell undergoes symmetric divisions generating copies of a specific set of cells.

Understanding Their Potency

Potency, in terms of stem cell biology refers to the differentiation potential of a certain group of stem cells in an organism. Based on the potency levels (Figure 2), stem cells can be classified to Totipotent, Pluripotent, Multipotent, and Unipotent stem cells.

Totipotent Stem Cells

Totipotent stem cells have the ability to differentiate into every type of cell in the body. With their uninhibited capability, they can form the placenta and entire foetus giving rise to a complete organism. These are formed by the fertilization of the sperm and an egg cell and are found in very early embryos that disappear later in fetal development. They are also referred to as omnipotent cells.

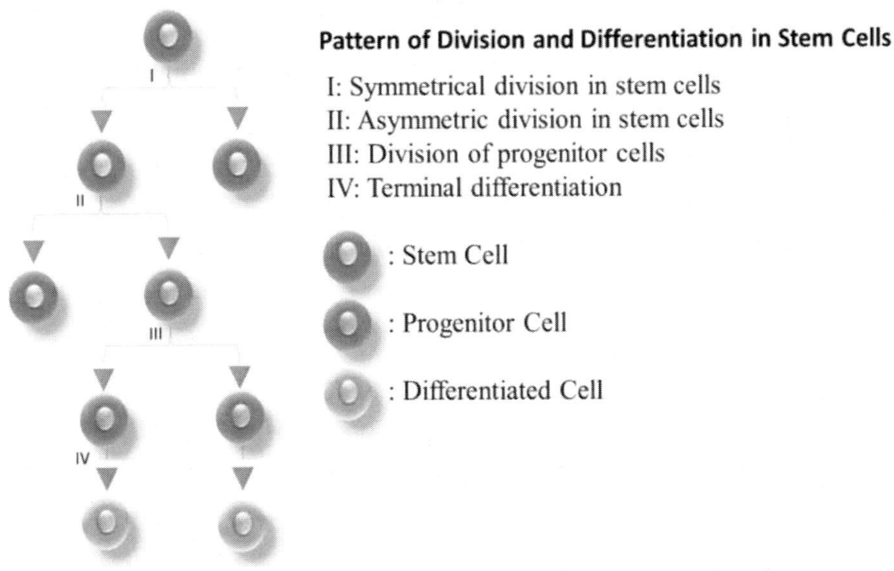

Figure 1. Image showing the pattern of division and differentiation in stem cells.

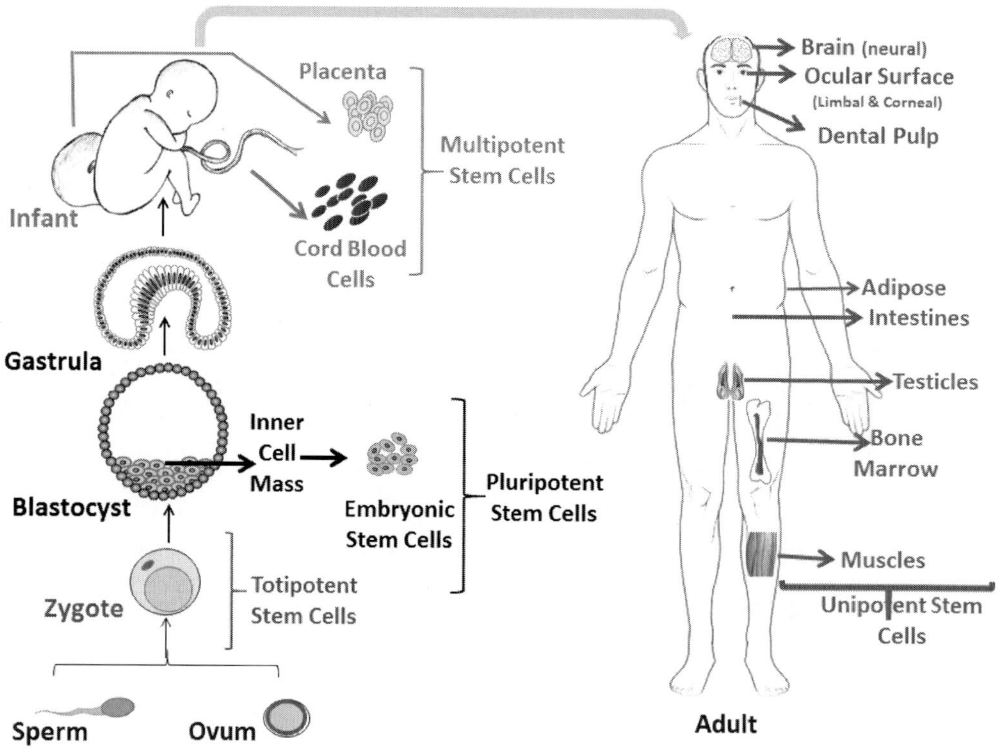

Figure 2. Types of stem cells: Illustrative image showing different types of stem cells based on potency.

The differentiation of a mammalian totipotent cell is unidirectional, which is not reversible, unlike plants. Plants have a special ability, where a terminal cell can dedifferentiate into a new and complete plant. A fertilized oocyte though has the capacity of forming an entire organism, it cannot be considered a stem cell because of the inability to generate itself i.e., it has no self-renewal (Bongso and Lee 2005). Eg: Zygote and 4-cell stage of embryo.

Pluripotent Stem Cells

Pluripotent stem cells are also found only in embryos and are able to form majority (but not all) cell types in the body. They have the ability to form all three germ layers but unlike the totipotent cells, they cannot form the whole organism (Bongso and Richards 2004). Embryonic stem cells are pluripotent, as they can give rise to every cell type in the fully formed body, but not the placenta and umbilical cord. They serve as a good source to study developmental stages of the body. E.g., Blastocyst

MULTIPOTENT STEM CELLS

They are able to form several types of specialized cells that are in closely related families of cells, generally in a specific tissue or organ. They exist in later stages of foetal development and adults as well. E.g., Mesenchymal stem cells (MSCs) and hematopoietic stem cells.

UNIPOTENT STEM CELLS

They can differentiate in to only one cell type and are found in all adult tissues. Arising from multipotent cells, a unipotent stem cell has the lowest differentiation potential that could differentiate into only one type of cell or tissue. Hence, they are referred to as lineage-restricted cells as well. E.g., Dermal skin cells.

Table 1. Table showing the chronology of various events of Stem Cell Research

Year	Event
1868	Ernst Haeckel uses the phrase Stem Cell.
1955	First Successful Stem Cell Transplantation. An indication that all blood cell types could be derived from a single precursor cell.
1958	Cloning of tadpoles.
1961	Existence of stem cells was proved by Dr. James Till and Dr. Ernest McCulloch (Till, J. E., and McCulloch, E. A. 1961)
1974	Creation of first transgenic mammals (mice), by inserting virus genes into embryos (Jaenisch, R., and Mintz, B. 1974)
1975	Mouse ES cells' cultivation
1978	Discovery of blood stem cells in human umbilical cord blood.
1980	Derivation of human embryonic carcinoma cell lines.
1981	Isolation of stem cells from mice embryos. (Martin 1981)
1985	Discovery of OCT4
1989	Discovery of Pre-implantation Genetic Diagnosis (PGD) – where a single stem cell can be removed from an IVF embryo and tested for inherited diseases.
1995	Non-human primates' ES cell line established.
1996	Birth of Dolly (sheep), the first mammal to be generated via cloning from an adult stem cell
1998	Derivation of human ES cells
2005	Discovery of iPS cells
2006	Creation of rodent induced pluripotent cells (iPS cells) by Shinya Yamanaka and Kazutoshi Takahashi. (Takahashi and Yamanaka 2006). Nobel prize in physiology and medicine awarded to Mario Capecchi, Martin Evans and Oliver Smithies for making gene modifications in mice, using ES cells.

Table 1. (Continued)

Year	Event
2007	Discovery of epiblast stem cells in rats and mice.
2009	First ES clinical trial in the world, was approved by FDA, US
2010	Mice skin cells modified to functional neurons *in vitro* (Vierbuchen, Thomas, et al., 2010). Mouse non-muscle cells reprogrammed into beating heart cells.(Leda M, Srivastava D et al., 2010). Human skins cells converted to blood cells (Szabo, Eva, et al. 2010).
2012	Nobel prize in Physiology and Medicine is awarded jointly to Shinya Yamanaka and John Gordon for reprogramming mature cells to become pluripotent. World's first iPS cell clinical study for retinal regeneration, approved by Japan. Adult stem cells extracted from patients following a heart attack, grown in a petri dish, were transplanted into patient's heart; helps to decrease scarring and lead to regrowth of heart tissue.
2013	Human skin cells were reprogrammed into ES cells, using a technique called Somatic Cell Nuclear Transfer (SCNT).
2014	ES cells improve the sight of patients suffering from macular degeneration, showing that transplantation of ES cells is safe in the long-term.
2016	Human Embryonic cells were successfully re-programmed to a fully functional anterior pituitary gland.

Types of Stem Cells

Stem cells are not only units of the biological organization but are also the actual units of regeneration. The human body has around 200 types of cells that emerge from zygote which is formed by the fusion of sperm and egg. Tenth-day post-fertilization, zygote turns in to blastocyst after a set of divisions which is the site of origin for pluripotent stem cells that at later stages develop into various organs and organ systems.

Stem cells are broadly classified into embryonic, foetal, cord blood and adult stem cells, based on their origin (Bongso and Lee 2005).

Embryonic Stem Cells

Embryonic stem cells (ESCs), make up the inner cell mass (ICM) of the blastocyst before implantation. They are the most studied stem cells, whose research remained most controversial till date. ESCs exist only for shorter duration i.e., during the stage of embryo development prior to implantation and tend to lose their pluripotency after commitment. They grow as round, compact colonies of small cells which are dependent on leukemia inhibitory factor (LIF), a cytokine produced by the endometrium that allows blastocyst implantation (Figure 3). LIF receptor (Gp130) is expressed by ESCs and maintains self-renewal and pluripotency by phosphorylating STAT3. ESCs are obtained by extracting cells from early embryos, at the blastocyst (hollow ball) stage and growing them in laboratory dishes. Human ES cells are generated mainly from blastocysts that are the result of *in vitro* fertilization for assisted reproduction but are not needed for implantation into the mother which the parents can donate the 'spare' blastocysts for medical research (Zare et al. 2014).

Adult Stem Cells

Adult stem cells are also called tissue-specific stem cells because each type of adult stem cell produces only a limited set of specialized cells characteristic of a particular tissue-epidermis, blood, etc. In adults, tissue-specific stem cells are located throughout the body. Other types of tissue-specific stem cells are usually found deep within tissues and are difficult to obtain and study, especially in humans. Familiar examples are the epidermal stem cells which continually renew the outer layer of the skin as it gets worn away, and the epithelial stem cells in the gut, that similarly replenish the gut lining. More recent discoveries are of bronchoalveolar stem cells from the lungs of adult humans that are thought to renew the lining of the lungs. Although new types of cells are discovered that appear to behave like stem cells when grown *in vitro*, it is difficult to know whether they act the same *in vivo*, because their natural behavior could not be monitored thoroughly.

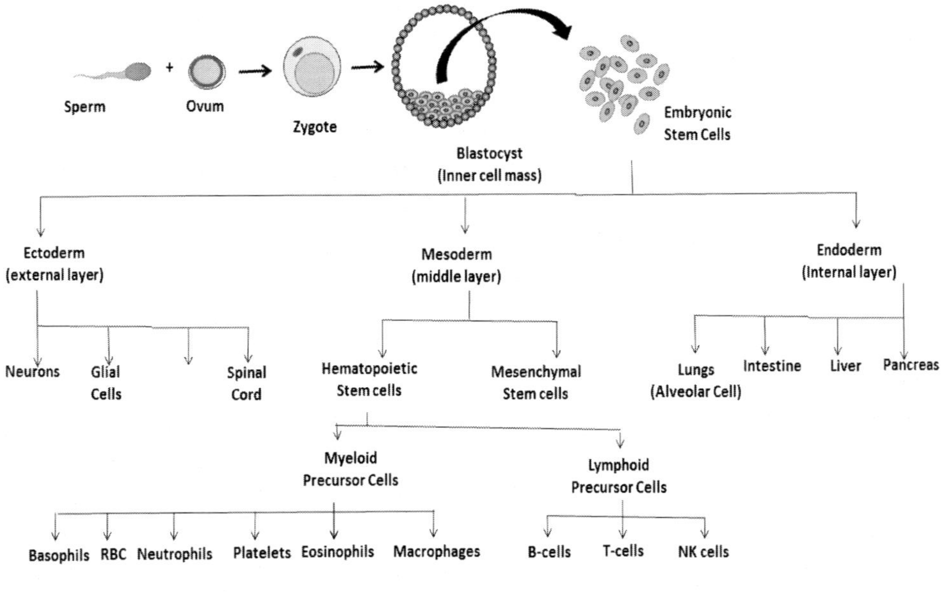

Figure 3. Flowchart showing different types cells of differentiated from embryonic stem cells.

Adult stem cells can be majorly classified into parenchymal and mesenchymal stem cells. Parenchymal stem cells generate new somatic non-connective tissues and the mesenchymal stem cells generate new somatic connective tissue cells such as fibroblasts, adipocytes, chondrocytes, osteoblasts, smooth muscle cells, skeletal muscle satellite cells and trabecular bone (Silvani et al. 2011). Given their regenerative and anti-inflammatory properties, MSCs have their role in reducing inflammation in disease models of autoimmune arthritis and diabetes, multiple sclerosis and lupus (Meregalli et al. 2011). Isolated from the bone marrow, adipocytes (Razmkhah et al. 2011) can be cultured *in vitro* because of their property of adherence to plastic. All these cells can be characterized by markers depending on their location. E.g., Muscle stem cells (MuSCs), small intestinal stem cells, mesenchymal stem cells (MSCs).

Foetal Stem Cells

Foetal stem cells are the primitive cells found in the foetal tissues or organs that were obtained after abortions. They are tissue-specific stem cells and give rise to that particular organ in the foetus. They are obtained from blood, bone marrow; liver and kidney. Foetal blood serves as a rich pool of hematopoietic cell population that has higher proliferative potential than those of bone marrow and cord blood cells.

Umbilical Cord Stem Cells

Umbilical cord stem cells are cord blood stem cells collected from the blood in the umbilical cord, soon after the baby's birth. They are tissue-specific and are multipotent stem cells. They are valuable source of stem cells, whose banking has become prevalent at a large scale. The blood collected from the umbilical cord is a good source of blood-forming stem cells and are currently banked and stored. These cells at a later point of time are used in treating blood related disorders such as leukaemia and to restore the circulatory system. Cord blood stem cells are not only used for autologous treatments but also for allogenic treatments as well. Cells collected from the cord during the birth of a new-born can be later used to treat his/her siblings.

They provide an opportunity for research when found to be non-useful for clinical applications. Despite their advantages, the cord blood cells have the drawbacks of low cell count that could be insufficient for adult patients and the risks of transmitting undiagnosed diseases.

Induced Pluripotent Stem Cells

Induced pluripotent stem cells (iPSCs) are genetically reprogrammed stem cells generated by inducing the differentiated cells to express stem cell specific genes that are expressed in embryonic stem cells. They have similar properties like the ESC, in transforming to cells of all the tissues and organs, but they are not identical to ESC. iPS cells from mouse skin fibroblasts were obtained by inducing four genes, OCT4, KLF4, SOX2, and cMYC (Takahashi and Yamanaka 2006). Human iPS (hiPS) cells were generated using LIN28 as the fourth gene omitting cMYC (Takahashi et al. 2007, Yu et al. 2007). Having these cells generated from almost all the somatic parts of humans, they now serve as models for human diseases *in vitro*. Induced pluripotent stem cells are genetically reprogrammed in the lab by converting adult cells such as skin cells into cells that behave like embryonic stem cells. They also help in drug testing and development, where they can generate disease-specific cell lines. In the near future, iPSCs can be an ultimate source of cells or tissues for medical treatments.

iPSCs had overcome the drawbacks of ethical issues surrounding the ESCs as well as the easiness of culture unlike the adult stem cells playing a huge role in stem cell therapy. Currently, 33 studies that involve iPSCs were being actively conducted, as per the clinical trails in the National Institute of Health (NIH) database. Yet, iPSCs were proven to induce an immune response in few cases (Zhao et al. 2011).

A Note on "Turn Back"

The theory of cellular plasticity says that a somatic cell can be converted back to stem cell, just the way a pluripotent cell differentiates into a somatic cell. This is often referred to as "Turn Back" mechanism.

Applications of Stem Cells

Regenerative medicine has made an enormous entry into various fields of human biology and its medicine. Stem cells, from their ability of proliferation and differentiation, have a wide variety of clinical implications in Ocular, Neural disorders, Cancer, Diabetes, and some autoimmune diseases (Figure 4) as well (Kumar 2011).

Currently, stem cells are a major solution for various unanswered questions in medicine and biology. They have the potential of revolutionizing, improving the quality and increasing the span of human lives. Till date, around 35019 studies were registered as "Cell Therapy" at clinical trails of the NIH database, of which around 8553 are active.

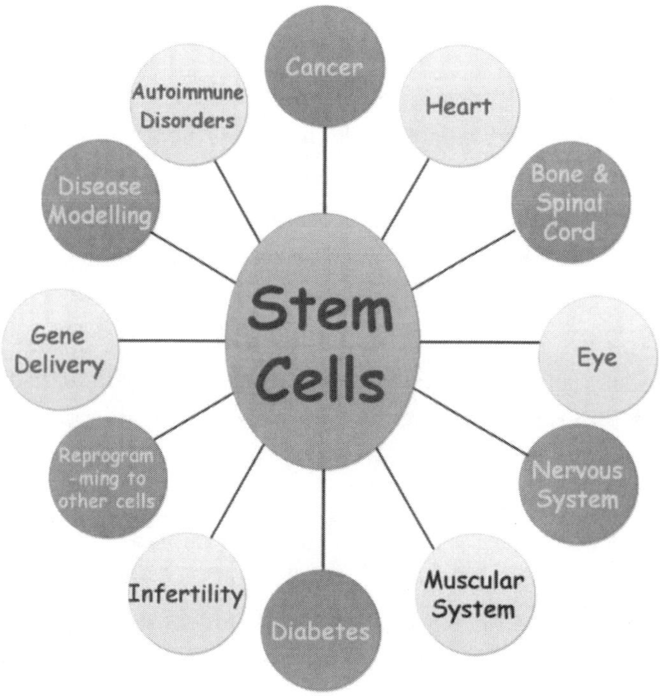

Figure 4. Image illustrating applications of stem cells in various fields of medicine.

Hematopoietic stem cells, collected from blood and bone marrow are used in treating cancers like leukaemia, lymphomas, multiple myeloma, etc. In most cases of the stem cell transplantation for leukaemia, myeloma and certain bone disorders like myelodysplasia, the transplantation is allogeneic. Source of these stem cells, would be histocompatible relatives of the patients, placental or cord blood cells. In cases of tumour, the T-cells of the donor graft may help in defending the host malignant cells thereby eliminating them which is referred to as graft versus tumour effect (GvT).

Allogeneic stem cell transplant (bone marrow-derived) is also done in the cases of sickle cell disease which would enhance the production of normal haemoglobin at a rapid rate (Fruchtman 2003).

MSCs help in cartilage regeneration (Pei, M., He, F., and Wei, L. 2011) and several other bone disorders. It is also reported that bone marrow-derived MSC cells are used to treat lung fibrosis and spinal cord injury (Meregalli et al. 2011). Autologous stem cell transplant in cases of coronary artery related diseases have shown promising results by reducing myocardial infarction (Duran and George 2011) and improving myocardial function.

Cord blood stem cells find their application in various malignancies, blood related and metabolic disorders. They are used in treating autoimmune diabetes type 1 in children. Cord blood stem cells sometimes do replace the bone-marrow transplants (Duran and George 2011) in case of the occurrence of mismatch in HLA.

STEM CELLS IN THE EYE

Prevalence of Blindness

Blindness extends to be one of the global challenges in the world. World Health Organization (WHO) estimates about 285 million people who are visually impaired of which 39 million are totally blind around the world (WHO August 2014). Similarly, in Indian scenario, approximately 8.07 million people are blind contributing to 20% of world blindness (Mariotti 2010). Out of this, National Programme for Control of Blindness approximates 6.8 million people having visual outcome less than 6/60 in at least one eye (NPCB December 2015) due to corneal disorders. Opacity of the cornea is the commonest form of visual impairment after cataract and glaucoma (Robaei and Watson 2014) and usually scarring or vascularization of cornea would lead to an impaired vision for rest of the individual's life (Whitcher et al. 2001).

Corneal Epithelium and Its Regeneration

Cornea is structurally complex with a transparent surface contributing to 2/3rd of the refractive eye power (Katzir and Howland 2003). Its transparency is a necessary requirement for optimal vision for focusing the incoming light. Being the outermost surface of the eye, corneal epithelium is susceptible to dryness and abrasion injuries and undergoes continuous stress due to dust, pollution, photo-damage, and infection. However, just as the surface of the skin does, cornea undergoes continuous regeneration and renewal through cell proliferation and migration (Thoft and Friend 1983) to preserve its integrity, thus keeping the ocular outer surface stable and functional. Source for continuous replenishment of cells in the cornea is identified from a specified region known as limbus (Cotsarelis et al. v1989) which surrounds the periphery of cornea demarcating cornea from conjunctiva (Buskirk 1989). Limbus microscopically is identified by distinct pigmented palisades of Vogt where stem cells are known to reside and this region acts as a barrier for conjunctival cells invading the cornea (Dua and Azuara-Blanco 2000). Limbal cells are reported to have slow cycling stem cells (Cotsarelis et al. 1989) and are known to reside in the deep limbal epithelial crypts which invaginate into the limbal stroma (Dua and Shanmuganathan 2005). These stem cells divide symmetrically to produce two identical daughter cells or asymmetrically to limbal stem cell (LSC) and a transiently amplifying cell (TAC) (German and Pollock 2006). TACs further give rise to terminally differentiated cells of which the corneal epithelial apical layers are made of (Sun and Lavker 2004). Injuries to the ocular surface such as alkali or thermal burns and ocular surface disorders like Steven Johnson's syndrome or Aniridia could cause damage to limbus leading to a condition known as limbal stem cell deficiency (LSCD) (Rossignol and Neumann 1998, Sangwan and Jain 2014).

Therefore, epithelial regeneration capacity of the cornea is either regressed or lost in cases of LSCD which still poses a challenge for visual rehabilitation. On the basis of the number of eyes involved, it could be unilateral or bilateral and severity could be partial or total LSCD.

Stem Cell Therapies in Cornea

It has been almost eight decades cell therapy has begun in European countries (Ebert and Kagen 1946) which went in hand to hand with tissue transplantation therapies (Gibson and Medawar 1943). However, until two decades earlier cell therapy has been not well accepted and faced several criticisms (Ledingham 1989, Last 1990). Eventually, cell along with the immunosuppressive therapies have gained popularity in treating cancers (Thompson and Shulman 1992) and successful bone marrow transplantations (Bach, Albertini et al. 1968). In recent years, cell therapy has been widely accepted as a promising approach in the treatment for various disorders (Starzl 2000). Nevertheless, immune related issues relating to the graft versus host diseases (GVHD) (Billingham 1966, Cerottini and Brunner 1974, Ferrara and Levine 2009) and recurring melanomas (Schwartz and Beldotti 1965, Giralt and Garderet 2015) had been reported since the beginning of the transplantation techniques. On the flip side, corneal transplantation is considered to be the most successful form of solid tissue transplantation (Amouzegar and Chauhan 2016) which was dated back to October 25th, 1928 where Ben Witt Key reported a case study bearing stable graft of transplanted cornea with nineteen months of follow up (Thomas 1933). Several refined measures have been reported in recent years for penetrating keratoplasty (PK) (Dandona and Ragu 1997; Thompson and Price 2003; Basu and Reddy 2012; Elbaz and Mireskandari 2016) which treated corneal disorders and providing successful outcomes in restoring vision. On the contrary, the corneal disorder relating to limbus deficiency is a two-step treatment procedure. This includes management of LSCD which would inhibit conjunctivalization of the cornea and further, a penetrating keratoplasty to correct the scarred cornea if any (Basu and Mohamed 2011, Sangwan and Jain 2014).

Various methods have been described in the literature for treating LSCD which has a common goal of reconstructing the corneal epithelium thereby restoring the ocular vision. The ocular surface restoration procedures have been classified based on the type of donor tissue, substrate material and either limbal or conjunctival graft used (Holland and Schwartz 1996, Holland and Schwartz 2004). The major clinical procedures include Conjunctival limbal autograft (CLAu) which was first reported by Thoft RA in 1977 (Thoft 1977) and was further improvised by Kenyon and Tseng in 1989. This involves excision of 5-6 mm of conjunctiva followed by superficial limbus along with 0.5mm into the cornea from the contralateral donor eye. The whole graft is placed on the recipient's eye after removal of the pannus (Kenyon and Tseng 1989). CLAu was reported to be successful in 94% (n = 39) in case of larger grafts (i.e., >120°) and 60% (n = 22) in smaller grafts used (Liang and Sheha 2009). Living-related conjunctival-limbal allograft (lr-CLAL) is yet another modification of CLAu which involves any living relative donor of the host whose Human Leukocyte Antigen (HLA) typing is considered to be closer (Kwitko, Marinho et al. 1995). In order to prevent jeopardizing the donor's eye, Keratolimbal allograft (KLAL) technique was used which involves 360^0 corneoscleral rims from a cadaveric donor which would be transplanted on to the host ocular surface (Thoft 1984). More recent and less invasive techniques such as *ex vivo* expansion of limbal cells or the oral mucosal cells have been appealing and more successful in reconstructing the ocular surface. Pelligrini et al. (1997) reported the first successful *ex vivo* expanded limbal stem cells transplantation in a patient with unilateral LSCD which is known cultivated limbal epithelial transplantation (CLET). Moreover, allogenic CLET is less preferred and is reported to have lower success rates due to the immunological rejection episodes compared to the autologous CLET (Pauklin et al. 2010, Sangwan et al. 2011, Qi et al. 2013). But in cases of bilateral LSCD, instead of the allogeneic transplant, an autologous cultivated oral mucosal epithelium transplant (COMET) could be performed which was reported to be successful but with persistent epithelial defects (PED) in few cases (Nakamura and Takeda 2011, Burillon and Huot 2012, Sotozono and Inatomi 2013).

From the above studies, limbal epithelial cell transplantation serves as an appropriate redressal for severe ocular surface disorders like burns, chemical injuries (Sangwan et al. 2011, Basu et al. 2016, Vazirani et al. 2016) or diseases like Steven-Johnson syndrome (Omoto and Shimmura 2010). In 2011, Sangwan et al. came up with a novel technique called Simple limbal epithelial transplantation (SLET) (Sangwan and Basu 2012) (Figure 5) which is a simplified form of Cultivated limbal epithelial transplantation (CLET) that was first conducted at LV Prasad Eye Institute, Hyderabad. This technique reduces cost as well as visits for the unilateral LSCD patients. Additionally, it completely eliminates the need for laboratory based processing as in case of CLET or the use of other cells to generate limbal epithelium *in vitro* thereby making it possible for it to be done by any well trained surgeon anywhere. While only a small portion of the limbal tissue is removed from the healthy eye, the stem cell expansion takes place not in the lab but in the damaged eye itself on the amniotic membrane. This has been already ensured that the healthy eye is never damaged. The procedure has less risk of contamination (as the expansion does not take place in a lab) and is very economical compared to CLET. Simple limbal epithelial transplantation (SLET) or autologous cultivated limbal epithelial transplantation (CLET) can be performed in cases of unilateral LSCD which has a success rate of 71%. Besides, our group as well other groups in the world are also working on the possibilities for developing alternative, structurally simple, synthetic biodegradable scaffold based to replace AM (Deshpande and Ramachandran 2013).

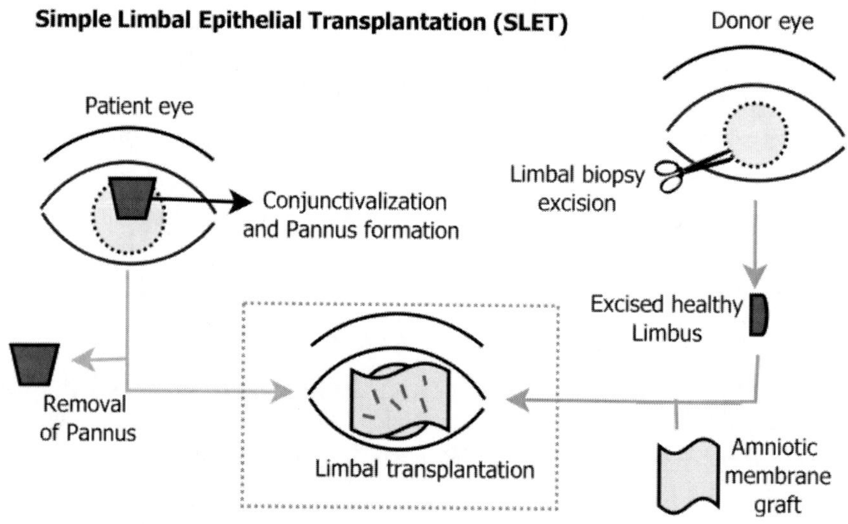

Figure 5. Simple limbal epithelial transplantation (SLET): Illustrative image depicting the limbal tissue transplantation therapy in treating limbal stem cell deficiency (LSCD), a corneal disease.

SLET for unilateral LSCD treatment is an attractive option as opposed to laboratory cultured limbal epithelium cells and it has been reported in a larger group where the results were promising (Singh and Shukla 2015, Basu and Sureka 2016). Even though clinical outcomes are outstanding in these cases, biological status post-transplantation is unknown. Hence, there is a need for studying limbal explant changes, pannus formation which only occurs in the living eye and the interaction between the recipient damaged limbal niche, corneal epithelium and the stromal cells. This can only be done in the intact animal model. Moreover, the effect of various carrier materials (amniotic membrane, polylactide-co-glycolide (PLGA) or fibrin glue) for limbal explant transplantation in unilateral LSCD generated rabbit model and the rate of healing of the eye after SLET followed by corneal transplant i.e., Penetrating keratoplasty (PK) can be assessed. These wound healing events in the cornea are mediated by cytokine–mediated interactions between the corneal/limbal epithelium and the underlying limbal niche and bone marrow-derived cells (Spandau and Toksoy 2003). There are no models currently available to study these complex interactions *in vitro* which would otherwise enhance the standards of transplantation in SLET. Yet, other groups have investigated the rabbit models for reconstructing the corneal surface in cases of limbal stem cell deficiency by culturing limbal cells of fetal and adult human on the amniotic membrane (Du and Chen 2003). Similarly, in other study using 16 rabbits as LSCD model, after transplantation with autologous ex-vivo expanded limbal epithelium on amniotic membrane (AM), restoration of a clear and transparent cornea was associated with a normal corneal epithelium and completes wound remodeling. In contrast, wound healing remains active and incomplete in conjunctivalised corneas, which remain opaque with myofibroblasts (Espana and Grueterich 2003).

Stem Cell Therapies in Retina

Degenerative diseases of the retina such as age-related macular degeneration (AMD), Stargardt's disease or retinitis pigmentosa (RP) possess challenging treatment methods since they involve the loss of cells in the complex retina structure. Retina involves neural networks which are usually the part of the central nervous system that does not tend to regenerate (Ramon and Cajal 1928) or is self-sufficient for its repair. Earlier experiments of retinal transplantation have begun in the rats at the early nineties which provided evidence of retinal tissues can be transplanted which could provide minimal responses in the eye (Klassen and Lund 1987). Later after a decade, human trials on retinal transplantation were conducted by Algvere et al. (1997) using human fetal retinal pigment epithelium (RPE) in patients suffering from AMD. The results had shown a small subset of transplanted graft survival with no improvement in vision.

Embryonic stem cells (ESCs) or the induced Pluripotent stem cells (iPSC) are the most commonly used cells in retinal transplantation. This involves differentiating them into their respective lineage i.e., the retinal cell types and transplanting the same in the sub-retinal space of the affected eye (Figure 6). Post transplantation, the donor cells may interact with the host cell types to establish connections thereby improving the visual response. Recent studies demonstrate the safety of the retinal graft post-transplantation without much of immunological rejections in animal models which led in encouraging human clinical trials in several parts of the world. The first transplant using autologous iPSC derived RPE cells was performed by RIKEN Centre for Developmental Biology, Japan in a 70-year-old female patient suffering from AMD (Cyranoski 2014). One year follow-up of the patient showed no significant improvement in vision but, the transplantation is deemed to be safe without rejection. This would be a milestone in the progress of developing effective regenerative treatments for retinal dystrophies.

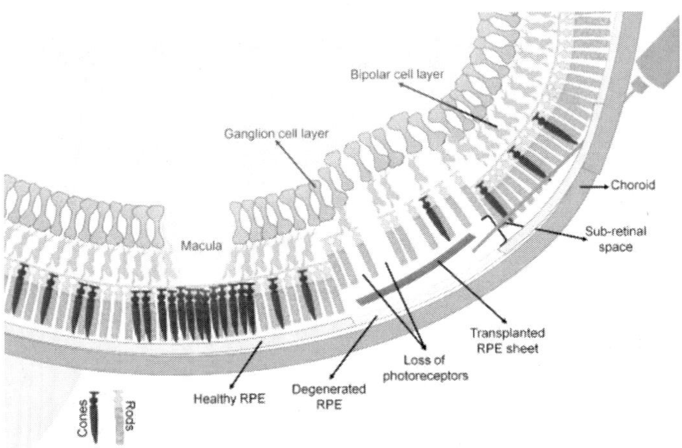

Figure 6. Retinal pigment epithelium (RPE) transplantation: Image representing the RPE sheet transplantation in sub-retinal space of the diseased retina.

ETHICS

"*Morality is largely constituted by categorical imperatives: non-negotiable rules of behavior to which every human being must conform.*"-Immanuel Kant

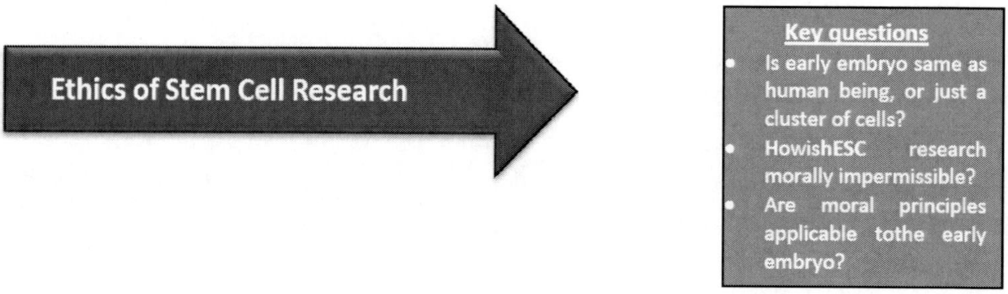

Human Embryonic Stem Cells (hESC) are very important and they play a vital role in the growth and development of the human body. They are capable of division and self-renewal (unspecialized cells) and are also capable of differentiating into specialized cell types for specific functions of the human body. These properties of hESC offer a ground to understand the basic mechanisms of the human body development on which extensive methods can be developed for the treatment of various diseases. Despite the tremendous promises that the hESC offers, it also raises sharp ethical and political controversies. The success in the conversion of adult human skin cells to have the properties of hESC by activating four genes in the adult cells (Takahashi et al. 2007; Yu et al. 2007) have provided us with "induced pluripotent stem cells" which could ultimately terminate the need of hESC. But as the potential of these iPSCs and the success of its transplantation into the human body being unknown, hESC research will continue at least in the near-term and thereby force us to choose between two moral principles:

- The principle of preventing or reducing suffering.
- The principle of respecting the value of human life.

Hence, in order to deal with this issue we will look into this in detail as follows:

I. The moral status and ethics of destroying human embryos.
  1. Commencement of the existence of human being.
  2. Do human embryos deserve moral status?
  3. Ethics of destroying human embryos.

II. Ethics and guidelines for using embryonic stem cells for research.
  1. Guidelines for the donor.
  2. Guidelines for derivation and its use.

III. What are the ethics of creating embryos for research?

I. The Moral Status and Ethics of Destroying Human Embryos

1. Commencement of the Existence of Human Being

When it is that human beings begin to exist? Can the early stage human embryo be considered a human being? These are the questions which makes it more difficult. Dianne N. Irving in her paper "*When do human beings begin? Scientific myths and scientific facts*" (Irving 1999) quotes that "The change from a simple part of one human being (i.e., a sperm) and a simple part of another human being (i.e., an egg) which simply have human life, to a new, genetically unique, newly existing, individual, whole living human being (a single-cell embryonic human zygote)" claims that a zygote is a whole new living human being and it is the beginning of a new human being. There is, however, another basis on which this statement was rejected. One such view which contradicts the above statement says that "The cells that make up the early embryo are a bundle of homogeneous cells that exist in the same membrane, but do not form a human organism because the cells do not function in a coordinated way to regulate and keep a single life." (Smith and Brogaard 2003; McMahon 2002).

According to The Warnock Committee, ethically significant biological changes or the primitive streak begins to appear after 14 days, i.e., the initiation of the formation of three germ layers (ectoderm, endoderm and mesoderm) begins after 14 days, which also include the cells that will become the central nervous system, which is responsible for consciousness and sensory feelings. After this stage it is indeed an individual which cannot be terminated.

2. Do Human Embryos Deserve Moral Status?

The moral status of the human embryo is one of the key pressure points about the ethics of the human embryonic stem cell research. Some accept that the human embryo is a human being but argue that the human embryo does not have the moral status required for the right to life (Siegel 2008). While there are others who claim that the human embryo should instead be given the benefit of any right regardless of its status, and then there are others who claim that embryo's lack the requisite for the right to life but have an intrinsic value that demands respect and some moral constraints on their use. An intermediate place stands for this, one with which many would agree the embryo deserves respect as a human life form, but not the same level of respect deserved to a human.

3. Ethics of Destroying Human Embryos

Stem cells derived from doomed embryos can be used for research purposes rather than being simply discarded, as it is claimed that, "It is morally permissible to kill an individual to help others who is about to be killed by someone else, where killing the individual will help others" (Curzer 2004) or as some others claim, the derivation of hESC from embryos which were to be discarded does not cause its death, rather, the decision to discard the embryo causes it.

II. Ethics and Guidelines For Using Embryonic Stem Cells for Research

1. Guidelines for a Donor

It is not ethical to collect embryonic stem cells without donor's consent and without providing complete information to the donor. The details on purpose of the derivation of the embryo have to be addressed with complete transparency to the donor. A set of guidelines needs to be followed before the derivation of embryonic stem cells.

The table below describes the details that need to be provided to the donor before donation of the embryo. A proper informed consent must be taken after the agreement of the donor.

2. Guidelines for the Derivation of the Embryonic Stem Cells

The debate about whether the use of embryonic stem cells be treated differently from its derivation is significant. Even though many individuals wouldn't want to use embryonic cells because of its source, but the process and its derivation and uses are different (NIH Human Embryo research panel 1994, National Bioethics Advisory Committee 1999). It can never be claimed that derivation of hESC are impermissible in all the cases, especially when it is a benefit from moral wrongs of others (Robertson 1988). For example, the derivation of hESC from a murdered victim or an accidental death due to drunken driving, etc. is permissible. Secondly, the derivation of embryonic stem cells is not always from a donor. It is also from a large pool of existing cell lines too. Hence, the primary investigator working with the hESC cell lines need not be worried about its source of derivation and bears no moral responsibility for the destruction of embryos.

Table 2.The table describes the information that needs to be provided to the embryo donor

• The risk involved and lack of the absolute guarantee of confidentiality.	• All women/couple should be offered the opportunity to cryopreserve and must be informed that all embryos may increase the chance of undergoing *in vitro* fertilization for reproductive purpose.
• No personal benefits are involved for the donors.	• The consent can be withdrawn any time before thawing for research, or before stem line is created in the case of stem cell research.
• The decision to discard/donation to another couple/to research of the cryopreserved embryo should not be made until the women/couple decide that they no longer need it.	• Medical care is not affected by the decision of embryo donation and patient safety is a priority.

By understanding the guidelines about derivation and usage of embryonic stem cells, it can be concluded that a set of rules must be implemented in the human embryonic stem-cell research. Rules such as, the use of embryos should be restricted to specific areas of research (where it is an absolute necessity), informed consent must be acquired from the donor, regulatory system for the guidelines for its use must be followed, systemic control on the usage of the human embryonic stem cells in research must be implemented, etc.

III. What are the Ethics of Creating Embryos for Research?

If the creation of embryos is for a non-reproductive purpose, then the method used for it maybe cloning or *in vitro* fertilization, it stands morally questionable. This is because each embryo is viewed originally as a potential child, whereas embryos used for research it is viewed as mere tools (Annas et al. 1996, President's Council on Bioethics 2002) and embryos created for research are produced with the intent of its destruction (Fitz Patrick 2003). This argument can be resolved as most of the hESC are derived from the leftover embryos, which were created for infertility treatment which if the patient agrees to donate for research purposes (Figure 7). When the donation is from the side of the patient in case of spare embryos, it cannot be considered immoral even when it is donated for research purposes.

Figure 7. Cartoon depicting the usage of leftover blastocysts in case of infertility cases.

These leftover embryos serve as a beneficial tool for research but, limit the range of research as it provides incomplete details on genetic diseases and limit the chances of the genetic matching. Hence, to overcome these hurdles, induced pluripotent stem cells will serve as a great tool from a therapeutic point of view as it can be used for specific genetic mutations and can offer an exact genetic match for the recipient. As for now, the best way to address this problem is by creating public stem cell banks (Faden *et al.* 2003, Lott & Savulescu 2007). This will provide a large pool of diverse stem-cell lines which can serve as a genetic match for the recipient and can open up a larger portal for research purpose when it comes to the study of genetic mutations.

By this the primary cause of the donation of embryos is reproduction; the leftover embryos of the infertility treatment can be used for research purposes than being discarded.

ACKNOWLEDGMENTS

We acknowledge Ms. Swatilekha Hazra for her contribution in few figures (2, 7) cited in this chapter. We thank Science and Engineering Research Board, New Delhi, India (Fast Track-SB/YS/LS-263/2013) for providing a search grant to Dr. Vivek Singh.

REFERENCES

Algvere, P. V., Berglin, L., Gouras, P., Sheng, Y. & Kopp, E. D. (1997) Transplantation of RPE in age-related macular degeneration: observations in disciform lesions and dry RPE atrophy. *Graefes Arch Clin Exp Ophthalmol 235*: 149-158.

Amouzegar, A., Chauhan, S. K. & Dana, R. (2016) Alloimmunity and tolerance in corneal transplantation. *J Immunol 196:* 3983-3991.

Annas, G., Caplan, A. & Elias, S. (1996) "The Politics of Human-Embryo Research—Avoiding Ethical Gridlock," *New England J Med 334:* 1329–32.

Arun Kumar, R., Sathish Kumar, D. & Nishanth, T. (2011) A Review on Stem cell Research & Their Role in Various Diseases. *J Stem Cell Res Ther 1:* 1-5.

Bach, F., Albertini, R., Joo, P., Anderson, J. & Bortin, M. (1968) Bone-marrow transplantation in a patient with the Wiskott-Aldrich syndrome. *The Lancet 292:* 1364-1366.

Barfoot, J., Kemp, E., Doherty, K., Blackburn, C., Sengoku, S., van Servellen, A., Gavai, A. & Karlsson, A (2013) Stem cell research: trends and perspectives on the evolving international landscape. *World Stem Cell Summit.* San Diego: SciVal.

Basu, S., Mohamed, A., Chaurasia, S., Sejpal, K., Vemuganti, G. K. & Sangwan, V. S. (2011) Clinical outcomes of penetrating keratoplasty after autologous cultivated limbal epithelial transplantation for ocular surface burns. *Am J Ophthalmol 152:* 917-924.

Basu, S., Reddy, J. C., Vaddavalli, P. K., Vemuganti, G. K. & Sangwan, V. S. (2012) Long-term outcomes of penetrating keratoplasty for keratoconus with resolved corneal hydrops. *Cornea 31:* 615-620.

Basu, S., Sureka, S. P., Shanbhag, S. S., Kethiri, A. R., Singh, V. & Sangwan, V. S. (2016) Simple limbal epithelial transplantation: long-term clinical outcomes in 125 cases of unilateral chronic ocular surface burns. *Ophthalmol 123:* 1000-1010.

Bernard, Lo. & Lindsay, Parham. (2009) Ethical issue in Stem Cell Research. *The Endocrine Reviews.*

Billingham, R. E. (1966) The biology of graft-versus-host reactions. *Harvey Lect 62:* 21-78.,

Bongso, A. & EngHin, L. E. (2005) Stem Cells: Their Definition, Classification and Sources. *Stem Cells: From Bench to Bedside.*

Bongso, A. & Richards, M. (2004) History and perspective of stem cell research. Best Pract. Res. Clin. Obstet. *Gynaecol 18:* 827-842.

Burillon, C., Huot, L., Justin, V., Nataf, S., Chapuis, F., Decullier, E. & Damour, O. (2012) Cultured autologous oral mucosal epithelial cell sheet (CAOMECS) transplantation for the treatment of corneal limbal epithelial stem cell deficiency. *Invest Ophthalmol Vis Sci 53:* 1325-1331.

Canelles, M. (2011) The Universality of Asymmetric Division. *J Stem Cell Res Ther, 1,* e103.

Cerottini, J. C. & Brunner, K. T. (1974) Cell-mediated cytotoxicity, allograft rejection, and tumour immunity. *Adv Immunol 18:* 67-132.

Cotsarelis, G., Cheng, S. Z., Dong, G., Sun, T. T. & Lavker, R. M. (1989) Existence of slow-cycling limbal epithelial basal cells that can be preferentially stimulated to proliferate: implications on epithelial stem cells. *Cell 57:* 201-209.

Cyranoski, D. (2014) Japanese woman is first recipient of next-generation stem cells. *Nature*

Dandona, L., Ragu, K., Janarthanan, M., Naduvilath, T. J., Shenoy, R. & Rao, G. N. (1997) Indications for penetrating keratoplasty in India. *Indian J Ophthalmol 45:* 163.

Deshpande, P., Ramachandran, C., Sangwan, V. S. & MacNeil, S. (2013) Cultivation of limbal epithelial cells on electrospun poly (lactide-co-glycolide) scaffolds for delivery to the cornea. *Corneal Regenerative Medicine: Methods and Protocols* 179-185.

Dianne, N Irving. (1999) "When do human beings begin? and Scientific myths and Scientific Facts". *International Journal of Sociology and Social Policy 19*: 22-36.

Du, Y., Chen, J., Funderburgh, J. L., Zhu, X. & Li, L. (2003) Functional reconstruction of rabbit corneal epithelium by human limbal cells cultured on amniotic membrane. *Mol Vis 9:* 635.

Dua, H. S. & Azuara-Blanco, A. (2000) Limbal stem cells of the corneal epithelium. *Surv Ophthalmol 44:* 415-425.

Dua, H. S., Shanmuganathan, V. A., Powell-Richards, A. O., Tighe, P. J. & Joseph, A. (2005) Limbal epithelial crypts: a novel anatomical structure and a putative limbal stem cell niche. *Br J Ophthalmol 89:* 529-532.

Duran, J. M. & George, J. C. (2011) A Review of the Basis of Autologous Stem Cel l Therapy for Coronary Artery Disease. *J Clin Exp Cardiol 2:* 130.

Ebert, M. & Kagen, M. S. (1946) Sickle cell ulcers treated with red blood cells. *Arch Derm Syphilol: 54*: 627.

Elbaz, U., Ali, A. & Mireskandari, K. (2016) Long-Term Corneal Endothelial Cell Counts After Penetrating Keratoplasty in Infants. *Cornea 35:* 784-788.

Espana, E. M., Ti, S. E., Grueterich, M., Touhami, A. & Tseng, S. C. G. (2003) Corneal stromal changes following reconstruction by ex vivo expanded limbal epithelial cells in rabbits with total limbal stem cell deficiency. *Br J ophthalmol 87:* 1509-1514.

Faden, R. R., *et al.*, (2003) Public Stem Cell Banks: Considerations of Justice in Stem Cell Therapy, *Hastings Center Report 33:* 13–27

Ferrara, J. L., Levine, J. E., Reddy, P. & Holler, E. (2009) Graft-versus-host disease. *The Lancet 373:* 1550-1561.

FitzPatrick, W. (2003) "Surplus Embryos, Nonreproductive Cloning, and the Intend/Foresee Distinction," *Hastings Center Report 33:* 29-36.

Fruchtman, S. (2003) Stem cell transplantation. *Mt Sinai J Med 70:* 166-170.

German, M. J., Pollock, H. M., Zhao, B., Tobin, M. J., Hammiche, A., Bentley, A., Cooper, L. J., Martin, F. L. & Fullwood, N. J. (2006) Characterization of putative stem cell populations in the cornea using synchrotron infrared microspectroscopy. *Invest Ophthalmol Vis Sci, 47:* 2417-2421.

Gibson, T. & Medawar, P. B. (1943) The fate of skin homografts in man. *J Anat 77:* 299.

Giralt, S., Garderet, L., Durie, B., Cook, G., Gahrton, G., Bruno, B., Hari, P., Lokhorst, H., McCarthy, P., Krishnan, A. & Sonneveld, P. (2015) American society of blood and marrow transplantation, European society of blood and marrow transplantation, blood and marrow transplant clinical trials network, and international myeloma working group consensus conference on salvage hematopoietic cell transplantation in patients with relapsed multiple myeloma. *Biol Blood Marrow Transplant 21*: 2039-2051.

Harold, T., et al. (1999) "Ethical issues in human stem cell research; Volume 1" *The Embryo Project Encyclopedia.*

Holland, E. J. & Schwartz, G. S. (1996) The evolution of epithelial transplantation for severe ocular surface disease and a proposed classification system. *Cornea 15:* 549-556.

Holland, E. J. & Schwartz, G. S. (2004) The Paton lecture: Ocular surface transplantation: 10 years' experience. *Cornea 23:* 425-431.

Holm, S. (2005) Embryonic stem cell research and the moral status of human embryos. *Reprod Biomed Online 10:* 63-67.

Hug, K. & Hermerén, G. (2011) Which patient groups should be asked to participate in first-in-human trials of stem-cell-based therapies? *J Clin Ethics 23:* 256-271.

Irving, D. N. (1999) When do human beings begin? Scientific myths and scientific facts. *Int J Sociol Soc Policy 19:* 22-46.

Jaenisch, R. & Mintz, B. (1974) Simian virus 40 DNA sequences in DNA of healthy adult mice derived from preimplantation blastocysts injected with viral DNA. *Proc Natl Acad Sci 71:* 1250–1254.

Jeff McMahan (2007) "Killing embryos for stem cell research." *Metaphilosophy 38*(2-3): 170-189.

Julian, Savulescu., et al. (2015) "The moral imperative to continue gene editing research on human embryos" *Protein Cell 6*(7): 476-479.

Kamani, N., Spellman, S., Hurley, C. K., Barker, J. N., Smith, F. O., Oudshoorn, M., Bray, R., Smith, A., Williams, T. M., Logan, B. & Eapen, M. (2008) *State-of-the-art review: HLA matching and outcome of unrelated donor umbilical cord blood transplants*. 1-6.

Katzir, G. & Howland, H. C. (2003) Corneal power and underwater accommodation in great cormorants (Phalacrocoraxcarbosinensis). *J Exp Biol 206:* 833-841.

Keivyon, K. R. & Tseng, S. C. (1989). Limbalautograft transplantation for ocular surface disorders. *Ophthalmol 96:* 709-723.

Klassen, H. & Lund, R. D. (1987) Retinal transplants can drive a pupillary reflex in host rat brains. *Proc Natl Acad Sci 84:* 6958-6960.

Kwitko, S., Marinho, D., Barcaro, S., Bocaccio, F., Rymer, S., Fernandes, S. & Neumann, J. (1995) Allograft conjunctival transplantation for bilateral ocular surface disorders. *Ophthalmol 102:* 1020-1025.

Last, P. M. (1990) Cell therapy: a cruel and dangerous deception. A drama in three acts. *J Paediatr Child Health 26:* 197-199.

Leda, M., Fu, J., Delgado-Olguin, P., Vedantham, V., Hayashi, Y., Bruneau, B. & Srivastava, D. (2010) Direct reprogramming of fibroblasts into functional cardiomyocytes by defined factors. *Cell 142:* 375-386.

Ledingham, I. M. (1989) Cell therapy in shock: a reasonable prospect or a lost cause? *Resuscitation 18:* S85-S99.

Liang, L., Sheha, H., Li, J. & Tseng, S. C. G. (2009) Limbal stem cell transplantation: new progress and challenges. *Eye 23:* 1946-1953.

Lo, B., & Parham, L. (2009) Ethical issues in Stem Cell Research. *Endocr Rev 30* (3): 204-13.

Lockwood, M. (2005) The moral status of the human embryo: implications for IVF. *Reprod Biomed Online 10:* 17-20.

Mariotti, S. P. W. H. O. (2010) *Global data on visual impairments.*

Martin, G. R. (1981) Isolation of a pluripotent cell line from early mouse embryos cultured in medium conditioned by teratocarcinoma stem cells. *Proc Natl Acad Sci 78:* 7634-7638.

McMahan, J. (2007) Killing embryos for stem cell research. *Metaphilosophy 38:* 170-189.

Meregalli, M., Farini, A. & Torrente, Y. (2011) Mesenchymal stem cells as muscle reservoir. *J Stem Cell Res Ther 1:* 1-9.

Michael, Lockwood (2005) "The moral status of the human embryo: implications for IVF" *Reproductive BioMedicine Online.*

Morrison, S. J., Shah, N. M. & Anderson, D. J. (1997) Regulatory mechanisms in stem cell biology. *Cell 88:* 287-298.

Nakamura, T., Takeda, K., Inatomi, T., Sotozono, C. & Kinoshita, S. (2011) Long-term results of autologous cultivated oral mucosal epithelial transplantation in the scar phase of severe ocular surface disorders. *Br J Ophthalmol 95:* 942-946.

Nelson, E., et al. (2008) "Informed consent to donate embryos for research purposes" *J Obstet Gynaecol Can. 30*(9): 824-36.

National Programme for Control of Blindness (December 2015) *Controlling Corneal Blindness in India: Facts and Figures, Dreams and Realities.*

Omoto, M., Shimmura, S., Hatou, S., Ichihashi, Y., Kawakita, T. & Tsubota, K. (2010) Simultaneous deep anterior lamellar keratoplasty and limbal allograft in bilateral limbal stem cell deficiency. *Jpn J Ophthalmol 54* (6): 537-543.

Ozone, C., Suga, H., Eiraku, M., Kadoshima, T., Yonemura, S., Takata, N., Yutaka, O., Takashi, T. & Sasai, Y. (2016) Functional anterior pituitary generated in self-organizing culture of human embryonic stem cells. *Nat Commun 7.*

Pauklin, M., Fuchsluger, T. A., Westekemper, H., Steuhl, K. P. & Meller, D. (2010) Midterm results of cultivated autologous and allogeneic limbal epithelial transplantation in limbal stem cell deficiency *In: Research Projects in Dry Eye Syndrome*, (Vol. *45*). Karger Publishers, Basel, Switzerland. pp. 57-70.

Pei, M., He, F. & Wei, L. (2011) Three-dimensional cell expansion substrate for cartilage tissue engineering and regeneration: a comparison in decellularized matrix deposited by synovium-derived stem cells and chondrocytes. *J Tissue Sci Eng 2:* 104.

Pellegrini, G., Traverso, C. E., Franzi, A. T., Zingirian, M., Cancedda, R. & De Luca, M. (1997) Long-term restoration of damaged corneal surfaces with autologous cultivated corneal epithelium. *The Lancet 349:* 990-993.

Pera, M. F. & Patrick, P. L. T. Extrinsic regulations of pluripotent stem cells. *Nature 465:* 713-720.

Qi, X., Xie, L., Chen, J., Zhai, H. & Zhou, Q. (2013) Characteristics of immune rejection after allogeneic cultivated limbal epithelial transplantation. *Ophthalmol 120:* 931-936.

Ramalho-Santos, M. & Willenbring, H. On the origin of the term "stem cell." *Cell Stem Cell 1*: 35-38.

Ramon, Y. & Cajal, S. (1928) Degeneration and regeneration of white matter. *Cajal's Degeneration and Regeneration of the Nervous System.* Oxford University Press, New York, USA. pp. 57-198.

Razmkhah, M., Jaberipour, M. & Ghaderi, A. (2011) Chemokines and Chemokin e Receptors Expression in the Adipose Derived Stem Cells (ASCs), Breast Tissues and in Peripheral Blood of Patients with Breast Cancer. *J Carcinog Mutagen 2:* 120.

Robaei, D. & Watson, S. (2014). Corneal blindness: a global problem. *Clin Exp. Ophthalmol. 42:* 213-214.

Robertson, J. (1988) "Fetal Tissue Transplant Research is Ethical," *IRB: A Review of Human Subjects Research, 6*(10): 5–8.

Rossignol, A. M. & Neumann, C. (1998), International environmental health: priorities from Huairou. *JPublic Health Policy 19*: 319-330.

Sangwan, V. S., Basu, S., MacNeil, S. & Balasubramanian, D. (2012) Simple limbal epithelial transplantation (SLET): a novel surgical technique for the treatment of unilateral limbal stem cell deficiency. *Brit J Ophthalmol 96*: 931-934.

Sangwan, V. S., Basu, S., Vemuganti, G. K., Sejpal, K., Subramaniam, S. V., Bandyopadhyay, S., Krishnaiah, S., Gaddipati, S., Tiwari, S. & Balasubramanian, D. (2011) Clinical outcomes of xeno-free autologous cultivated limbal epithelial transplantation: a 10-year study. *Brit J Ophthalmol 95* (11): 1525-1529.

Sangwan, V. S., Jain, R., Basu, S., Bagadi, A. B., Sureka, S., Mariappan, I. & MacNeil, S. (2014) Transforming ocular surface stem cell research into successful clinical practice. *Indian J Ophthalmol 62*: 29.

Savulescu, J., Pugh, J., Douglas, T. & Gyngell, C. (2015) The moral imperative to continue gene editing research on human embryos. *Protein Cell 6*: 476-479.

Schwartz, R. S. & Beldotti, L. (1965) Malignant lymphomas following allogenic disease: transition from an immunological to a neoplastic disorder. *Science 149*: 1511-1514.

Siegel, A. (2008). *Ethics of stem cell research.*

Silvani, G., Galli, D., Benedetti, L., Ceccarelli, G. & Crosetto, N. (2011) Peculiar characteristics of human mesenchymal stem cell clones suitable as tissue engineering models. *J Tissue Sci Eng 2*: 1000106.

Singh, V., Shukla, S., Ramachandran, C., Mishra, D. K., Katikireddy, K. R., Lal, I., Chauhan, S. K. & Sangwan, V. S. (2015) Chapter Two-Science and Art of Cell-Based Ocular Surface Regeneration. *Int Rev Cell Mol Biol 319*: 45-106.

Soren, Holm. (2005) "Embryonic stem cell research and the moral status of human embryo" *Reproductive BioMedicine Online*.

Sotozono, C., Inatomi, T., Nakamura, T., Koizumi, N., Yokoi, N., Ueta, M., Matsuyama, K., Miyakoda, K., Kaneda, H., Fukushima, M. & Kinoshita, S. (2013) Visual improvement after cultivated oral mucosal epithelial transplantation. *Ophthalmol 120* (1): 193-200.

Spandau, U. H., Toksoy, A., Verhaart, S., Gillitzer, R. & Kruse, F. E. (2003) High expression of chemokinesGro-α (CXCL-1), IL-8 (CXCL-8), and MCP-1 (CCL-2) in inflamed human corneas *in vivo*. *Arch Ophthalmol 121*: 825-831.

Starzl, T. E. (2000) History of clinical transplantation. *World J Surg 24*: 759-782.

Sun, T. T. & Lavker, R. M. (2004). Corneal epithelial stem cells: past, present, and future. *J Investig Dermatol Symp Proc 9*: 202-207.

Szabo, E., Rampalli, S., Risueno, R., Schnerch, A., Mitchell, R., Fiebig-Comyn, A., Levadoux-Martin, M. & Bhatia, M. (2010) Direct conversion of human fibroblasts to multi-lineage blood progenitors. *Nature 468*: 521-526.

Takahashi, K., Tanabe, K., Ohnuki, M., Narita, M., Ichisaka, T., Tomoda, K. & Yamanaka, S. (2007) Induction of pluripotent stem cells from adult human fibroblasts by defined factors. *Cell* 131: 861-872.

Takahashi, K. & Yamanaka, S. (2006) Induction of pluripotent stem cells from mouse embryonic and adult fibroblast cultures by defined factors. *Cell 126*: 663–676.

Thoft, R. A. (1977). Conjunctival transplantation. *Arch Ophthalmol 95*: 1425-1427.

Thoft, R. A. (1984). Keratoepithelioplasty. *Am J Ophthalmol*, *97*, 1-6.

Thoft, R. A. & Friend, J. (1983) The X, Y, Z hypothesis of corneal epithelial maintenance. *Invest Ophthalmol Vis Sci*, *24*: 1442-1443.

Thomas, J. W. (1933) Corneal Transplantation on an Opaque Cornea. *Proc R Soc Med 26*: 597-605.

Thompson, J. A., Shulman, K. L., Benyunes, M. C., Lindgren, C. G., Collins, C., Lange, P. H., Bush, W. H., Benz, L. A. & Fefer, A. (1992) Prolonged continuous intravenous infusion interleukin-2 and lymphokine-activated killer-cell therapy for metastatic renal cell carcinoma. *J Clin Oncol 10*: 960-968.

Thompson, R. W., Price, M. O., Bowers, P. J. & Price, F. W. (2003) Long-term graft survival after penetrating keratoplasty. *Ophthalmol 110*: 1396-1402.

Till, J. & McCulloch, E. A. (1961) direct measurement of the radiation sensitivity of normal mouse bone marrow cells. *Radiat Res 14*: 213-222.

Van Buskirk, E. M. (1989) The anatomy of the limbus. *Eye 3*: 101-108.

Vazirani, J., Ali, M. H., Sharma, N., Gupta, N., Mittal, V., Atallah, M., Amescua, G., Chowdhury, T., Abdala-Figuerola, A., Ramirez-Miranda, A. & Navas, A. (2016) Autologous simple limbal epithelial transplantation for unilateral limbal stem cell deficiency: multicentre results. *Br J Ophthalmol 100*: 1416-1420.

Vierbuchen, T., Ostermeier, A., Pang, Z. P., Kokubu, Y., Sudhof, T. C. & Wernig, M. Direct conversion of fibroblasts to functional neurons by defined factors. *Nature 463*: 1035-41.

Whitcher, J. P., Srinivasan, M. & Upadhyay, M. P. (2001) Corneal blindness: a global perspective. *Bull World Health Organ 79*: 214-221.

World Health Organization (August 2014) *Visual impairment and blindness*.

Yu, J., Vodyanik, M. A., Smuga-Otto, K., Antosiewicz-Bourget, J., Frane, J. L., Tian, S., Nie, J., Jonsdottir, G. A., Ruotti, V., Stewart, R. & Slukvin, I. I. (2007) Induced pluripotent stem cell lines derived from human somatic cells. *Science 318*: 1917-1920.

Zare, S., Kurd, S., Rostamzadeh, A. & Nilforoushzadeh, M. A. (2014) Types of Stem Cells in Regenerative Medicine: A Review. *J Skin Stem Cell 1*(3).

Zhao, T., Zhang, Z. N., Rong, Z. & Xu, Y. (2011) Immunogenicity of induced pluripotent stem cells. *Nature 474*: 212-215.

In: Trends in Life Science Research
Editors: R. P. Sinha and U. P. Shrivastava

ISBN: 978-1-53613-241-0

Chapter 5

TUBERCULOSIS: PATHOGENESIS, DIAGNOSIS AND TREATMENT

Anil Kumar[1,*], Vivek Kumar Kashyap[2] and Vikram Pal Gandhi[1]

[1]Centre for Life Science, School of Natural Sciences,
Central University of Jharkhand, Brambe, Ranchi, Jharkhand, India
[2]Department of Pharmaceutical Sciences,
University of Tennessee Health Science Center, Memphis, TN, US

ABSTRACT

Tuberculosis is one of the most deadly diseases caused by a non-spore forming acid fast aerobic bacillus, *Mycobacterium tuberculosis*, which is slender, straight or slightly curved and measures about 0.3-0.6 x 1-4 µm in size. This chapter incorporates brief information about the history of the tuberculosis (TB) epidemics, routes and steps of disease initiation, the strategies adopted by the highly adopted pathogen to escape the host defence system, different diagnostics for detecting TB, and the drugs and major strategies developed and under development for the cure of the disease. Chemotherapy for the disease with first-line anti-tuberculosis drugs provides better treatment for the drug susceptible TB, however it fails to protect from drug resistant form of TB (m/XDR-TB). Addition of second and third line drugs accompanied by prolonged therapy provides control over the disease to some extent. The control over the disease has been achieved by implementation of the various TB control programmes at country level which are coordinated by the World Health Organization. A brief outline of principles and aims of the Global tuberculosis control programmes has been given under DOTS and Stop TB Strategies. Besides, unstopped efforts, we still need new tool to efficiently detect TB, and new vaccines and chemotherapeutics to keep in check the pathogens. Research is going on to find the agents to protect the humans from this deadly pathogen, and hope seems to be positive.

* Corresponding Author Email: kumaranilvp@rediffmail.com; anil.kumar@cuj.ac.in.

Keywords: tuberculosis, pathogenesis, diagnosis and treatment, MDR, DOTS, Stop-TB, trend, granuloma, Xpert-MTB/RIF

INTRODUCTION

Right from birth we are exposed to a world of microbes that invade us in various ways such as breath inhalation, oral feeding, blood transfusion, injuries etc. Many of the invading microbes become commensal to different organs of the body providing benefits to the systems. Pathogens, on the other hand, pose detrimental health effects. Almost all the invading microbes are well taken care of by our innate or acquired immune system; there are few species that have developed systems which enables them to escape the immune system leading to pathogenesis, disease and deaths, if left uncured. Tuberculosis is amongst one of the most dreadful diseases known and is caused by the intracellular pathogen *Mycobacterium tuberculosis*. *M. tuberculosis* was discovered on March 24, 1882 by Robert Koch who was awarded the Nobel Prize for this discovery in 1905 (The Nobel Prize in Physiology or Medicine 1905). The pathogen resides inside the macrophages, one of the constituent of our immune system which recruits and inhibits the invading pathogens. The ability of the pathogen to survive within the macrophages and persist within humans for long periods without any symptomatic overt disease is the key to the success of *M. tuberculosis* (Pieters 2008). Tuberculosis is characterised by symptoms that an individual exhibits during the acute phase of infection by tubercle bacilli *M. tuberculosis* such as: persistent fatigue, anorexia, progressive loss of weight, persistent moderate fever, and production of a chronic, often contagious cough (World Health Organization-WHO 2009, Zahrt 2003). On the basis of the organ that is colonized and affected by the tubercle bacilli, tuberculosis can be divided in pulmonary (tuberculosis of lung) and extra-pulmonary (tuberculosis of kidney, intestine, bone, uterus, fallopian tube etc. other than lung) tuberculosis. However, on the basis of time of infection and type of infection response, pulmonary tuberculosis can be classified in primary and post-Primary (secondary) tuberculosis (WHO 2009, Zahrt 2003).

About one third of the human population is infected with *M. tuberculosis* resulting in an annual casualty case of ~2 million people worldwide (Yew and Leung 2008). According to WHO (2010), the global burden of tuberculosis in 2009 was estimated as a total of 9.4 million incident cases, 14 million prevalent cases, 1.3 million deaths among HIV-negative cases and 0.3 million deaths among HIV-positive cases. Most of the cases were from South-East Asia (35%), African (30%) and Western Pacific (20%) regions. 11-13% of the incident cases were HIV-positive out of which 80% cases belonged to the African Region only (WHO 2010). Although, anti-mycobacterial therapy is currently available, the drugs are only partially effective because of the impermeable nature of the mycobacterial cell wall and the propensity of *M. tuberculosis* to develop resistance

(Warner and Mizrahi 2006). The improper treatment of tuberculosis patients due to their poverty and non-adherence to the treatment strategies has also lead to the emergence of drug resistant *M. tuberculosis* causing multi-drug resistant (MDR)-TB and XDR (extensively drug-resistant)-TB. WHO has reported 3.3% MDR-TB cases among the all new TB cases, and at least one XDR-TB case has been confirmed in 58 counties by March, 2010 (WHO 2010).

History of Tuberculosis

TB has been reported to be present in the form that attacks bone and causes skeletal deformities. Bone can be preserved for thousands of years, allowing identification of individuals having bone TB who died even more than 4,000 years ago. The evidence from skeletons (bones) with apparent tubercular deformities in ancient Egypt, Neolithic sites in Italy, Denmark, and countries in the Middle East indicate that TB was present throughout the world up to 4,000 years ago. The origin of *M. tuberculosis*, the causative agent of TB, has been recently investigated, and it is thought that the bacteria in *Mycobacterium* genus, like other actimomycetes, were initially found in soil and that some species evolved somehow to live and populate in mammals. The evolution of the soil mycobacteria to the closely related human mycobacterial pathogen, *M. tuberculosis,* has been proposed to occur due to adaptation which passed through the domesticated livestock. Characterisation of human and animal pathogens in *M. tuberculosis* complex i.e., *M. tuberculosis, M. bovis, M. africanum*, *M. microti* and *M. canetti* by DNA sequencing, distribution of deletions and insertions in the genomes and synonymous single-nucleotide polymorphisms (sSNP) provides strong evidence for the contemporary and independent evolution of both *M. tuberculosis* and *M. bovis* from precursor species, possibly related to *M. canetti*. The description of the patients with TB-like symptoms recorded in seventh century B.C. and fifth century B.C. indicates that the disease was already well established. It has been suggested that Indo-European herders, who were carrying the infection by virtue of their exposure to cattle, spread the disease to Europe and Asia during their migrations to these regions. Europe, then, became the epicentre for many TB epidemics during 16th and 17th centuries. This disease reached to its peak in Europe and Paris in the first half of the 19th century, and about one-quarter Europeans and 250 of 696 cadavers examined in Paris hospital died of TB. In one study in a Paris hospital at this time, 250 of 696 cadavers examined showed that the individuals had died of this disease. In the last half of the 19th century, mortality due to TB decreased, largely due to improved sanitation, housing and public health measures. Morbidity and mortality rates due to TB steadily dropped during the 20th century in the developed world, aided by better public health practices and widespread use of vaccination by *M. bovis* BCG, as well as the development of anti-tubercular antibiotics in the 1950s. Increased homelessness, poverty in the developed world and the emergence of AIDS again ended the downward trend of the disease in the mid-1980s and new cases of TB were seen. This "mini-epidemic" of new TB cases has been reversed in Europe and the United States by massive expenditures of funds and human resources, mainly by directly monitored antibiotic delivery. However, the underdeveloped world is still suffering from TB and a global crisis is always maintained essentially in Third World due to poverty, lack of healthy living conditions/adequate medical care, and the emergence of multi-drug resistance in countries like the former Soviet Union, South Africa, and India (Smith 2003).

Earlier notions about the disease that it develops in poorer (working) classes from external causes like malnutrition, poor sanitation, and overwork, have been resolved by the efforts made by Jean-Antoine Villemin (1865), Robert Koch (1882) and Edward Trudeau (late 19th and early 20th century) who proved that the TB is caused by a bacterium *M. tuberculosis* and the environmental conditions greatly influence the course of disease development. Their works not only resolved the earlier controversies but also founded the basis of the TB sanatorium in 1800 and paved a way for the treatment of TB. In 1952 Rene Dubos clearly rearticulated an idea that TB is caused by a bacterium, but environmental factors also play a major role in the development of the disease. Dubos stated that pure medical solutions alone would not be sufficient to cure and prevent TB. With the discovery of streptomycin by Schatz and Waksman in the 1940s and its use to treat TB along with introduction of many other antibiotics like isoniazid, rifampicin, pyrazinamide and ethambutol that are useful against TB form the first line of anti-TB drugs. An attenuated vaccine against TB (BCG) has been developed by Calmette and Guerin in Paris in the 1920s, and is widespread in its use to provide acquired immunity to TB infections in childhood (Andersen 2002). Despite, new cases of TB have increased showing the inefficacy of the vaccine in adults. Poverty, malnutrition, improper treatment, patient non-adherence and emergence of drug resistance have further amplified the problem. Emergence of MDR-TB and XDR-TB are the today's havoc that is pretty difficult to treat today (WHO 2009). The present scenario of TB demands development of new efficient vaccines, new drugs, efficient-rapid-affordable diagnostics and coordinated approaches to control, treat and end TB (WHO 2016).

TAXONOMIC STATUS AND CLASSIFICATION OF *M. TUBERCULOSIS*

According to Bergey's Manual of Systemic Bacteriology, 2nd ed., vol.1, mycobacteria have been classified under taxonomic hierarchy: Kingdom-Prokaryotes, Order-Actinomycetales, Family- Mycobacteriaceae and Genus- *Mycobacterium*. *M. tb*. belongs to the group *M. tuberculosis* complex of the genus mycobacterium, which comprises 177 species and 13 subspecies as on 31 January 2017 (Euzeby 2017). About one third of these species are associated with human diseases (Katoch 2004). The first requisite feature of an unknown microorganism to be classified under mycobacterium is that it should be "acid-fast." Acid-fast is the staining feature of bacteria used to describe those bacteria that resist decolourization with acidified alcohol once they have been stained with a dye carbol fuchsin. The technique used to test acid-fastness of the bacterium is called "Acid-Fast" staining. Certain other bacterial species, "partially acid-fast" i.e., *Nocardia* and *Rhodococcus* species, are also able to retain carbolfuchsin but resist decolourization only by low-concentration inorganic acids such as 1% H_2SO_4 or HCl.

Classification of mycobacteria has changed from time to time to incorporate the newly discovered mycobacteria and the additional features confronted by different working groups. Mycobacteria are usually divided into *M. tuberculosis* complex and atypical mycobacteria. The most medically significant classification, given by Runyon (1959), divides atypical mycobacteria into four groups: Group I (photochromogens), Group II (scotochromogens), Group III (nonphotochromogens) and Group IV (rapid growers) (Table 1).

PATHOGENESIS

The host receives infection of *M. tuberculosis* following the inhalation of aerosols containing a small number of bacilli (Kaufmann 2001). When the bacilli reach to lungs, they are phagocytosed by the alveolar macrophages. Phagocytosis of *M. tuberculosis* by macrophages proceeds through a series of membrane invagination, budding, and fusion events, resulting in the formation of the phagosome (Aderem and Underhill 1999).

Infection by *M. tuberculosis* can be divided in three separate but interrelated stages (Figure 1) (Zahrt 2003). During the first stage of infection, individuals become infected by the inhalation of air-borne droplet nuclei containing *M. tuberculosis*. Within the lungs, *M. tuberculosis* primarily infects and resides in resident phagocytic cells such as alveolar macrophages and dendritic cells, or in the monocytes recruited from peripheral blood. Although the macrophages provide an effective initial check against infection by most bacterial pathogens, *M. tuberculosis* has evolved a number of survival strategies that allow it to survive and establish an initial infection within these cells of the immune system. *M.tb.* has evolved various mechanisms that: alters the recruitment or association of various proteins to the phagosomal membrane, modifies the normal phagosomal trafficking pathway of the host, and prevents acidification of *M. tuberculosis*-containing phagosome. These strategies help the organism to escape destruction by immune cells and allow multiplication during the initial stages of infection (Pieters 2008, Zahrt 2003).

Table 1. Group characteristic of the mycobacteria and the representative species

Group	Representative species	Growth	Pigmentation
Tuberculosis complex	*M. tuberculosis* *M. bovis* *M. africanum* *M. microti* *M. canetti*	Slow	Pigment (intensely yellow carotenoids)
Runyon I	*M. kansasii,* *M. marinum,* *M. simiae,* *M. asiaticum*	Slow	Photochromogens (Light–induced)
Runyon II	*M. szulgai* *M. xenopi* *M. lenyiflavium*	Slow	Schotochromogens
Runyon III	*M. avium* *M. intracelluare* *M. haemophilum* *M. celatum* *M. bronderi* *M. shimoide*	Slow	Nonchromogenic
Runyon IV	*M. fortuitum* *M. peregrinum* *M. abscessus* *M. chelonae* *M. mucogenum*	Rapid Growers	

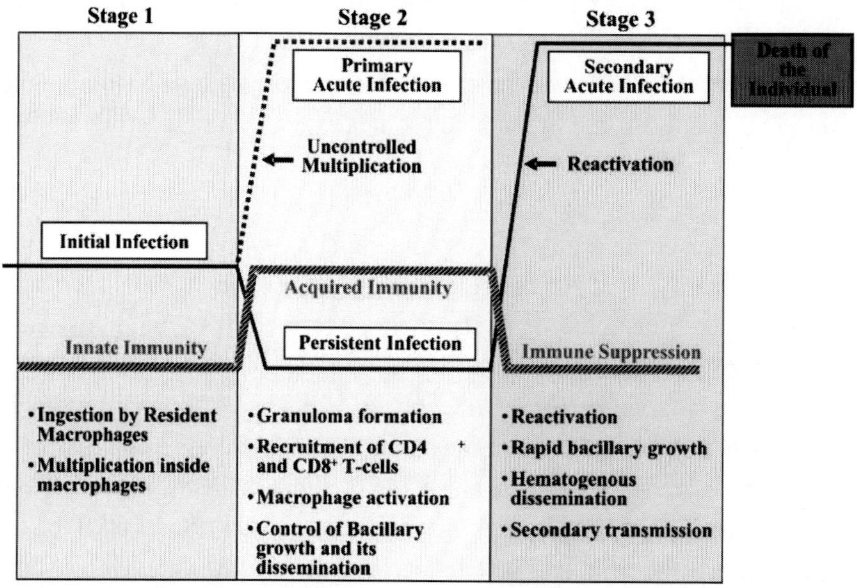

Figure 1. Three stages of infection by *M. tuberculosis*. Solid/square dot line represents progression of infection and disease. Upward diagonal line represents immunity status of the infected host.

The course of infection by *M. tb* during the second stage primarily depends on the status of the host's immune response. For example, an immunocompromised host develops an acute infection characterised by uncontrolled bacillary proliferation and dissemination of the organism to distal sites. Individuals having the acute infection/disease exhibit the symptoms of persistent fatigue, anorexia, progressive loss of weight, persistent moderate fever, and production of a chronic, often contagious cough. Alternatively, in the immunocompetent host, the immune system will clear the initial infection by delivering the mycobacterial phagosomes to lysosomes, or it will hold the infection in check using mechanisms (granuloma formation) that prevent further bacillary proliferation, limit the dissemination of the organism, and concentrate the immune response directly to sites of infection. These individuals receive persistent infection by *M. tuberculosis* and harbour latent carriers of the organism, they neither exhibit overt signs of disease symptoms nor are they infectious. However, they show positive test for a delayed-type hypersensitivity response (Zahrt 2003).

The third and final stage of infection is exhibited by the immunocompetent individual, having persistent and or latent infection in granuloma, when their immune system is suppressed due to HIV infection, malnutrition, age and steroid therapy. The infection is characterised by reactivation of the tubercle bacillus from latency, and subsequent initiation of secondary acute infection in the host due to resumed growth of bacilli (Zahrt 2003).

Mechanisms-Intracellular Survival of *M. tuberculosis*

A schematic diagram of mechanisms adopted by mycobacteria to escape the destructive methods exercised by macrophages has been shown in Figure 2. The phagocytosis of *M. tuberculosis* by macrophages may involve different receptor molecules such as: mannose receptor, complement receptors, dendritic cell-specific intercellular adhesion molecule (ICAM)-3-grabbing nonintegrin (DC-SIGN), and Fc receptors (Pieters 2008). The specific receptor involved in phagocytic entry, may have a considerable impact on the survival chance of *M. tuberculosis* inside the macrophages. For example, internalization of particles *via* complement receptor type 3 (CR3) prevents the activation of the macrophages, whereas ingestion of particles through Fc receptors results in a respiratory burst and an inflammatory response in the macrophages. CR3-mediated uptake of *M. tuberculosis* is strictly dependent on the presence of host plasma membrane cholesterol. The presence of cholesterol at phagosomes harbouring mycobacterium also prevents the delivery of phagosomes to the lysosomes enabling the mycobacteria to utilize its cell wall cholesterol to survive within chronically infected macrophages (Kaufmann 2001, Pandey and Sassetti 2008, Pieters 2008). Thus, internalization *via* certain precise receptors, such as CR3, is beneficial to *M. tuberculosis* to help escape immediate and long-term lysosomal destruction inside macrophages. Once inside the non-activated macrophages, the *M. tuberculosis* interferes with the intracellular trafficking *via* its secretary proteins such as- acid phosphatase (SapM), serine/threonine protein kinase G (PknG), protein phosphatases (PtpA, PtpB) and its lipids: the mycobacterial cell-wall component lipoarabinomannan (LAM) thus preventing the phagosome maturation. These proteins and lipids are known to inhibit: a) the rise in cytosolic Ca^{2+} level of the macrophages and b) retention of early endosome autoantigen-1 (EEA1) and hepatocyte growth factor-regulated tyrosine kinase substrate (Hrs) on the phagosomes (Pieters 2008).

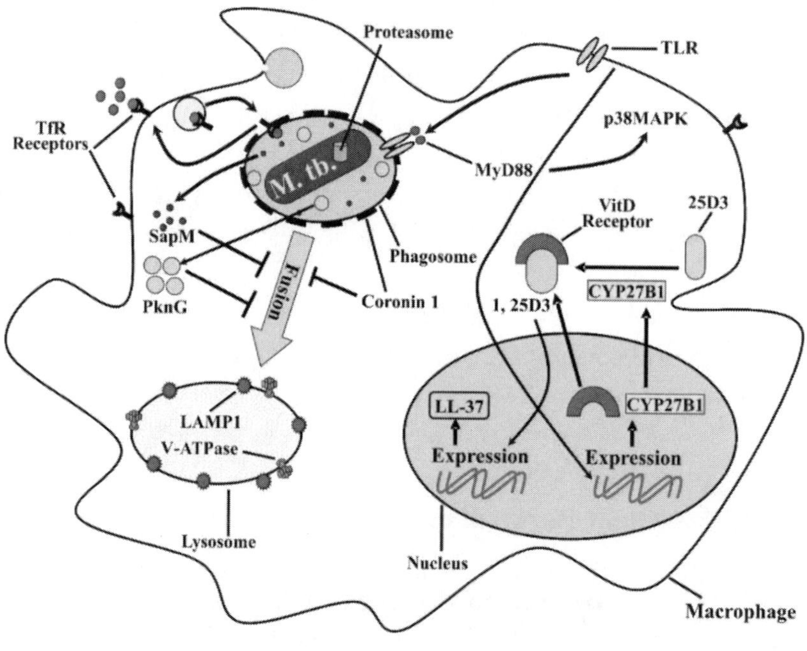

Figure 2. A schematic diagram: mycobacterial escape mechanism against destructive methods exercised by macrophages.

Mycobacteria produces not only virulence factors such as SapM and PknG, but also has evolved the mechanisms to utilize the molecules of host origin for their own survival (Pieters 2008). Seven isoforms of coronin has been identified in mammalian cells and are thought to regulate the F-actin-dependent dynamics of cytoskeleton. The coronin 1-dependent cytosolic influx of calcium activates the calcium-dependent phosphatase calcineurin. This coronin 1 dependent activation of calcineurin has been shown to be a requirement for blocking phagosome-lysosome fusion. Following internalization of mycobacteria into macrophage phagosomes, coronin 1 (earlier TACO for Tryptophan aspartate containing coat protein, also known as P57) is recruited to the mycobacterial phagosome. The macrophages infected with mycobacteria respond with a sustained cytosolic influx of calcium that has been shown dependent of the presence of coronin 1. Retention of Coronin 1 in macrophage phagosome leads to the activation of the

Ca^{2+} dependent phosphatase calcineurin, which is responsible for preventing phagosome-lysosome fusion, thereby allowing *M. tuberculosis* survival (Kaufmann 2001, Pieters 2008). Other mechanism involve production of katG (a catalase-peroxidase) and mycobacterial proteasomes for neutralizing the destructive effect of reactive oxygen and nitrogen respectively produced by macrophage activation through cytokine signaling by IFNγ and TNFα. The cell-wall components LAM as well as its glycosylated forms (manLAM) can modulate signaling pathways that can be initiated to induce activation of macrophages, interferon-γ-mediated gene expression, TLR activation, and phagosome-lysosome fusion. LAM can also block the activation of an important kinase (mitogen-activated protein kinase) that is downstream of various stimuli that cause macrophage activation (Houben et al. 2006, Knutson et al. 1998, Pieters 2008).

Granuloma - A Balance between *M. tuberculosis* and Macrophage

As we have mentioned above infection of *M. tuberculosis* in the immunocompetent host does not lead to an acute infection because the pathogen is either cleared by immune system or the infection is held in check within a structure known as granuloma that prevent further bacillary proliferation. This is how the disease is not developed by most individuals, yet they do harbour certain viable mycobacteria in their macrophages. The precise biology of granuloma induction, formation and maturation is still not clearly understood however, the studies so far reveal that granulomas are structured clusters comprising of *M. tuberculosis* infected macrophages in the centre, surrounded by different types of immune cells i.e., activated macrophages, nonactivated macrophages and T-lymphocytes (Pieters 2008). There exists a mutual interaction between activated macrophages and the T lymphocytes. Activated macrophages, after processing the mycobacterial components, present the mycobacterial antigens to the surrounding T-lymphocytes. The T lymphocytes bind *via* its T-cell receptors (TCRs) to the mycobacterial antigens presented by macrophages, and become activated. Upon activation, T-cells secrete cytokines and chemokines keeping the macrophages in an activated state and ensuring the recruitment of other immune cells to the site of infection. The precise status of *M. tuberculosis* within granulomas is not clear; however, they may exist as actively dividing bacilli or in a dormant state. Depending on the stage of disease, both actively dividing and dormant bacilli can exist within the same infected individual. Irrespective of the fact that *M. tuberculosis* can persist in granulomas for a long time, even up to the lifetime of the host and as long as host immunity is effective, adverse effects of the *M. tuberculosis* on the host's health has usually not been seen. Thus, the granuloma probably represents a balance between a potentially dangerous pathogen and the effective host immune system. This balance is mutually maintained by both macrophages and *M. tuberculosis via* utilizing various mechanisms. Disruption of this delicate balance by the deterioration of host immunity due to co-infection of HIV, malnutrition or steroid chemotherapy results in a potentially life-threatening condition of an individual harbouring latent live *M. tuberculosis* (Pieters 2008).

The Diagnosis of TB

Early diagnosis of TB is crucial in the TB treatment as well as control of its spread. Radiology of Chest/lung can detect TB only in an advanced stage so we need an early diagnosis to detect TB infection. The diagnostics pipeline in use and under development can be divided into two tests, one that require culture and others are those that do not require culture. Culture methods are slow, cumbersome and need expertise and safety parameters however, remains the gold standard and often the most sensitive test for TB. Drug sensitivity testing is dealt with in both, culture and non-culture sections (Martin et al. 2006).

Conventional culture methods using Lowenstein-Jensen (LJ) or 7H11 medium, while cheap and simple, take 20 to 56 and 17 to 21 days respectively for diagnosis. However, both the media require four to six weeks for drug sensitivity testing. Although rapid culture methods have been developed and are commercially available, most of them are difficult to implement in the field due to the complexity of the technique or the required equipment. There are also some emerging simplified culture techniques that can reduce time to diagnosis or drug sensitivity testing (DST) that seem more appropriate for use in resource-limited settings. However, the culture technique faces some limitations due to the following: HIV positive patients and children have difficulty in producing sputum, extra-pulmonary (EP) forms of TB can not be detected by sputum culture, EPTB is very common in HIV positive patients and is more fatal compared to EPTB in HIV negative patients, bacilli may lie protected in lung cavities of a patient with active TB, or the bacilli may be absent in the particular sputum sample or they may be lost during decontamination treatment applied to process the sputum for mycobacterial culture. Thus it invites non-culture based diagnoses which make use of radiation X-ray (useful only in advanced stage of active TB) and various biomarkers from pathogen and host sides. Limitations of culture based diagnosis have been overcome with the efforts made toward whole genome sequencing of *M. tuberculosis.* The knowledge implementation of whole genome sequence information has improved resolution, speed and automation of non-culture and nucleic acid based TB-diagnosis. In turn it offers more valuable data for further management of TB (Skenders et al. 2016). These biomarker based diagnostic may be used to diagnose active TB, monitor the efficacy of TB therapy and identify latent TB (Figure 3) (Goletti et al. 2016, Hanifa et al. 2016, Shah et al. 2016). Various methods for TB diagnosis and drug sensitivity testing are presented in Table 2 (Martin et al. 2006, Ryu 2015).

Table 2. Various methods, commercialised and under development, for the diagnosis of TB

Type Detection	Methods	Company	Commercialisation stage	Remarks
A. Culture Methods				
	BACTEC 460-TB®	Becton Dickinson (US)	No longer commercialised, but still available in some laboratories	Culture on radioactive reagent, detects the presence of mycobacteria based on their metabolism rather than visible colonial growth
	BACTEC MGIT 960® (automated)	Becton Dickinson (US)	Commercialised	Fluorescence quenching-based oxygen sensor, O_2 consumption due to growth produces fluorescence upon UV illumination
	BACTEC MGIT 960® (manual)	Becton Dickinson (US)	Field evaluated by Médecins Sans Frontières (MSF)	Make use of Hand-held UV lamp
	MB/Bact T® system	Organon Teknika (Netherlands)	Commercialised	For diagnosis and DST of first and second-line drugs, fully automated non-radiometric technique, detects CO_2 in closed and continuous monitoring system
	Versa Trek system (Earlier ESP culture system II)	AccuMed International (USA)	Commercialised	For diagnosis and DST for first-line drugs, fully automated, nonradioactive, continuous monitoring based on detection of gases released by bacteria
	E Test	AB Biodisk (Sweden)	Commercialised	The E Test is strip based; require high bacterial concentration and culture on solid medium
	MB redox®	Biotest (Germany)	Commercialised	Based on reduction of a tetrazolium salt indicator broth, 96 well plate format, read by eye/spectrophotometer, not for DST
	TK medium	Salubris (Turkey)	Under evaluation	Based on colour change, Red- no growth, Yellow- TB, Green- other bacterial contamination
	Phage-based tests	Biotec (UK and South Africa)	Under evaluation	Based on counting of plaques formed by mycobacteriophage
	Luciferase reporter phages	Sequella Inc. (USA)	Under development, not commercialised.	Expression of luciferase gene incorporated in recombinant mycobacteriophage when it infects mycobacterium; detection by luminometer or by photosensitive film in a Polaroid film box (Bronx Box); used for DST

Table 2. (Continued)

Type Detection	Methods	Company	Commercialisation stage	Remarks
A. Culture Methods				
	MODS	Not commercialised	Field evaluations have been published	Microscopic-observation drug-susceptibility (MODS) assay is performed in 7H9 broth with/without drug incorporated in the medium
	Nitrate reductase assay	Not commercialised	Used from culture, needs validation for sputum/other clinical samples	The assay is based on the ability of *M. tuberculosis* to reduce nitrate to nitrite *via* nitrate reductase; Used for DST of First line drugs
	Thin Layer Agar (microcolony detection)	Not commercialised	Validated for diagnosis. Under evaluation for DST. Need more validation.	The technique is based on growth and microcolony formation of *M. tuberculosis* on thin layer of 7H11 agar in plates and its microscopic examination at alternate days for first two weeks and then less frequently
	"Resa" Colorimetric method	Not commercialised	Validated in large field studies	Colorimetric method (known as the resazurin assay) is based on the ability of live bacteria to reduce a coloured indicator upon growth producing a change of colour from blue to pink; Used for DST of first and second-line drugs
B. Non-culture Methods				
Antibody detection	Many but some are presented here: ABP Diagnostics Focus Sure Check TB Advanced Diagnostics Tuberculosis Rapid Test American Bionostica Rapid Test for TB Ameritek dBest One Step TB Test CTK Biotech TB Antibody onsite Rapid Screening Test Kit Standard Diagnostics SD Rapid TB		Fast and Commercialised; sensitivity below 40% for methods showing 80-90% specificity. So none are suitable to replace microscopy	All tests detect antibodies in serum

Type Detection	Methods	Company	Commercialisation stage	Remarks
Antigen detection	LAM urine test	Chemogen (USA).	Published study conducted on 200 healthy and 242 TB suspect patients	The test detects lipoarabinomannan (LAM: a component TB bacterial cell wall) in urine as a surrogate marker for TB infection
	Antigen-based detection test	Proteome Systems (Australia)	Still at the research stage	Proteome Systems have been used to identify *M. tuberculosis* specific antigens that could potentially be employed in development of a rapid test and diagnostics, or to monitor treatment efficacy. Biological fluids such as sputum, saliva, plasma or whole blood can be used
	Flow-through filter tests	ANDA (France) for *patho-tb* and BioMed (India) for *diagnos-tb*.	ANDA under field evaluation. No information about BioMed test	Polyclonal antibody based detection of *M. tuberculosis* in sputum or body fluids using a flow-through device
Nucleic Acid Amplification (NAA)	Amplified *Mycobacterium Tuberculosis* Direct Test	Gen-Probe Inc. (USA)	Commercially available	Isothermal NAA. Isothermal (42°C) transcription-mediated amplification of *M. tuberculosis* target 16S rRNA. The test is specific for *M. tb.* complex
	AMPLICOR® MTB tests	Roche Diagnostic Systems (USA)	Commercialised	PCR test targeting the 16S rRNA gene. The process can be automatically performed on the Roche COBAS AMPLICOR machine
	BD ProbeTec ET® assay	Becton Dickinson Biosciences (USA)	Commercialised	Isothermal NAA. Isothermal co-amplification of target sequences of IS*6110* and 16S rRNA gene by SDA (strand displacement amplification) process, and simultaneous detection in the ProbeTec instrument
	In-house PCR	N/A	Non-commercial	"Home brew" NAA assays. Amplification of *M. Tb.* complex-specific target, the insertion sequence IS*6110* repeats, which is present in multiple copies in the *M. tuberculosis* genome.
	Real-time PCR	N/A	Non-commercial	The techniques make use of different probes: TaqMan probe, fluorescence resonance energy transfer (FRET) probes, molecular beacons and bioprobes for detection. Used for rapid detection of rifampicin resistance.

Table 2. (Continued)

Type Detection	Methods	Company	Commercialisation stage	Remarks
Nucleic Acid Amplification (NAA)				It can be applied to *M. tuberculosis* strains and directly in clinical samples
	Simplified NAT test, TB-LAMP test	Eiken Chemical Co Ltd (Japan).	Under development	LAMP (loop mediated isothermal amplification) utilizes amplification of TB DNA directly from clinical samples. A positive result is signalled by a visible chromogenic reaction
	INNO-LiPa assays	Innogenetics (Belgium).	Commercialized	Two Inno-LiPa assays are commercialised: INNO-LiPA Mycobacteria Assay for tuberculosis diagnosis and INNO-LiPA Rif TB Assay for detection of rifampicin resistance. The assay is based on hybridization of amplified mycobacterial 16S-23S rRNA spacer region DNA from cultured strains or clinical samples to the 10 probes (covering the core region of the *rpo*B gene of *M. tb.*) immobilized on a nitrocellulose strip
	GenoType Assays	Hain Lifesciences (Germany).	Commercialised	Two assays are commercialised. GenoType Mycobacteria Assay for tuberculosis diagnosis and GenoType MTBDR Assay for detection of rifampicin and isoniazid resistance. 16S-23S ribosomal DNA spacer region is isolated by PCR amplification; the product is biotinylated and then hybridized with 16 specific oligonucleotide probes immobilized as parallel lines on a membrane strip. Detection of resistance to isoniazid and rifampicin in culture samples is based on detection of most common mutations in *kat*G and *rpo*B genes respectively
	PCR sequencing	N/A.	A gold standard for monitoring resistance	Amplification and sequencing of specific *M. tuberculosis* genetic material, mostly repeat insertion sequence IS*6110*, allowing the DNA to be "read". This is most widely used method for defining genetic resistance for drug sensitivity testing and is the gold standard method. It has been used for characterising mutations in the *rpo*B gene in rifampicin-resistant strains and to detect mutations responsible for other anti-tuberculosis drugs

Type Detection	Methods	Company	Commercialisation stage	Remarks
Nucleic Acid Amplification (NAA)	PCR directly from AFB slides	N/A.	Early research stage.	The basic technique consists of conducting PCR directly on samples washed off sputum slides. Different methods have been used
	T-cell based tests Quantiferon TB Elispot	Cellestis (Australia/USA) Oxford Immunotec (UK)	Several versions have been released. These tests are being validated in children and HIV-positive patients.	Blood from a patient is incubated with tuberculin antigen for 12 hours in sterile culture. The release of a protein (TB patient with previously mounted immune response to TB, will have "primed" cells in their blood that produce immune response to tuberculin antigen) that is associated with immune activation is measured in the test. A replacement to Mantoux test (tuberculin skin testing)
	Transdermal patch test for active TB	Sequella (USA).	An evaluation study was conducted in South Africa showed 65% sensitivity and 96% specificity.	A transdermal patch containing a TB protein, when applied on the forearm, leaves a red spot after a few hours in the patient exposed to TB. A replacement to Mantoux test. It allows both, the distinction between active and latent infection, and the monitoring of treatment success
	DNA Microarrays	Not commercialised, research tool	At research laboratories	The method is based on hybridization of DNA obtained from clinical samples to oligonucleotides immobilised on a solid support, such as miniaturized glass slides. Used for detecting drug resistance in *M. tb.*: Microarrays have been mainly used to detect resistance to rifampicin
	Xpert MTB/RIF	Cepheid, Sunnyvale, CA, USA	Commercialised	Rapid and Cartridge based NAA test from sputum automated from sample processing to detection of amplified *rpo*B gene target. It simultaneously detects rifampicin resistance directly from sputum
Non-NAA based DNA detection	-	Various companies and researchers identified	Under development	Various methods to detect DNA sequences without amplification by PCR. For example, a probe binds to the specific target, thereby exposing the catalytic site buried in the probe molecule. The exposed catalyst acts to change the colour of the buffer
Breath detection	-	Various companies and researchers identified.	Currently at the research stage. The technology is not very specific yet	The method is based on the detection of volatile organic compounds in the breath of TB patients relative to controls

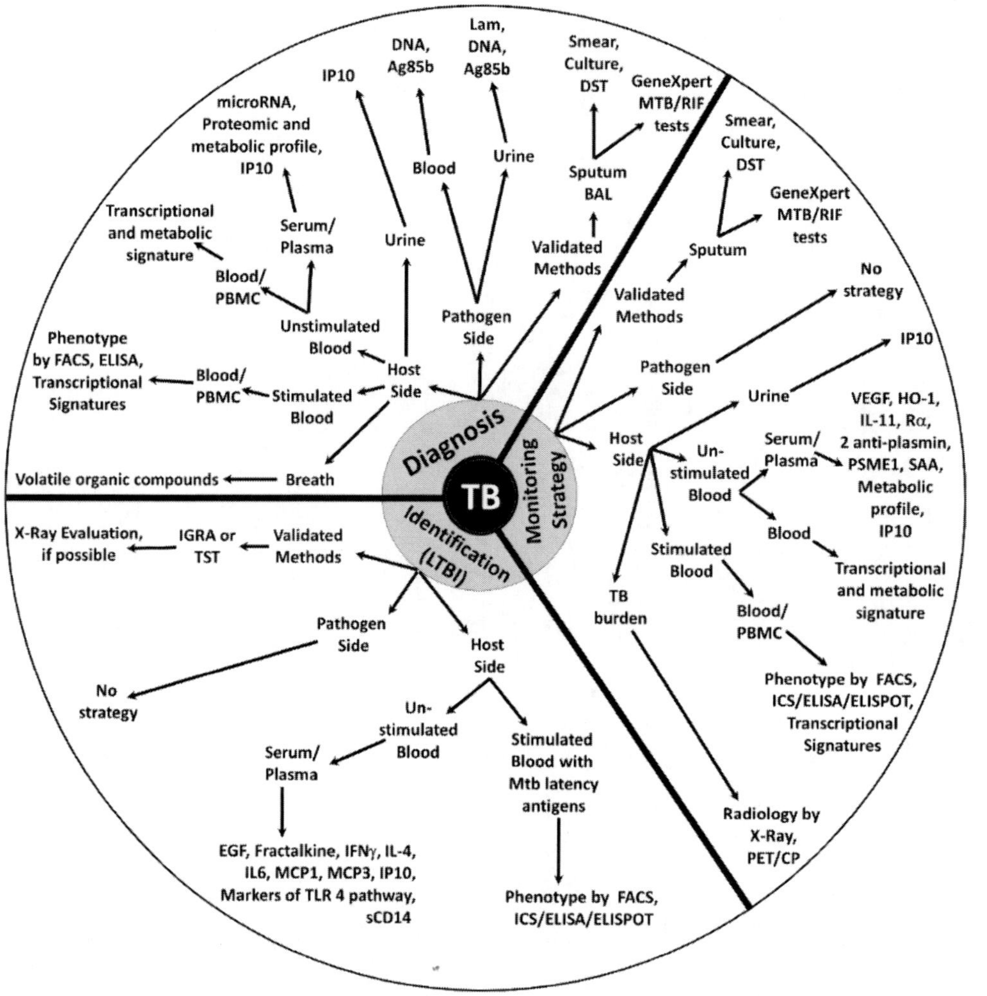

Figure 3. Flow chart representation of biomarkers for diagnosis and monitor therapy efficacy of active and latent TB. TB: tuberculosis, Ag: antigen, LAM: lipoarabinomannan, BAL: broncholavage, IP: interferon-γ inducible protein, FACS: Fluorescence-activated cell sorting, VEGF: vascular endothelial growth factor; IL: interleukin; PSME1: proteasome activator complex subunit 1; SAA: serum amyloid A; PET/CP: positron emission tomography/computed tomography, LTBI: latent tuberculosis infection; TST: tuberculin skin test; IGRA: IFNγ release assay; EGF: endothelial growth factor; MCP: monocyte chemo attractant protein; TLR: toll like receptors; sCD14: soluble CD14; ICS: intracellular staining.

THERAPY OF TUBERCULOSIS

Case Definition

It is crucial step to define a TB case before the treatment is started. The case definition refers to the diagnosis (recognition) of an active disease (symptomatic) case and assigning the type of TB case. There are four determinants of case definition: site of

TB disease (PTB, EPTB), bacteriology (result of sputum smear), severity of TB disease (Bacillary load, extent of disease and anatomical site) and history of previous treatment of TB. Miliary, also known as disseminated TB is considered to be severe. The EPTB forms are: meningeal, pericardial, peritoneal, bilateral or extensive pleural effusive, spinal, intestinal and genitourinary are classified as severe, whereas lymph node, pleural effusion (unilateral), bone (except spine), peripheral joints and skin tuberculosis are classified as less severe. History of previous treatment of TB can be distinguished as new, relapse, treatment after failure, treatment after default, transfer in and other-chronic cases. Although rare, smear-negative pulmonary and extra-pulmonary cases may also be relapses, failures, returns after default or chronic cases. This case defining step applies to all age categories of TB patients, i.e., adults and children. The case definition is important for: proper patient registration and case notification; prioritized treatment of sputum smear-positive cases, which are the main sources of infection transmission in the community; allocation of cases to appropriate standardized treatment regimens; appropriate evaluation of the proportion of cases according to site of TB, bacteriology and treatment history; and cohort analysis of treatment outcomes. A cohort refers to a group of patients who have been diagnosed and registered for treatment during a specific time period (usually is one-quarter of a year) (WHO 2009).

Chemotherapy

Discovery of streptomycin in 1944 has made chemotherapy for tuberculosis (TB) possible but had some limitations. Later, the introduction of isoniazid formed the basis of anti-TB chemotherapy in the 1950s to 1960s. The anti-TB drug regimen at that time usually comprised a total duration of 18 months with initial months of streptomycin, isoniazid and para-aminosalicylic acid followed by isoniazid and para-aminosalicylic acid. Discovery of rifampicin in 1965 shortened the duration of TB chemotherapy, and a short-course treatment for TB was commenced in the 1970s. Short-course chemotherapy given to new TB (presumed, or known, to have drug susceptible TB) cases includes intensive phase of 2 months with rifampicin, isoniazid, pyrazinamide and ethambutol followed by continuation phase of 4 months with rifampicin and isoniazid (WHO 2009). Short-course chemotherapy for TB, if administered properly, confers bactericidal and sterilizing actions against *M. tuberculosis* and prevents emergence of resistance to the anti-TB drugs (Nuermberger et al. 2010). However, the chemotherapies of the tuberculosis suffer today the major problem of emerging of drug resistance (WHO 2009).

The emergence of resistance to the anti-tuberculosis drugs has been attributed to spontaneous mutation. Emergence of drug resistance has necessitated the incorporation of multiple drugs for first line therapy. Drug resistance in *M. tuberculosis* may arise from man-made selection of mutants that result from spontaneous mutation. Non-adherence of the patients, inappropriately prescribed regimens, poor drug quality and erratic drug supply are the key elements likely to facilitate selective amplification of these drug-resistant mutants over the years. Infection by drug resistant *M. tuberculosis* reduces the efficacy of treatment and lead to failures in the implementation of TB control programmes (Nuermberger et al. 2010). This drug resistance may be "primary" (pre-treatment) when the patient is infected for the first time by a tubercle bacilli which has already got resistance, or "secondary" (acquired, post-treatment), when the sensitive infecting bacilli becomes resistant due to inadequate treatment. Emergence of resistance to one drug increases the probability of resistance development to susceptible agent. These bacilli, resistant at least to rifampicin and isoniazid, pose a very serious situation in patients leading to multidrug resistant tuberculosis (MDR-TB) which do not respond to first line drugs. Prolonged/revised treatment of MDR-TB with first-line drugs might lead to amplification of resistance. Therefore prompt identification of MDR and initiation of MDR treatment with second-line drugs is indispensable, leading to better cures as well as prevent the development and spread of further resistance (WHO 2009). The health problem of TB patients has further been complicated by the emergence of XDR-TB, and an increasing list of XDR-TB cases are being reported by various countries (WHO 2010a). XDR-TB is defined as MDR-TB plus resistance to a fluoroquinolone and at least one second-line injectable agent (amikacin, kanamycin and/or capreomycin) (WHO 2010b).

Different standardized treatment regimens are recommended for the treatment of different categories, as sorted by case definition, of TB cases (WHO 2009). Treatment of TB cases aims to: cure the patient; prevent late effect and death from active TB; prevent relapse of TB; minimise transmission of TB; and to prevent development of resistance to the anti-tuberculosis drugs. A number of essential anti-tuberculosis drugs have been recommended which possess three main properties; bactericidal activity, sterilizing activity and ability to prevent resistance. Different anti-tuberculosis drugs possess varying extents of these three properties. Isoniazid and rifampicin are most powerful bactericidal (active against all population of TB bacilli) and rifampicin is the most potent sterilizing drug. However, ethambutol and thiocetazone are used in combination with the powerful drugs to prevent emergence of resistant bacilli. The anti-tuberculosis drugs can further be grouped in different groups as shown in Table 3. These drugs have been used to formulate different treatment regimens for the treatment of various forms of TB. WHO have proposed different standardised treatment regimens for the treatment of TB cases classified by its case definition (WHO 2003, 2009). WHO has also proposed the guidelines and treatment regimens for the treatment of TB cases in special situation of pregnancy and breast feeding, liver disorders, oral contraception, acute hepatitis, HIV infection, and renal failure and severe renal in-sufficiency. In addition to these treatment regimens, other national and international strategies for tuberculosis control have been implemented such as DOTS and vaccination with Bacille Calmette-Guérin (BCG).

Besides regular BCG, other new vaccines/subunit vaccines are being developed and tested in clinical trials which may boost BCG efficacy (Orme 2005, Rowland and McShane 2011).

Directly Observed Treatment, Short Course Chemotherapy (DOTS)

DOTS strategy is a well programmed strategy which ensures 'regular intake' of the prescribed medication and 'targeted cure' by providing effective medicines. DOTS strategy does not require patient admission to the hospital however, health workers and the trained persons are employed to issue medicines and monitor the patients for uninterrupted medication and routine health check-ups to monitor the treatment outcomes. The deputed personnel keep track of the registered patients until they are cured. The essential principles of DOTS were first demonstrated in India at the Tuberculosis Research Centre (NIRT) in Chennai in the 1950s and 1960s. The world-wide success of the DOTS strategy lies in the central "Mantra" that "curing the TB patients is the responsibility of the health worker and not the patients." DOTS strategy has been proven most successful in China with cure rates of more than 90% whereas cure rate in India is 80% (Elzinga et al. 2004).

In 1991, World Health Assembly (WHA) resolution recognized TB as a major global public health problem and launched DOTS in 1994 as the internationally recommended tuberculosis control strategy. It was started with its five components: 1- sustained government commitment; 2- case diagnosis by quality-assured sputum-smear microscopy; 3- standardized short-course chemotherapy to all confirmed sputum smear-positive cases under proper case management conditions; 4- a system for uninterrupted supply of effective and quality assured drugs; and 5- a standard recording and reporting system for programme monitoring by systematic assessment of treatment outcomes of all the registered patients (Elzinga et al. 2004, WHO 2003). Global tuberculosis monitoring and assessment project are coordinated by WHO to which the involved countries report their annual progress in implementation of the DOTS strategy at their national tuberculosis control programmes. WHO intermittently monitors and evaluates the outcomes of national TB control programmes, and accordingly set the revised guidelines of DOTS. DOTS forms the core of all national and international TB control programmes. Intermittent expansion of DOTS framework and its association with other TB control partners help countries to improve their national TB control programmes (NTPs) and make major progress in TB control i.e., the DOTS Plus strategy for MDR-TB and the Stop TB strategy for 'a world free of TB' (Revised National tuberculosis Control Programme 2010, Verma et al. 2013, WHO 2006).

Table 3. Anti-tuberculosis drugs and their mode of action

Category of drugs and Drugs	Biological activity	Mode of Action
Category A: First-line oral drugs*		
Isoniazid (H)	Bactericidal on actively dividing cells Bacteriostatic on non-replicating bacteria	Inhibition of mycolic acid biosynthesis Inhibition of catalase-peroxidase
Rifampicin (R)	Bactericidal	Inhibition of DNA dependent RNA polymerase, transcription
Pyrazinamide (Z)	Bactericidal (Active in acidic environment)	Unknown
Ethambutol (E)	Bacteriostatic	Inhibition of arabinosyl transferase (arabinogalactan biosynthesis)
Category B: Injectable agents#		
Streptomycin *	Bactericidal	Inhibition of protein synthesis
Amikacin	Bactericidal on actively dividing extracellular organisms	Disruption of protein synthesis by inhibiting ribosome function
Kanamycin		
Capreomycin	Bactericidal on non-replicating bacteria	Inhibition of protein synthesis
Category C: Fluoroquinolones#		
Ofloxacin	Bactericidal	Inhibition DNA gyrase
Levofloxacin	Bactericidal	
Moxifloxacin	Bactericidal	
Category D: Oral Bacteriostatic second-line agents#		
Ethionamide	Bacteriostatic	Inhibition of mycolic acid biosynthesis
Prothionamide	Bacteriostatic	Inhibition of mycolic acid biosynthesis
Cycloserine	Bacteriostatic	Inhibition of cell wall biosynthesis
Terizidone	Bacteriostatic	Inhibits cell wall biosynthesis by inhibiting L-alanine racemase and D-alanine ligase
p-Aminosalicylic acid	Bacteriostatic	Interferes with bacterial folic acid biosynthesis
Linezolid (MDR-TB) (WHO 2016)	Bactericidal	Inhibition of ribosomal protein synthesis
Clofazimine (MDR-TB) (WHO 2016)	Bacteriostatic	Binds to mycobacterial DNA
Category E: Third line drugs		
Rifabutin	Bactericidal	Blocks DNA dependent RNA polymerase, transcription
Macrolides	Bactericidal	binds to 50S, inhibits ribosomal translocation, peptidyl transferase & protein biosynthesis
Thioridazine	Bactericidal against XDR *M. tb.*	Influence bacterial DNA metabolism, Inhibits efflux pumps and Ca^{++} uptake, alters bacterial cell membrane's lateral organization and fluidity (Kopec and Khandelia 2014)

Category of drugs and Drugs	Biological activity	Mode of Action
Arginin/Vitamin-D	Improves patient physiology	Minimises side effects of powerful drugs, Isoniazid, Ethambutol etc.
R207910	-	Affects proton pump for ATP synthase
Category F: Approved Anti-TB activity but efficacy not certain		
Metronidazole	Bactericidal	Forms reactive radicals that damage DNA
Ampicillin-sulbactam	Bactericidal	Inhibition of mucopeptide biosynthesis, inhibits beta-lactamases
Thiacetazone**	Bacteriostatic	Targets cyclopropane mycolic acid synthases inhibiting cyclopropanation of mycolic acids in mycobacteria
Imipenem/cilastatin	Bacteriostatic	Imipeneme: Inhibits Cell wall biosynthesis. Cilastatin: Inhibits imipeneme metabolism by inhibiting renal dehydropeptidase (Pastel 1986).
Clarithromycin	Bactericidal	As the macrolides

* First line drug;
# Second line drug;
** Third line drug.

The Stop TB Strategy

The Stop TB strategy has six principal components: 1- pursue high-quality DOTS expansion and enhancement; 2- address TB/HIV and MDR-TB and other special challenges; 3- contribute to health system strengthening; 4- engage all care providers; 5- empower people with TB, and communities; 6- enable and promote research. The first component, DOTS expansion and enhancement, forms the cornerstone of the Stop TB strategy and provides foundation for the rest five. The basic features of the Stop TB strategy have been summarized below (Uplekar et al. 2006):

Vision of the strategy: 'a world free of TB' incidences.

Goal: dramatic reduction in the global burden of TB by the end of 2015 in line with the Millennium Development Goals (MDGs) and the Stop TB Partnership targets.

Objectives:

- Universal access to high-quality diagnosis and patient-centred treatment.
- To alleviate suffering and socioeconomic burden due to TB.
- Protection of poor and vulnerable populations from TB, TB/HIV and MDR-TB.
- Promote/support development of efficient tools, and its timely and effective use.

Targets:

- To make TB cases halt by 2015 and then to reverse the incidence.
- Targets aimed by MDGs and endorsed by the Stop TB Partnership.

- Detection, at least 70%, of new sputum smear-positive TB cases and cure, at least 85%, of these cases by year 2005.
- 50% reduction in TB prevalence and associated death rates, caused in 1990, by year 2015; and finally.
- The eradication (<1 case per million population) of TB by year 2050.

Current Trends in TB Research

Including proper diagnosis at the right time, the development of granuloma and hypoxia during TB infections and cases of reactivation/post-treatment-relapse with difficult pathogens (MDR-XDR) poses a major hurdle to the treatment and management of TB cases. Besides the strive to look for new anti-tubercular drugs in the arsenal to fight end TB, the current trend in TB-research encompass global aspects of TB pathogenesis, diagnosis and treatment. Progress is being made towards development of granuloma model which potentially can be employed in a mechanistic pathophysiology model of development/progression of infection or disease (Parasa et al. 2014). Routine and traditional TB diagnostics (culture based, AFB staining etc.), though cheaper, are slow and inadequate, and are being either replaced or fortified with new, fast, versatile, throughput, multi-parametric diagnostics but are costly. Some of these developing techniques have to be used at three different levels: a- Reference Level Laboratory, b- Intermediate level Laboratory and c- Peripheral level laboratories (WHO 2016). The current diagnostics developments aim to automatize detection of TB/drug resistance/combined TB and drug-resistance, and so are mostly making use of molecular markers (Goletti et al. 2016, WHO-2016). One stop online suite "Molecular Property Diagnostic Suite (MPDS^{TB})" is dedicated to tuberculosis and it has been developed for the advancement of drug discovery process (Gaur et al. 2017). To add, emphasis are given to detect TB, drug resistance, TB and drug resistance, identify and treating LTB, drug-sensitivity test to existing and new drug-regimen for MDR/XDR-TB, investigation of TB, discovery of new efficient drugs and formulation of drug-regimen for paediatric treatment of TB cases (Gaur et al. 2017, Goletti et al. 2016, WHO 2016).

Conclusion

The History of TB epidemics is too old but still the epidemic of TB is not a history. Unlike diabetes, TB is the disease of poor people because of its high prevalence among the poor and developing countries. The pathogen consumes the lung of immunologically weak or immunosuppressed individuals. However, it remains hidden inside alveolar macrophages of the immunocompetent people until the immune system weakens.

Extended co-evolution of *M. tuberculosis* with its human host within macrophages has provided the bacterium with a number of strategies for its intracellular survival. They produce an array of effector molecules that interferes with host trafficking systems, macrophage activation and maturation, and thus allow intracellular survival. Emergence of MDR-TB and XDR-TB, as a consequence of development of resistance to the first-line and second-line anti-tuberculosis drugs, has further complicated the treatment and cure of TB. The epidemics of TB has to some extent been taken under control through improved living conditions, discovery of effective antibiotics and formulation of standardised drug regimens, and worldwide implementation of global as well as national tuberculosis control programmes (DOTS, DOTS-Plus, MDGs, Stop TB partnership). It has been assumed, the goal of Stop TB Strategy, that the TB would be eliminated by year 2050. To achieve the goal of Stop TB strategy "a world free of TB" further developments is efficient diagnostics, therapeutics, new efficient vaccines, honest implementation of coordinative tuberculosis control programmes, education and compliance of the patients are highly admitted. Development of new anti-TB drugs are needed to: (i) shorten and simplify the treatment of drug-susceptible TB cases as well as latent TB infections; (ii) introduce shorter, safer, cheaper and more effective treatment alternatives for M/XDR-TB); (iii) eliminate hurdles to provide effective treatment of TB in special situations of HIV and others; and (iv) shorten treatment of LTBI. However the WHO (2014) global tuberculosis report is indicative of inadequate success, especially to MDR cases, to Stop TB strategy probably due to poor treatment outcomes/capacity (growing waiting list) and patient noncompliance. So, the End-TB strategy has been proposed with three high level indicators: TB incidence rate, absolute number of TB death and percentage of TB patients, and their households costs. This strategy targets at these indicators for year 2030 and 2035 including milestone for year 2020 and 2025 by addressing health system barriers, diagnostic and treatment challenges, and inadequate funding for care and research (Cheon et al. 2016, WHO 2014). Beside all efforts the worldwide decline in rate of TB incidence was only 1.5% during 2014-2015 which needs to be accelerated to 5% (WHO 2016).

References

Aderem A, Underhill DM (1999) Mechanisms of phagocytosis in macrophages. *Annu Rev Immunol* 17: 593-623.

Andersen P (2002) TB vaccines: progress and problems. *Trends Immunol* 22: 160-168.

Cheon SA, Cho HH, Kim J, Lee J, Kim HJ, Park TJ (2016) Recent tuberculosis diagnosis toward the end TB strategy. *J microbial methods* 123: 51-61.

Elzinga G, Raviglione MC, Maher D (2004) Scale up: meeting targets in global tuberculosis control. *Lancet* 363: 814-819.

Euzeby JP (2017) *List of prokaryotic names with standing in nomenclature-Genus Mycobacteriaum*. (http://www.bacterio.net/mycobacterium.html).

Gaur AS, Bhardwaj A, Sharma A, John L, Vivek MR, Tripathi N, Bharatam PV, Kumar R, Janardhan S, Mori A, Banerji A, Lynn AM, Hemrom AJ, Passi A, Singh A, Kumar A, Muvva C, Madhuri C, Choudhury C, Kumar AD, Pandit D, Bharti DR, Kumar D, Singam AEr, Raghava GPS, Sailaja H, Jangra H, Raithatha K, Tanneeru K, Chaudhary K, Karthikeyan M, Prasanthi M, Kumar N, Yedukondalu N, Rajput NK, Saranya PS, Narang P, Dutta P, Krishnan RV, Sharma R, Srinithi R, Mishra R, Hemasri S, Singh S, Venkatesan S, Kumar S, Jaleel UCA, Khedkar V, Joshi Y, Sastry GN (2017) Assessing Therapeutic Potential of Molecules: Molecular Property Diagnostic Suite for Tuberculosis ($MPDS^{TB}$). *J Chem Sci* 129: 515-531.

Goletti D, Petruccioli E, Joosten SA, Ottenhoff TH (2016) Tuberculosis biomarkers: from diagnosis to protection. *Infect Dis Rep* 8: 2.

Hanifa Y, Fielding KL, Chihota VN, Adonis L, Charalambous S, Karstaedt A, Churchyard GJ (2016) Diagnostic Accuracy of Lateral Flow Urine LAM Assay for TB Screening of Adults with Advanced Immunosuppression Attending Routine HIV Care in South Africa. *PloS one* 11: e0156866.

Houben ENG, Nguyen L, Pieters J (2006) Interaction of pathogenic mycobacteria with the host immune system. *Curr Opin Microbiol* 9: 76-85.

Katoch VM (2004) Infections due to non-tuberculous mycobacteria (NTM). *Indian J Med Res* 120: 290-304.

Kaufmann SH (2001) How can immunology contribute to the control of tuberculosis? *Nat Rev Immunol* 1: 20-30.

Knutson KL, Hmama Z, Herrera-Velit P, Rochford R, Reiner NE (1998) Lipoarabinomannan of Mycobacterium tuberculosis promotes protein tyrosine dephosphorylation and inhibition of mitogen-activated protein kinase in human mononuclear phagocytes. Role of the Src homology 2 containing tyrosine phosphatase 1. *J Biol Chem* 273: 645-652.

Kopec W, Khandelia H (2014) Reinforcing the membrane-mediated mechanism of action of the anti-tuberculosis candidate drug thioridazine with molecular simulations. *J Comput Aided Mol Des* 28: 123-134.

Martin G, Martin U, James A (2006) Tuberculosis diagnosis and drug sensitivity testing: an overview of the current diagnostic pipeline. *Campaign for Access to Essential Medicines*, Médecins Sans Frontières. Geneva, Switzerland.

Nuermberger EL, Spigelman MK, Yew WW (2010) Current development and future prospects in chemotherapy of tuberculosis. *Respirology* 15: 764-778.

Orme IM (2005) Current progress in tuberculosis vaccine development. *Vaccine* 23: 2105-2108.

Pandey AK, Sassetti CM (2008) Mycobacterial persistence requires the utilization of host cholesterol. *Proc Natl Acad Sci* USA 105: 4376-4380.

Parasa VR, Rahman MJ, Hoang ATN, Svensson M, Brighenti S, Lerm M (2014) Modeling *Mycobacterium tuberculosis* early granuloma formation in experimental human lung tissue. *Dis Model Mech* 7: 281-288.

Pastel DA (1986) Imipenem-cilastatin sodium, a broad-spectrum carbapenem antibiotic combination. *Clin Pharm* 5: 719-736.

Pieters J (2008) Mycobacterium tuberculosis and the macrophage: maintaining a balance. *Cell Host Microbe* 3: 339-407.

Revised national tuberculosis control programme: DOTS-Plus guideline-2010. Central TB Division, Directorate General of Health Services, Ministry of Health & Family Welfare, New Delhi, India. http://www.tbcindia.org/pdfs/DOTS_Plus_Guidelines_Jan2010.

Rowland R, McShane H (2011) Tuberculosis vaccines in clinical trials. *Expert Rev Vaccines* 10: 645-658.

Ryu YJ (2015) Diagnosis of pulmonary tuberculosis: recent advances and diagnostic algorithms. *Tuberc Respir Dis* 78: 64-71.

Shah M, Hanrahan C, Wang ZY, Dendukuri N, Lawn SD, Denkinger CM, Steingart KR (2016) Lateral flow urine lipoarabinomannan assay for detecting active tuberculosis in HIV-positive adults. *Cochrane Database Syst Rev* 5: 1-111.

Skenders G, Soolingen D (2016) Molecular typing of *Mycobacterium tuberculosis* complex isolates. In: *Handbook on TB laboratory diagnostic methods for the European Union.* European Centre for Disease Prevention and Control (*ECDC*), Stockholm, Sweden. pp. 83-91.

Smith I (2003) Mycobacterium tuberculosis pathogenesis and molecular determinants of virulence. *Clin Microbiol Rev* 16: 463-496.

Stop TB (2009) *Partnership Working Group on New TB Vaccines*. Tuberculosis Vaccine Candidates.

The Nobel Prize in Physiology or Medicine (1905) Nobelprize.org. *Nobel Media AB* 2014. Web. 24 Mar 2017. http://www.nobelprize.org/nobel_prizes/medicine/laureates/1905/.

Uplekar M, Figueroa-Munoz J, Floyd K, Getahun H, Jaramillo E (2006) *The Stop TB Strategy: building on and enhancing DOTS to meet the TB-related Millennium Development Goals.*

Verma R, Khanna P, Mehta B (2013) Revised national tuberculosis control programme in India: the need to strengthen. *Int J Prev Med* 4: 1-5.

Warner DF, Mizrahi V (2006) Tuberculosis chemotherapy: the influence of bacillary stress and damage response pathways on drug efficacy. *Clin Microbiol Rev* 19:558-570.

WHO (1991) *World Health Assembly (WHA) resolution to eliminate leprosy*. Geneva, Switzerland: WHO.

WHO (1994) *Tuberculosis Programme: framework for effective tuberculosis control.* Geneva, Switzerland: WHO.

WHO (2003) *Treatment of Tuberculosis: guidelines for national programmes.* 3rd ed. Geneva, Switzerland: WHO.

WHO (2006) *Global Tuberculosis Report 2006.* Geneva, Switzerland: WHO.

WHO (2009) *Treatment of Tuberculosis: guidelines,* 4th ed. Geneva, Switzerland: WHO. (WHO/HTM/TB/2009.420).

WHO (2010a) *Stop TB partnership.* 2010/2011 tuberculosis global facts. Geneva, Switzerland: WHO.

WHO (2010b) *Multidrug and extensively drug-resistant TB (M/XDR-TB): 2010 global report on surveillance and response.* Geneva, Switzerland: WHO.

WHO (2014) *Drug-resistant TB: surveillance and response.*

WHO (2016) *Global Tuberculosis Report 2016.*

WHO and Stop TB Initiative (2010) *Treatment of tuberculosis: guidelines.*

Yew WW, Leung CC (2008) Update in tuberculosis 2007. *Am J Respir Crit Care Med* 177: 479-485.

Zahrt TC (2003) Molecular mechanisms regulating persistent *Mycobacterium tuberculosis* infection. *Microb Infect* 5: 159-167.

In: Trends in Life Science Research
Editors: R. P. Sinha and U. P. Shrivastava

ISBN: 978-1-53613-241-0

Chapter 6

INNATE IMMUNE RESPONSE DURING BACTERIAL KERATITIS

Sanhita Roy[1,2,*], Prashant Garg[2] and Mausita Karmakar[3]

[1]Prof. Brien Holden Eye Research Centre, LV Prasad Eye Institute, Hyderabad, India

[2]Tej Kohli Cornea Institute, Hyderabad, India

[3]Department of Physiology and Biophysics, Case Western Reserve University, Cleveland, US

ABSTRACT

Corneal ulcers caused by microbial infections are a major cause of preventable blindness worldwide. Bacterial keratitis of the cornea is extremely painful, blinding and debilitating disease that is often difficult to treat and might require surgical intervention. In this chapter, we review the ocular innate immune response towards the pathogen by corneal epithelial cells and about clinical characteristics and management. Nearly 30% of bacterial keratitis cases have been reported in India, with infection caused by Gram-positive bacteria being more prominent. The bacterial infection of the cornea causes activation of the Toll-like receptors, which leads to cytokines and chemokines production along with recruitment of the neutrophils that are essential for killing the bacteria. Along with the pathogen recognition receptors on the ocular surface, tears, mucins, inflammasomes and antimicrobial peptides play an important role to detect and combat bacterial invasion. Though antibiotics remain the first choice and main form of treatment for corneal infections, continuous efforts are being made to identify alternative targets for therapy for bacterial infection to overcome the resistance attained by the pathogens against the antibiotics. Further research will help us better understand the disease and might provide management beyond antibiotic therapy.

* Corresponding Author Email: sanhita@lvpei.org.

Keywords: antimicrobial peptides, cornea, keratitis, innate immune response, ocular surface, Toll like receptors

INTRODUCTION

Microbial keratitis is worldwide most leading cause of non-refractive blindness, and almost 30-35% is of bacterial etiology (Gopinathan et al. 2009). Bacterial keratitis results corneal opacity that might lead to vision loss. It is extremely painful and explores rapidly and can perforate of the cornea and causes blindness. The monocular vision loss resulting from opacity of cornea adversely affects vision-related quality of life. The ocular surface is protected from infection by the intact epithelium that provides the first line of defense. Gram-positive as well as Gram-negative bacteria can cause bacterial keratitis. However, Gram-positive seems to be the most prevalent cause. Figure 1 is representative corneas of patients infected with Gram-positive or Gram-negative bacteria. The common pathogens include *Staphylococcus aureus*, *Pseudomonas aeruginosa*, *Streptococcus pneumoniae*, *Serretiamarcenes*.

Identification of virulence factors and the host response to the invading pathogen are critical to understanding the disease process in order to develop effective treatment modalities. Diffusion of inflammatory products like cytokines elicits outpouring of inflammatory cells and may create hypopyon. In case of severe infection and inflammation, complications like corneal leukoma and irregular astigmatism sometimes occurs. Chronic infections result in epithelial regeneration, vascularization, edema, myofibroblatic transformation and stromal scarring. The pathogenesis of bacterial keratitis is often complex, involving both bacterial factors and host components.

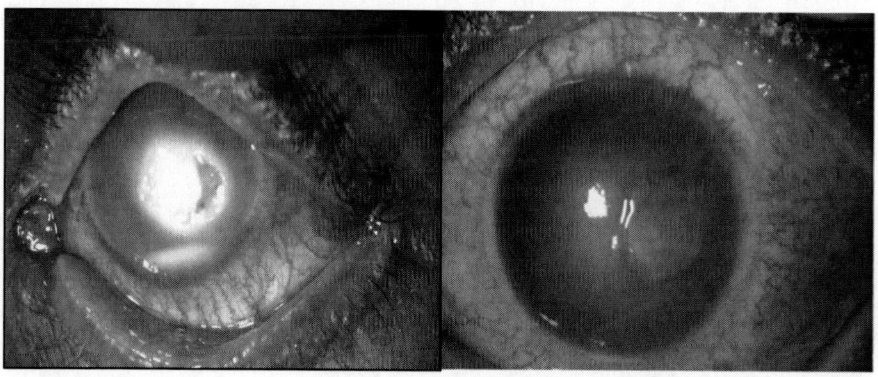

Figure 1. Clinical representation of eyes infected with Gram-positive (left) and Gram-negative bacteria (right). Corneal opacity is clearly visible with high level of inflammation.

EPIDEMIOLOGY

Lack of extensive epidemiological studies and serious risks of bacterial keratitis are nationwide and globally. Bacterial keratitis is one of the most prevalent causes of corneal infections and world widely second common cause of blindness after cataracts. The epidemiology for bacterial infection including causative microorganisms shows wide geographic variation. Therefore, a periodic region wide review in terms of etiology and epidemiology must be updated for diagnosis and management of the disease. It was reported that, 11 incidence of bacterial keratitis per 100,000 persons in USA (Erie et al. 1993), compared to about 800 per 100,000 persons in Nepal (Upadhyay et al. 2001), whereas it is as low as 6.3 per 100,000 in Hong Kong (Lam et al. 2002). Various studies across the world have reported bacterial keratitis to be accounted for over 30% of all cases of corneal infections. It has also reported 32.3% of bacterial keratitis cases in Madurai, India, as compared to almost 30% in Tiruchirapalli, India to about 25% in Ghana and 35% in south Florida (Kaliamurthy et al. 2013; Srinivasan et al. 1997).

An extensive study with almost 6000 patients at LV Prasad Eye Institute, a tertiary center at Hyderabad, reports bacterial keratitis to be present in over 50% of all the cases (Gopinathan et al. 2009). The Gram-positive bacteria were responsible for more than 60% of all the bacterial keratitis cases with *Staphylococcus epidermis* being the most prevalent agent, unlike reports from other parts of the world where *Streptococcus pneumoniae* is most dominant. Among the Gram-negative bacteria, *Pseudomonas aeruginosa* is the most common to cause infection. Prior ocular surgery and ocular trauma are the most pronounced predisposal factors for bacterial keratitis (Gopinathan et al. 2009). However, report based on the study done in western part of India, infection by Gram-positive bacteria was most prevalent, *S. aureus* being the causative agent, and ocular trauma being the most predisposing factor. A recent study from Brazil also reports more than 55% cases of microbial keratitis are of bacterial etiology. Apart from ocular trauma, contact lens wear is another most frequent predisposing factor for bacterial keratitis. The risk factor associated with corneal infection for contact lens wearers is poor storage case hygiene and infrequent case replacement. The risk of developing corneal infections is almost ten times greater with overnight use of contact lens. This is mainly as wearing lens over extended period of time induces hypoxia, increases corneal temperature and decreases tear flow over corneal epithelium. Unlike non-contact lenses related bacterial keratitis, where Gram-positive bacteria are predominant, *P. aeruginosa* is the most commonly recovered species. Severe microbial keratitis with vision loss in contact lens wearers is more likely to be caused by environmental pathogen. Genetic mutation in innate immune system may also be responsible for individual susceptibility to corneal infection in these cases. It was reported that in contact lens wearers with bacterial keratitis a link was identified between haplotypes of IL-10 and susceptibility of keratitis

(Keijser et al. 2009). The genotype of IL-6 has also been associated with severity of contact lens related keratitis (Carnt et al. 2011).

Innate Immune Response at the Ocular Surface

The surface of the cornea is intact and covered with thin layer of tear film and contains several antimicrobial peptides, lysozyme and mucins that prevent the adherence of microbial entities. The corneal epithelial cells constitute the first line of defense against all kinds of microbial pathogens. The epithelium is the outermost layer of the cornea which is about 50 μm thick and is stratified non-keratinized squamous layer. Figure 2 shows the image of the section of human cornea. Ocular surface epithelium selectively responds to microbial components and induces inflammation. Several studies have shown that corneal epithelium actively participates in host immune response to infection by secreting proinflammatory cytokines that recruit immune cells to fight pathogens, immune defense molecules like antimicrobial peptides, mucins that forms a dense glycocalyx and restricts bacterial adherence. Another unique feature of immune response at ocular surface epithelium is its co-existence with commensal bacteria.

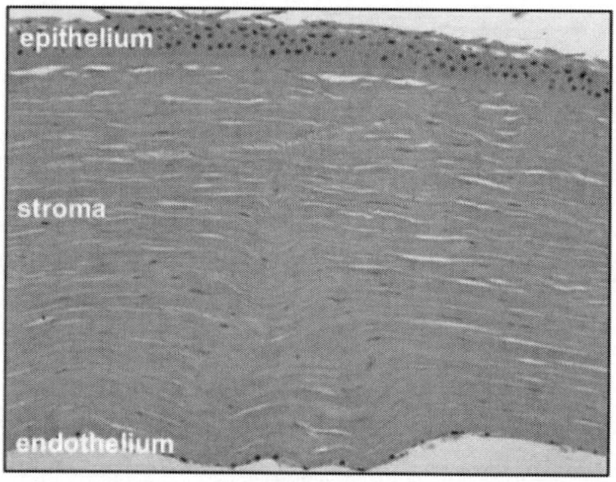

Figure 2. Representative image of section of human cornea. The three main layers, epithelium, stroma and endothelium are marked in the image. The epithelium is the outer most layer, whereas the endothelium is the innermost layer of the cornea.

Mucins (MUC) are hydrophilic O-glycosylated proteins associated with the glycocalyx of conjunctiva and corneal epithelium in secreted and membrane associated forms. It performs multiple functions as they prevent the drying of the ocular surface, hold the aqueous layer of the tear film, and protect the epithelium from microbial invasion. The ocular surface epithelium mainly consists of MUC1, 4, and 16. MUC 1 and MUC 16 are expressed by corneal and conjunctival epithelia whereas MUC4 is expressed exclusively by conjunctival epithelium. A gel forming mucin MUC5AC and a small soluble mucin MUC7 are also present at the ocular surface. Increase in corneal infection by *Corynebacteria* sp. and *Staphylococci* sp. has been reported in Muc $1^{-/-}$ mice (Kardon et al. 1999). Very recently it has also been reported that MUC 1 and 16 modulate inflammatory response at ocular surface by limiting TLR2 and TLR5 induced IL-6, IL-8 and TNFα expression (Menon et al. 2015).

Tears also form a main component of innate defense mechanism of ocular surface. It is often described as a three-layered structure comprising of lipid layer, an aqueous layer and a mucus layer over the corneal epithelium. The aqueous component accounts for 60% of the tear and mainly contains electrolytes and proteins. Tears help to keep the surface moist and flush foreign particles along with the transport of antimicrobial proteins like lysozyme, lactoferrin, lipocalin to the surface to prevent infection. The enzyme lysozyme mainly acts on the glycosidic bonds present in the cell wall of bacteria, whereas lactoferrin protects the ocular surface from the effects of free radicals. Tears have a high concentration of IgA that prevents bacterial adherence to the corneal epithelium. Increased level of IL-6, IL-8, and IL-1β are found in the tears of the patients with bacterial infection of the cornea, however there were no differences in IL-10 and IL-17A levels in tears from infected and uninfected eyes. Tear also participates in transportation of oxygen and carbon dioxide helping in the metabolism process.

Pathogen Recognition Receptors on Ocular Surface

A healthy ocular surface is free from any inflammatory responses. One of the first steps in innate immune response to bacterium is the activation of Toll like Receptors (TLR) family of pathogen recognition molecules. TLRs initially recognize different patterns that are intrinsic to microorganisms and boost an immune response in an attempt to clear the pathogens. Ten functional human TLRs have been identified along with their microbial ligands and deciphering of signaling pathways. TLR1, 2, 4, 5, 6 and 10 are surface expressed and recognize specific extracellular ligands. TLR3, 7, 8 and 9 are intracellular and are localized in the endosomal compartments. Once they activated by different ligands such as lipopolysaccharides or lipoteichoic acids, then the intracellular signals are relayed through signaling cascades mostly by activation of nuclear factor-kB (NfkB) and activator protein-1 (AP-1) through MyD88 dependent pathways. Alternatively, it also leads to activation of type I interferons in a TRIF dependent pathway that eventually lead to induction of several cytokines, chemokines, antimicrobial peptides and cell adhesion molecules. The TLR signaling in corneal epithelium is highly regulated, as it is in constant exposure to pathogens, both commensal and environmental. The signaling is controlled by several negative regulators and expression patterns of these TLRs. Ocular surface epithelial cells like corneal and conjunctival epithelial cells are known to express several TLRs (Pearlman et al. 2008). Several studies have also shown that TLRs are expressed on corneal macrophages, and depletion of macrophages and dendritic cells impairs bacterial killing. Gram-positive bacteria are mostly recognized by TLR2, whereas TLR4 recognizes lipopolysaccharide present in the Gram-negative bacteria. There are reports that show human corneal epithelial cells are poorly responsive to LPS due to low expression level of MD-2, an essential co-activator of TLR4 (Roy et al. 2011). However, MD-2 gets up regulated in presence of infection and initiates immune response in these cells. TLR4 is also expressed in corneal stromal fibroblasts, iris endothelial cells and retinal pigment epithelial cells. As mentioned above, activation of TLR signaling must be controlled and fine-tuned, otherwise prolonged inflammatory response causes tissue damage. Several negative regulator of TLR signaling like ST2 and Sigirr have been identified over past several years. ST2, a member of TLR/IL-1R superfamily, is expressed on the surface of fibroblasts in both soluble and transmembrane form and has anti-inflammatory properties (Yu and Hazlett 2006). Sigirr has the typical conserved TIR domain and has been shown to play a role in *P. aeruginosa* keratitis by down regulating IL-1R and TLR4 signaling (Huang et al. 2006). It was also demonstrated that, enhanced *P. aeruginosa* keratitis when Sigirr was blocked. Along with TLRs, Nod like receptors (NLR), expressed within immune cells, play a critical role in regulation of ocular innate immune responses. They function as important sensors of microbial derived and danger associated molecules in the host cell cytoplasm. Till date, NLRP3 and NLRC4 are the two inflammasomes that have been identified to assemble on the corneal surface. These inflammasomes forming NLRs mediate processing of proinflammatory cytokines that are then secreted by cells. Non-inflammasomes forming NLR members activate other major cell signaling pathways. *S. pneumoniae* and *P. aeruginosa* has been shown to activate NLRP3 and NLRC4 in human keratitis. Similarly, NLRP3 also gets induced in human corneal epithelial cells in response to *C. pseudodiphtheriticum* (Roy et al. 2015).

TLRS IN CORNEAL INFLAMMATION

Although TLRs play a major role in the innate immune response to bacteria, it also plays role in regulating ocular inflammation. Dysfunctional tear syndrome or commonly known as dry eye is the most common autoimmune condition affecting the eye, and results due to immune based inflammation in the tears and ocular surface tissues. The clinical manifestations are eye irritation, photophobia and blurred vision. TLRs are also found to be involved in dry eye and up-regulation of TLR 2, 3 and 5 were found in an experimental mouse model with dry eye along with the down regulation of TLR4 in the corneal epithelium (Redfern et al. 2013). It was also observed that there was decreased expression of AMPs like LL-37 and mBD-4 in these mice. Several recent reports also show that dry eye patients tend to have an altered/higher bacteria load at the ocular surface compared to healthy eye. Detailed research is needed to see if dry eye patients are more prone to corneal infection compared to their healthy counterparts.

TLRS IN CORNEAL INFECTION

Gram-positive bacteria are a major cause of corneal infections. *S. aureus* mediates cytokine production (Adhikary et al. 2008) and CCL20 expression (Heimer et al. 2010) by both HCEC and neutrophils in TLR-2, but not TLR-4, dependent manner. *S. pneumoniae* is a major cause of corneal infection in India and developing countries. It has been observed that expression of TLR2, 4, 5 and 9 get elevated in human corneas infection with these bacteria (Karthikeyan et al. 2011). Increased expression of TLR 2, 3, and 4 were also observed in patients with *C. pseudodiphtheriticum* infection, along with increased expression of antimicrobial peptides S100A8 and S100A9. Similar up-regulation of AMPs was found in human corneal epithelial cells mediated by activation of NfkB and p38 pathways (Roy et al. 2015). Ocular innate immune response to *P. aeruginosa* is mediated by TLR4, 5 and activation of both TRIF and MyD88 pathways take place in epithelial cells and resident macrophages. Increased IL-6, IL-8 and IL-1β expression mediated by TLRs is also observed in *P. aeruginosa* corneal infections. Along with TLR4, CD14 also regulates cytokine production, immune cells infiltration and bacterial clearance during *P. aeruginosa* keratitis (Roy et al. 2014). TLR9 on the other hand appears to negatively regulate *P. aeruginosa*-induced corneal inflammation, and TLR9 depletion in a mouse model of keratitis caused increased bacterial load compared to the control eye (Huang et al. 2005). Clinical isolate of *Serretia marcescens* has been shown to activate TLR4, 5 and induce chemokine production regulated by MyD88 and IL-1R1 in a murine model of keratitis (Zhou et al. 2012). The roles of TLRs are also been reported in sterile keratitis in a mouse model, where LPS from *P. aeruginosa* was used (Khatri et al. 2002).

Apart from the role of TLRs in infection and inflammation, it has been reported that injury to corneal epithelium activates TLRs and contribute to epithelial repair by inducing cell migration and proliferation (Eslani et al. 2014). TLRs are also known to play a role in bacterial endophthalmitis (Novosad et al. 2011).

INFLAMMASOMES AND BACTERIAL KERATITIS

Inflammasomes are innate immune sensory receptors that detect pathogenic microorganisms including bacteria, virus, and fungus as well as sterile stressors like ATP, uric acid crystals, high mobility group box-1 (HMGB1) that are released out of dead or dying cells at the site of inflammation. Activation of inflammasome in response to these PAMPs (Pathogen associated molecular patterns) or DAMPs (Danger associated molecular patterns) leads to a series of downstream signaling to generate active caspase-1 which can then produce highly proinflammatory cytokines like IL-1β and IL-18 (Broz and Dixit 2016; Guo et al. 2015; Latz et al. 2013). Another functional output of active caspase-1 is that it induces a proinflammatory form of cell death called pyroptosis in inflammasome activated cells. Recent studies have shown that active caspase-1 can cleave a predominantly cytosolic protein called gasdermin-D in macrophages to generate a small N-terminal fragment. This N-terminal cleaved gasdermin-D oligomerizes to create a pyroptotic pore in the host cell plasma membrane and cause rapid lytic form of cell death called pyroptosis(He et al. 2015; Man and Kanneganti 2015; Shi et al. 2015). Inflammasomes are classified based on their structural features into nucleotide-binding domain-like receptors (NLRs), absent in melanoma 2 (AIM-2)-like receptors (ALRs) and the recently identified pyrin. The NLR can be further subdivided into NLRP or NLRC inflammasome based on whether it contains a pyrin or caspase activation and recruitment domain (CARD), respectively (Latz et al. 2013). Among different types of inflammasomes, NLRP3 and NLRC4 are expressed in corneal ulcers of patients with *P. aeruginosa* and *S. pneumoniae* keratitis (Karthikeyan et al. 2013). Recent study has shown that in murine model of *S. pneumoniae* keratitis, neutrophils were the major source of IL-1β at the site of inflammation. Additionally, neutrophils were shown to express functional NLRP3 inflammasome which can be activated by pore forming toxin - pneumolysin expressed by *S. pneumonia* leading to K^+ efflux and formation of multiple NLRP3/ASC complexes together with active caspase-1, resulting in secretion of mature IL-1β by these cells (Karmakar et al. 2015). Neutrophil NLRP3 derived IL-1β in the cornea helps to orchestrate the inflammatory processes during *S. pneumoniae* keratitis. Furthermore, extracellular ATP that are released by dying cells in the cornea during *S. pneumoniae* keratitis can further bind to purinergic P2X7 receptors expressed on neutrophil plasma membrane and activate NLRP3 inflammasome to mediate additional IL-1β secretion by these cells (Karmakar et al. 2016). Since neutrophils are abundant in early stages of microbial keratitis where extracellular ATP is generated, activation of P2X7 receptor on these cells exacerbates inflammatory responses by mediating NLRP3 inflammasome assembly and IL-1β secretion. *Staphylococcus aureus*, another causative agent of bacterial keratitis, expresses Panton-Valentine leukocidin (PVL)- a pore-forming toxin (Archer 1998) that has been shown to activate NLRP3 inflammasome in human monocyte and macrophages and secrete proinflammatory IL-1β in caspase-1 dependent manner (Holzinger et al. 2012). This also correlates with the fact that PVL+ clinical isolates were associated with a trend to worse clinical outcome and more surgical interventions (Sueke et al. 2013). This suggests that *this pore forming toxin* may be an important virulence factor in *S. aureus* keratitis which drives robust inflammatory responses by activating NLRP3 inflammasome and IL-1β secretion. α-hemolysin is another pore-forming toxin produced by *S. aureus* that is capable of producing extensive corneal disease. α-toxin-producing strains have been shown to cause significantly greater ocular inflammation and corneal damage than α-toxin-deficient mutant strains in experimental rabbit model of keratitis (O'Callaghan et al. 1997). Since α-hemolysin can activate NLRP3 inflammasomes (Craven et al. 2009; Kebaier et al. 2012), it is possible that the severity of corneal infection by α-toxin-producing *S. aureus* are at least in part due to exacerbated IL-1β derived from inflammasomes. This is further supported by the fact that goblet cells in rat conjunctiva can activate NLRP3 inflammasomes and secrete proinflammatory IL-1β in a caspase-1 dependent fashion, when infected with toxin-expressing *S. aureus* (McGilligan et al. 2013).

Although inflammasome dependent IL-1β has been shown to mediate inflammatory responses in multiple bacterial keratitis, neutrophil derived serine proteases like elastase can also process pro IL-1β to mature form and contribute to inflammatory responses in murine model of *P. aeruginosa* keratitis (Karmakar et al. 2012). Since NLRC4 inflammasome expression was found to be elevated in corneal ulcers of patients with *P. aeruginosa* keratitis (Karthikeyan et al. 2013), it is possible that the bacterial virulence factors like Type III secretion system (T3SS) or flagellin-which are known activators of NLRC4 inflammasome and may contribute further to proinflammatory IL-1β at the site of infection.

Cellular Infiltrates in Corneal Infection

Although mammalian cornea for a long time has been thought to be immune privileged and being devoid of myeloid cells owing to its avascular nature, several studies have been demonstrated that corneal epithelium and stromal contain heterogeneous population of bone marrow derived cells including macrophages and dendritic cells. Earlier studies reported the presence of Langherhans cells in the peripheral of human cornea (Rodrigues et al. 1981). Once infection with bacteria takes place through epithelial ulceration, resident cells within tissue initiates production of proinflammatory cytokines and chemokines leading to destruction of Bowman's layer and Descemet's membrane. Neutrophils are the first cells that migrate from peripheral limbal vessels, as a principal host defense mechanism, since their primary role is to destroy the invading pathogens. Although macrophages play an important role in modulation of immune response, more than 70% of circulating leukocytes are comprised of neutrophils in human. Several steps are involved in recruitment of immune cells to the cornea that has been elegantly deciphered in several studies of corneal infection using animal models. Macrophages also play an important role in host defense in bacterial keratitis. The dendritic cells in cornea have been seen to communicate with the neighboring dendritic cells by forming highly sophisticated membrane nanotubes. It is now clear that human cornea is dwelled by resident myeloid cells along with MHV II+/CD45+ bone marrow derived cells (Sosnova et al. 2005). There are also reports regarding the essential role played by CD4+IFNg+T cells in murine model of bacterial keratitis (Kwon and Hazlett 1997). Very recently CD4+IL-17+T cells have also been shown to play a role in neutrophil influx and pathology during bacterial keratitis.

ANTIMICROBIAL PEPTIDES

Antimicrobial peptides are naturally occurring, ancient innate immune response against variety of organisms. The presence of defensins in human tears has been detected almost twenty years back (Haynes et al. 1999). Human β defensins has been shown to be expressed by human corneal epithelial cells in response to bacteria and several bacterial ligands in NfkB and MAPK dependent pathways, also mediated by TLRs (Froy 2005). A few years back, it was reported that the presence of novel DEF-109 AMP on ocular surface and was found to have constitutive expression (Abedin et al. 2008). Another novel antimicrobial peptide, known as keratin derived antimicrobial peptide has been reported to be expressed from cytokeratin-6A of ocular epithelial cells (Tam et al. 2012). Expression of LL-37 has also been reported in human neutrophils during *P. aeruginosa* keratitis (Alalwani et al. 2010). In our laboratory, we have seen expression of LL-37 by human corneal epithelial cells in response to *S. pneumoniae* as shown in Figure 3 (unpublished data). Several antimicrobial peptide like defensins, cathelecidins and S100 proteins expression has been also seen in patients infected with *C. pseudodiphtheriticum* (Roy et al. 2015).

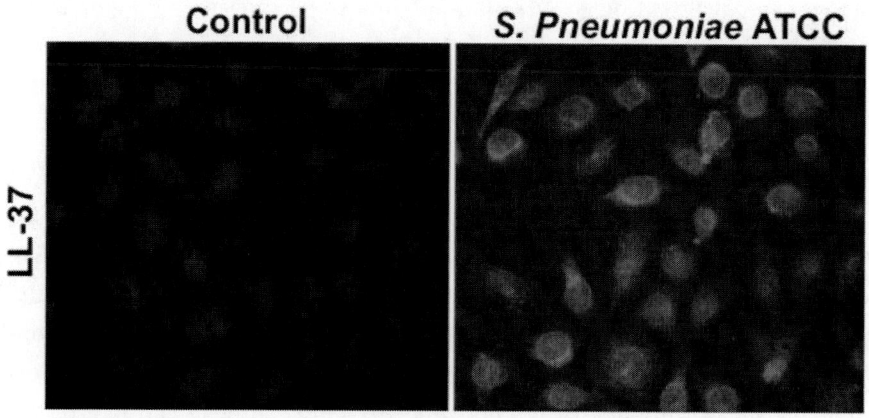

Figure 3. Expression of LL-37 by human corneal epithelial cells. Human corneal epithelial cells were exposed to *S. pneumoniae* for 2h, washed and fixed, and LL37 expression were determined by incubating the cells with LL-37 antibody followed by Alexafluor 488 secondary antibody and visualized and imaged by fluorescent microscopy using 20X objective.

Virulence Factors

A detailed knowledge about virulence factors is prerequisite needs to understand the pathogenesis of disease causing bacteria and to develop appropriate drugs to combat infections. *S. aureus* is the most common cause of bacterial keratitis, and virulence for this bacteria has been attributed to alpha-toxin, a lytic cytotoxin produced by all *S. aureus* strains. Among the other toxins responsible for the virulence of *S. aureus* are gamma-toxins and setnm-1 gene, which is a part of a cluster of genes for super-antigen-like proteins (Marquart and O'Callaghan 2013). The outer capsule of *S. pneumoniae* is the major virulence factor along with variety of proteins. One such protein is pneumolysin that binds to the host cell membranes, and forms pore that causes host cell lysis (Johnson et al. 1990). *P. aeruginosa* on the other hand has evolved a horde of virulence factors and mechanisms that allow it to survive in human cornea and causes disease. The major virulence factor is the type III secretion system in which ExoS, ExoU or ExoT are injected into host cells on contact with *P. aeruginosa*. ExoS and ExoT are highly homologous bifunctional enzymes whereas ExoU is a potent phospholipase that causes tissue destruction (Hauser 2009). Lipopolysaccharide is another important virulence factor for *P. aeruginosa* along with exotoxin A, elastases, alkaline protease, protease IV or phospholipase C (Heimer et al. 2013).

Bacterial Resistance

Antibiotic resistance among ocular isolates is on rise, and the primary factor responsible is the overuse of topical antibiotic in the eye. Other factors include improper dosage regimen, and misuse of antibiotics for viral and other nonbacterial infections. Different antibiotics that are used for treatment of bacterial infections include aminoglycosides, polymyxin B combinations and fluoroquinolones. In a recent study conducted in a tertiary center in South India, *S. pneumoniae* and *P. aeruginosa* were two most common pathogens with resistance. Antibiotic non-susceptibility was generally uncommon for these two organisms. In contrast, there was increase in fluoroquinolone non-susceptibility over the period of 12-year study period (Lalitha et al. 2016). In a study conducted in United Kingdom, the Gram-negative isolates were found to be highly resistance to chloramphenicol, and more sensitive to combination of gentamicin and cefuroxime, whereas ciprofloxacin resistance was found in about 17% of the Gram-positive isolates (Shalchi et al. 2011). There also has been an increase in percentage of *P. aeruginosa* isolates showing resistance to ciprofloxacin, and it is not uncommon to come across multidrug resistant *P. aeruginosa* isolates in keratitis. Multidrug resistant *P.*

aeruginosa keratitis is a major therapeutic challenge without the availability of alternative antibiotics. Certain level of antibiotic resistance *Enterobacteriaceae* has also been isolated from eye infections (Sharma 2011).

Clinical Characteristics and Management

The clinical picture in bacterial keratitis is typically characterized by sudden onset of pain, redness, watering from eyes and light sensitivity. Clinical examination shows cloudy cornea with epithelial defects, stromal infiltration, surrounding cellular infiltration and edema and anterior chamber reaction. The clinical picture though depends on the nature of microorganisms and the duration of disease occurrence, shows wide variability among different individuals. Antibiotics remain the main form of treatment for bacterial keratitis. In the absence of laboratory diagnosis, the initial therapy is usually broad-spectrum intensive therapy. Collagen cross linking of the cornea is becoming a new mode of treatment for multidrug resistant infectious keratitis. The combination of riboflavin and UVA has been extensively used in collagen cross-linking (Vazirani and Vaddavalli 2013).

Pathology

The pathology of bacterial ulcer comprises of breach in corneal epithelium with loss of corneal stromal associated with dense infiltration by polymorphonuclear leucocytes. Bacterial strains show presence of causative microorganisms in the corneal tissue. Previous studies showed that, in animal models of keratitis is difficult to induce infection in healthy eyes. The experience point out that healthy eye is equipped with defence mechanism to provide protection against invasion by pathogenic microorganisms. A disruption in the defence factors predisposes an individual to corneal infection. Therefore, the pathology of corneal ulcer represents interplay of agent factors and host responses both for the genesis and progression of the disease. While a lot of effort has gone into understanding agent factors and in fact the current management strategies comprise of drugs that eliminate invading microorganisms. It will be important to understand host factors. This will help us better understand the disease and might provide management beyond antibiotic therapy.

ACKNOWLEDGMENTS

Financial support from SERB-DST, Govt. of India (SB/YS/LS-180/2013) to S.R. and from Wellcome Trust and DBT, Govt. of India to P.G. is heartily acknowledged.

REFERENCES

Abedin A, Mohammed I, Hopkinson A, Dua, HS (2008) A novel antimicrobial peptide on the ocular surface shows decreased expression in inflammation and infection. *Invest Ophthalmol Vis Sci* 49: 28-33.

Adhikary G, Sun Y, Pearlman E (2008) C-Jun NH2 terminal kinase (JNK) is an essential mediator of Toll-like receptor 2-induced corneal inflammation. *J Leukoc Biol* 83: 991-997.

Alalwani SM, Sierigk J, Herr C, Pinkenburg O, Gallo R, Vogelmeier C, Bals R (2010) The antimicrobial peptide LL-37 modulates the inflammatory and host defense response of human neutrophils. *Eur J Immunol* 40: 1118-1126.

Archer GL (1998) *Staphylococcus aureus*: a well-armed pathogen. Clinical infectious diseases: an official publication of the *Infectious Diseases Society of America* 26: 1179-1181.

Broz P, Dixit VM (2016) Inflammasomes: mechanism of assembly, regulation and signalling. *Nat Rev Immunol* 16: 407-420.

Carnt N, Keay L, Willcox MD, Evans V, Stapleton F (2011) Pilot study of contact lens practitioner risk-taking propensity. Optometry and vision science: official publication of the *American Academy of Optometry* 88: E981-987.

Craven RR, Gao X, Allen IC, Gris D, BubeckWardenburg J, McElvania-Tekippe E, Ting JP, Duncan JA. (2009) *Staphylococcus aureus* alpha-hemolysin activates the NLRP3-inflammasome in human and mouse monocytic cells. *PloS ONE* 4: e7446.

Erie JC, Nevitt MP, Hodge DO, Ballard DJ (1993) Incidence of ulcerative keratitis in a defined population from 1950 through 1988. *Arch Ophthalmol* 111: 1665-1671.

Eslani M, Movahedan A, Afsharkhamseh N, Sroussi H, Djalilian AR (2014) The role of toll-like receptor 4 in corneal epithelial wound healing. *Invest Ophthalmol Vis Sci* 55: 6108-6115.

Froy O (2005) Regulation of mammalian defensin expression by Toll-like receptor-dependent and independent signalling pathways. *Cell Microbiol* 7: 1387-1397.

Gopinathan U, Sharma S, Garg P, Rao GN (2009) Review of epidemiological features, microbiological diagnosis and treatment outcome of microbial keratitis: experience of over a decade. *Indian J Ophthalmol* 57: 273-279.

Guo H, Callaway JB, Ting JP (2015) Inflammasomes: mechanism of action, role in disease, and therapeutics. *Nat Med* 21: 677-687.

Hauser AR (2009) The type III secretion system of Pseudomonas aeruginosa: infection by injection. Nature reviews. *Microbiology* 7: 654-665.

Haynes RJ, Tighe PJ, Dua HS (1999) Antimicrobial defensin peptides of the human ocular surface. *Br J Ophthalmol* 83: 737-741.

He WT, Wan H, Hu L, Chen P, Wang X, Huang Z, Yang ZH, Zhong CQ, Han J (2015) Gasdermin D is an executor of pyroptosis and required for interleukin-1beta secretion. *Cell Res* 25: 1285-1298.

Heimer SR, Evans DJ, Stern ME, Barbieri JT, Yahr T, Fleiszig SM (2013) *Pseudomonas aeruginosa* utilizes the type III secreted toxin ExoS to avoid acidified compartments within epithelial cells. *PLoS One* 8: e73111.

Heimer SR, Yamada A, Russell H, Gilmore M (2010) Response of corneal epithelial cells to *Staphylococcus aureus. Virulence* 1: 223-235.

Holzinger D, Gieldon L, Mysore V, Nippe N, Taxman DJ, Duncan JA, Broglie PM, Marketon K, Austermann J, Vogl T, Foell D (2012) *Staphylococcus aureus* Panton-Valentine leukocidin induces an inflammatory response in human phagocytes via the NLRP3 inflammasome. *J LeukocBiol* 92: 1069-1081.

Huang X, Barrett RP, McClellan SA, Hazlett LD (2005) Silencing Toll-like receptor-9 in *Pseudomonas aeruginosa* keratitis. *Invest Ophthalmol Vis Sci* 46: 4209-4216.

Huang X, Hazlett LD, Du W, Barrett RP (2006) SIGIRR promotes resistance against *Pseudomonas aeruginosa* keratitis by down-regulating type-1 immunity and IL-1R1 and TLR4 signaling. *J Immunol* 177: 548-556.

Johnson MK, Hobden JA, Hagenah M, O'Callaghan RJ, Hill JM, Chen S (1990) The role of pneumolysin in ocular infections with *Streptococcus pneumoniae*. *Curr Eye Res* 9: 1107-1114.

Kaliamurthy J, Kalavathy CM, Parmar P, Nelson Jesudasan CA, Thomas PA (2013) Spectrum of bacterial keratitis at a tertiary eye care centre in India. *Biomed Res Int* 2013: 181564.

Kardon R, Price RE, Julian J, Lagow E, Tseng SC, Gendler SJ, Carson DD (1999) Bacterial conjunctivitis in Muc1 null mice. *Invest Ophthalmol Vis Sci* 40: 1328-1335.

Karmakar M, Katsnelson M, Malak HA, Greene NG, Howell SJ, Hise AG, Camilli A, Kadioglu A, Dubyak GR, Pearlman E (2015) Neutrophil IL-1beta processing induced by pneumolysin is mediated by the NLRP3/ASC inflammasome and caspase-1 activation and is dependent on K+ efflux. *J Immunol* 194: 1763-1775.

Karmakar M, Katsnelson MA, Dubyak GR, Pearlman E (2016) Neutrophil P2X7 receptors mediate NLRP3 inflammasome-dependent IL-1beta secretion in response to ATP. *Nat Commun* 7: 10555.

Karmakar M, Sun Y, Hise AG, Rietsch A, Pearlman E (2012) Cutting edge: IL-1beta processing during *Pseudomonas aeruginosa* infection is mediated by neutrophil serine proteases and is independent of NLRC4 and caspase-1. *J Immunol* 189: 4231-4235.

Karthikeyan RS, Leal SM, Jr, Prajna NV, Dharmalingam K, Geiser DM, Pearlman E, Lalitha P (2011) Expression of innate and adaptive immune mediators in human corneal tissue infected with *Aspergillus* or *fusarium*. *J Infect Dis* 204: 942-950.

Karthikeyan RS, Priya JL, Leal SM, Jr, Toska J, Rietsch A, Prajna V, Pearlman E, Lalitha P (2013) Host response and bacterial virulence factor expression in *Pseudomonas aeruginosa* and *Streptococcus pneumoniae* corneal ulcers. *PloS one* 8: e64867.

Kebaier C, Chamberland RR, Allen IC, Gao X, Broglie PM, Hall JD, Jania C, Doerschuk CM, Tilley SL, Duncan JA (2012) *Staphylococcus aureus* alpha-hemolysin mediates virulence in a murine model of severe pneumonia through activation of the NLRP3 inflammasome. *J Infect Dis* 205: 807-817.

Keijser S, Kurreeman FA, de Keizer RJ, Dogterom-Ballering H, van der Lelij A, Jager MJ, Nibbering PH (2009) IL-10 promotor haplotypes associated with susceptibility to and severity of bacterial corneal ulcers. *Exp Eye Res* 88: 1124-1128.

Khatri S, Lass JH, Heinzel FP, Petroll WM, Gomez J, Diaconu E, Kalsow CM, Pearlman E (2002) Regulation of endotoxin-induced keratitis by PECAM-1, MIP-2, and toll-like receptor 4. *Invest Ophthalmol Vis Sci* 43: 2278-2284.

Kwon B, Hazlett LD (1997) Association of CD4+ T cell-dependent keratitis with genetic susceptibility to *Pseudomonas aeruginosa* ocular infection. *J Immunol* 159: 6283-6290.

Lalitha P, Manoharan G, Karpagam R, Prajna NV, Srinivasan M, Mascarenhas J, Das M, Porco TC, Lietman TM, Cevallos V, Keenan JD (2017) Trends in antibiotic resistance in bacterial keratitis isolates from South India. *Br J Ophthalmol* 101(2):108-113.

Lam DS, Houang E, Fan DS, Lyon D, Seal D, Wong E, Hong Kong Microbial Keratitis Study, G (2002) Incidence and risk factors for microbial keratitis in Hong Kong: comparison with Europe and North America. *Eye (Lond)*16: 608-618.

Latz E, Xiao TS, Stutz A (2013) Activation and regulation of the inflammasomes. *Nature reviews. Immunology* 13: 397-411.

Man SM, Kanneganti TD (2015) Gasdermin D: the long-awaited executioner of pyroptosis. *Cell Res* 25: 1183-1184.

Marquart ME, O'Callaghan RJ (2013) Infectious keratitis: secreted bacterial proteins that mediate corneal damage. *J Ophthalmol* 2013: 369094.

McGilligan VE, Gregory-Ksander MS, Li D, Moore JE, Hodges RR, Gilmore MS, Moore TC, Dartt DA (2013) *Staphylococcus aureus* activates the NLRP3 inflammasome in human and rat conjunctival goblet cells. *PloS One* 8: e74010.

Menon BB, Kaiser-Marko C, Spurr-Michaud S, Tisdale AS, Gipson IK (2015) Suppression of Toll-like receptor-mediated innate immune responses at the ocular surface by the membrane-associated mucins MUC1 and MUC16. *Mucosal Immunol* 8: 1000-1008.

Novosad BD, Astley RA, Callegan MC (2011) Role of Toll-like receptor (TLR) 2 in experimental *Bacillus cereus* endophthalmitis. *PLoS One* 6: e28619.

O'Callaghan RJ, Callegan MC, Moreau JM, Green LC, Foster TJ, Hartford OM, Engel LS, Hill JM (1997) Specific roles of alpha-toxin and beta-toxin during *Staphylococcus aureus* corneal infection. *Infect Immun* 65: 1571-1578.

Pearlman E, Johnson A, Adhikary G, Sun Y, Chinnery HR, Fox T, Kester M, McMenamin PG (2008) Toll-like receptors at the ocular surface. *Ocul Surf* 6: 108-116.

Redfern RL, Patel N, Hanlon S, Farley W, Gondo M, Pflugfelder SC, McDermott AM (2013) Toll-like receptor expression and activation in mice with experimental dry eye. *Invest Ophthalmol Vis Sci* 54: 1554-1563.

Rodrigues MM, Rowden G, Hackett J, Bakos I (1981) Langerhans cells in the normal conjunctiva and peripheral cornea of selected species. *Invest Ophthalmol Vis Sci* 21: 759-765.

Roy S, Karmakar M, Pearlman E (2014) CD14 mediates Toll-like Receptor 4 (TLR4) endocytosis and spleen tyrosine kinase (SYK) and IRF3 activation in epithelial cells and impairs neutrophil infiltration and *Pseudomonas aeruginosa* killing in vivo. *J Biol Chem* 289: 1174-1182.

RoyS, Marla S, Praneetha DC (2015) Recognition of *Corynebacterium pseudodiphtheriticum* by Toll-like receptors and up-regulation of antimicrobial peptides in human corneal epithelial cells. *Virulence* 6: 716-721.

Roy S, Sun Y, Pearlman E (2011) Interferon-gamma-induced MD-2 protein expression and lipopolysaccharide (LPS) responsiveness in corneal epithelial cells is mediated by Janus tyrosine kinase-2 activation and direct binding of STAT1 protein to the MD-2 promoter. *J Biol Chem* 286: 23753-23762.

Shalchi Z, Gurbaxani A, Baker M, Nash J (2011) Antibiotic resistance in microbial keratitis: ten-year experience of corneal scrapes in the United Kingdom. *Ophthalmology* 118: 2161-2165.

Sharma, S (2011) Antibiotic resistance in ocular bacterial pathogens. *Indian J Med Microbiol* 29: 218-222.

Shi J, Zhao Y, Wang K, Shi X, Wang Y, Huang H, Zhuang Y, Cai T, Wang F, Shao F (2015) Cleavage of GSDMD by inflammatory caspases determines pyroptotic cell death. *Nature* 526: 660-665.

Sosnova M, Bradl M, Forrester JV (2005) $CD34^{+}$ corneal stromal cells are bone marrow-derived and express hemopoietic stem cell markers. *Stem cells* 23: 507-515.

Srinivasan M, Gonzales CA, George C, Cevallos V, Mascarenhas JM, Asokan B, Wilkins J, Smolin G, Whitcher JP (1997) Epidemiology and aetiological diagnosis of corneal ulceration in Madurai, south India. *Br J Ophthalmol* 81: 965-971.

Sueke H, Shankar J, Neal T, Winstanley C, Tuft S, Coates R, Horsburgh MJ, Kaye S (2013) lukSF-PV in *Staphylococcus aureus* keratitis isolates and association with clinical outcome. *Invest Ophthalmol Vis Sci* 54: 3410-3416.

Tam C, Mun JJ, Evans DJ, Fleiszig SM (2012) Cytokeratins mediate epithelial innate defense through their antimicrobial properties. *J Clin Invest* 122: 3665-3677.

Upadhyay MP, Karmacharya PC, Koirala S, Shah DN, Shakya S, Shrestha JK, Bajracharya H, Gurung CK, Whitcher JP (2001) The Bhaktapur eye study: ocular trauma and antibiotic prophylaxis for the prevention of corneal ulceration in Nepal. *Br J Ophthalmol* 85: 388-392.

Vazirani J, Vaddavalli PK (2013) Cross-linking for microbial keratitis. *Indian J Ophthalmol* 61: 441-444.

Yu FS, Hazlett LD (2006) Toll-like receptors and the eye. *Invest Ophthalmol Vis Sci* 47: 1255-1263.

Zhou R, Zhang R, Sun Y, Platt S, Szczotka-Flynn L, Pearlman E (2012) Innate immune regulation of *Serratiamarcescens*-induced corneal inflammation and infection. *Invest Ophthalmol Vis Sci* 53: 7382-7388.

In: Trends in Life Science Research
Editors: R. P. Sinha and U. P. Shrivastava

ISBN: 978-1-53613-241-0

Chapter 7

ULTRAVIOLET RADIATION-INDUCED DNA DAMAGE AND MECHANISMS OF REPAIR IN CYANOBACTERIA: AN OVERVIEW

Rajneesh[1], Ananya Chatterjee[1], Jainendra Pathak[1], Haseen Ahmed[1], Vidya Singh[1], Deepak K. Singh[1], Abha Pandey[1], Shailendra P. Singh[2], Richa[1], Donat-P. Häder[3,†] and Rajeshwar P. Sinha[1,*]

[1]Laboratory of Photobiology and Molecular Microbiology,
Centre of Advanced Study in Botany, Institute of Science,
Banaras Hindu University, Varanasi, India
[2]Centre of Advanced Study in Botany, Institute of Science,
Banaras Hindu University, Varanasi, India
[3]Friedrich-Alexander University, Department of Biology,
Möhrendorf, Germany

ABSTRACT

For the proper functioning of all living organisms, including cyanobacteria, the stability and fidelity of a DNA molecule is very much important. DNA is easily affected by toxic chemicals and harmful radiations. Ultraviolet radiation (UVR), specifically UV-B promotes DNA damage and lesions, such as: cyclobutane-pyrimidine dimers (CPDs), 6-4 photoproducts (6-4PPs), their Dewar valence isomers and even DNA strand break. Cyanobacteria emerged on Earth's when the protective shield of stratospheric

[†] Donat-P. Häder, Emeritus from Friedrich-Alexander University.

[*] Corresponding Author Email: r.p.sinha@gmx.net.

ozone layer was absent and during the course of evolution, have developed a number of repair mechanisms to cope up with UV stress. The first and foremost mechanism is photoreactivation with the help of the enzyme "photolyase." The other mechanisms include base excision repair (BER) and nucleotide excision repair (NER) in which many enzymes with polymerase, glycosylase and nuclease activity have utmost role. Furthermore, other mechanisms such as recombinational repair, SOS repair are also important, and, if the damage is not repairable, apoptosis or programmed cell death is the final step to prevent the damaged DNA to get passed on in next generations. In this chapter, we have broadly discussed the types of DNA damage that can occur in cyanobacteria in response to UV stress and the various pathways to repair the damage or lesion. The objective will be a broad and clear understanding of these mechanisms and approaches to apply these in human cellular environment.

Keywords: UV stress, DNA damage, cyclobutane-pyrimidine dimers, photolyase, base excision repair, nucleotide excision repair

INTRODUCTION

Cyanobacteria are phylogenetically a primitive group of Gram negative prokaryotes having a cosmopolitan distribution ranging from hot springs to the Arctic and Antarctic regions (Stanier and Cohen-Bazire 1977). They were probably the first photosynthetic oxygen-evolving organisms that appeared during the Precambrian era (between 2.8 and 3.5×10^9 years ago) and provided a favourable condition for the evolution of current aerobic life (Fischer 2008). In addition, their inherent capacity to fix atmospheric nitrogen makes them ecologically important for rice-growing countries where they add to fertility of the rice fields as natural biofertilizer (Vaishampayan et al. 2001). They use the enzyme "nitrogenase" which directly converts the atmospheric nitrogen into ammonium (NH_4^+), a form through which nitrogen enters the food web. It has been calculated that cyanobacteria fix over 35 million tons of nitrogen annually and thus occupy a central position in nutrient cycling of ecosystems (Häder et al. 1989; Rajneesh et al. 2017a).

The stratospheric ozone layer is continuously depleting due to the release of atmospheric pollutants such as chlorofluorocarbons (CFCs), chlorocarbons (CCs), and organobromides (OBs). Consequently there is an increase in the incidence of ultraviolet radiation (UVR) on the Earth's surface (Lubin and Jensen 1995; Newman et al. 2006). In all the groups of UVR (UV-A: 315-400 nm; UVB: 280-315 nm; UV-C: <280 nm) UV-B radiation produces adverse effects on diverse habitats, even though most of the extraterrestrial UV-B is absorbed by the stratospheric ozone (Mckenzie et al. 2003). The depletion of stratospheric ozone increases deleterious UVR at the Earth's surface. These biologically effective doses of UV-B radiation can penetrate deep into the water column, and cause leakage of reactive oxygen species (ROS) from the photosynthetic system in plant cells (Häder and Sinha 2005, Häder et al. 2003). These increased ROS can easily destroy proteins, DNA and other biological molecules, and subsequently affect growth and reproduction, survival, photosynthetic energy harvesting enzymes, and the content of photosynthetic pigments (Wang and Zhang 2001; He and Häder 2002; Apel and Hirt 2004; Holzinger and Lütz 2006). Solar UVR is responsible for a wide range of biological effects including alteration in the structure of protein, DNA, and many other biologically important molecules, chronic depression of key physiological processes, and acute physiological stress leading to either reduction in growth and cell division, pigment bleaching, N_2 fixation, energy production, or photoinhibition of photosynthesis in several organisms (Sinha et al. 2008; Zeeshan and Prasad 2009; Llabrés et al. 2010, Kumar 2004, Björn 2007).

Special attention has been given to the Antarctic and more recently to the Arctic (Camus and Gulliksen 2005), where the autocatalytic destruction of stratospheric ozone (ozone hole) has resulted in an enhanced flux of UV-B radiation (Häder et al. 2007) and a decrease in primary productivity (Smith et al. 1992). Ozone depletion at rates between -6.6% to -7.6% and -2.6% to -3.0% per decade (Madronich et al. 1998) have also resulted in an increase in UV-B radiation at temperate latitudes (30-50°N) of the northern hemisphere (Meyer-Rochow 2000).

One of the most prominent targets of solar UVR is cellular DNA, which absorbs UV-B radiation and causes adverse effects on living systems such as bacteria and cyanobacteria (Sinha et al. 2001). Structural changes in DNA are considered to be one of the primary consequences of the deleterious effects of UV-B on the cellular level (Karentz et al. 1991; Buma et al. 1995). The detrimental effects of UVR in sunlight is that it can cause DNA damage by inducing the formation of a cyclobutane pyrimidine dimer (CPD; ~80% of the total damage) or a pyrimidine-pyrimidone (6-4) photoproduct (6-4PP; ~20% of the total damage) (Johns et al. 1964; Taylor 1994). CPDs, especially thymine dimers (T<>T) are the predominant lesions induced by UV-B (Tyrrell 1986). These photoproducts block transcription and replication of the DNA and cause mutations or even cell death (Setlow et al. 1963; Swenson and Setlow 1966; Boelen 2000).

However, cyanobacteria have also developed adaptation strategies to counteract the damaging effects of UVR, which includes avoidance, scavenging, screening, repair and programmed cell death (Quesada and Vincent 1997; Sinha et al. 1998; Sinha and Häder 2008; Singh et al. 2010). DNA damage can be repaired by two mechanisms: photoenzymatic repair (photoreactivation), which reverses cyclobutane dimers in DNA, and nucleotide excision repair (dark repair). The detrimental effects of UVR on biological systems (CPD accumulation) result from kinetic processes that involve a balance between damage and repair (Vetter et al. 1999) and it is assumed that UVR must have been a very significant evolutionary force during their earliest evolutionary time (Castenholz and Garcia-Pichel 2000). DNA lesions constitute ubiquitous threat to the accurate maintenance of the genome, and their incidence and importance have motivated the development of precise cellular repair functions, some of which have been highly conserved during evolution. Although repair mechanisms of photoproducts have generally focused on plants, bacterial, yeast and mammalian systems (Britt 1996; Dany et al. 2001; Bray and West 2005; Kunz et al. 2005; Friedberg et al. 2006; Roldán-Arjona and Ariza 2009), informations are available on DNA repair processes in cyanobacteria (Levine and Thiel 1987; Mühlenhoff 2000; Ng and Pakrasi 2001; Rastogi et al. 2010a). Cyanobacteria have the ability to revert their damaged DNA by applying the strategies such as photoreactivation, excision repair, recombinational repair, SOS (save our soul) response and programmed cell death (PCD) or apoptosis.

The protective responses either at cellular or biochemical level were found to be stimulated under UV-B exposure only. With the advances in manipulation of UV-B radiation as low fluence rate (Singh et al. 2012), it was discovered that rather than damaging DNA they were evoking protective responses which were photomorphogenic and a loophole was there in understanding these responses as neither phytochrome nor cryptochrome could absorb UV-B and thus explaining these responses with known photoreceptors was not possible (Parihar 2015).

Damage is mainly through direct absorption by DNA. DNA absorbs in three spectral regions: I (with a peak at 260-264 nm), II (with a peak at 192 nm), and III (below 125 nm). Spectral regions II and III would only have relevance in outer space (Zalar et al. 2007a,b; Olsson-Francis et al. 2010).

With whole genome profiling of some phototrophic prokaryotes now possible, transcripts from genes involved in various light (UVR) intensity transitions can be probed with full genome DNA microarrays (Gill et al. 2002; Los et al. 2008). CPD induction is mainly caused by UV-B, whereas photorepair of DNA damage depends on the intensity of UV-A and/or photosynthetically active radiation (PAR) (Friedberg et al. 1995). The effect of UV-C (254 nm max) negatively affected short-term photosynthetic rate and survival in *Agmenellum quadruplicatum* (*Synechococcus* sp. PCC 7309) (Van Baalen, 1968) though the damage to DNA was cured by photoreactivation at wavelengths from ~395 to 450 nm (Van Baalen and O'Donnell 1972).

Some of the negative effects on DNA (formation of CPDs and 6-4 photoproducts that block DNA and RNA polymerases) may be repaired by photoreactivation, utilizing UV-A and/or violet/ blue wavelengths that are absorbed by chromophores of photolyase, a DNA-bound enzyme that restores the intact bases, and also by nucleotide excision repair (Jägger 1985; Sancar 1994;1996). Increases in temperature have been shown to enhance the DNA photoreactivation repair rates in algae (Pakker et al. 2000).

UV-Induced Generation of ROS

In photosynthetic organisms such as cyanobacteria, ROS are produced by means of photosynthetic electron transport chain (Latifi et al. 2009). Production of ROS can also be seen as a result of normal metabolic activity. The high-energetic UV-B radiation may affect normal state of life either through direct effects on cellular DNA and proteins or indirectly by the production of ROS. ROS such as singlet oxygen ($^1O_2$), superoxide anion (O_2^-), hydrogen peroxide (H_2O_2) and the hydroxyl radical (OH^-) are potent and highly active oxidizing agents. UV-induced production of ROS has been reported in some cyanobacteria (He and Häder 2002; Rastogi et al. 2010b; Rajneesh et al. 2017b). Rastogi et al. (2010b) for the first time detected UV-induced generation of ROS by confocal image analysis in *Anabaena variabilis* PCC 7937 using a ROS sensing probe 2',7'-dichlorodihydrofluorescein diacetate (DCFH-DA). UV-induced ROS can damage the components of DNA such as purine/pyrimidine bases and deoxyribose sugar backbone leading to single/double DNA strand breaks. It has been established that exposure of DNA to UVR led to the formation of a number of oxidation products of purine such as 8-oxo-7,8-dihydroguanyl (8-oxoGua), 8-oxo-Ade, 2,6-diamino-4-hydroxy-5-formamidoguanine (FapyGua), FapyAde, and oxazolone (Doetsch et al. 1995; Hall et al. 1996). UVR can have direct effects on biological macromolecules, including DNA, by photosensitization, or indirect effects on macromolecules by photomodification (Lampi et al. 2005). Photosensitization generally leads to the production of singlet oxygen or other ROS that are highly damaging to biomolecules (Chen et al. 2006). Photomodification results in the formation of new compounds that exhibit greater toxicity than the parent phototoxicant (Brack et al. 2003).

DNA Damage

The effects of UVR on matter are disseminated in two ways. First, if the UVR is absorbable, molecules in matter are excited leading to their photochemical alteration. Second, if UVR cannot be directly absorbed, energy transfer from nearby molecules called photosensitizers indirectly affects matter. UVR damage DNA by both pathways (Chatterjee and Walker 2017). Kumar et al. (2004) have investigated the impacts of UVR on the filamentous cyanobacterium, *Anabaena* strain BT2 by adopting the PCR based assay and found that UVR drastically affects the genomic DNA.

The increase in solar UVR reaching the Earth's surface has become an important issue in the last few decades. Although very small proportions of solar UVR contribute to the total irradiance on the Earth's surface (UV-C; 0%, UV-B; <1% and UV-A; <7%), this part of the solar spectrum is energetically highly active (Kirk, 1994). There are a number of direct targets for the deleterious UV-B radiation in cyanobacteria such as proteins and DNA which have absorption maxima in this region while UV-A irradiation has indirect effects via energy transfer from UV-A stimulated chromophores to the DNA target or *via* photosensitized (chlorophylls and phycobilins can act as photosensitizers) production of ROS (Hargreaves et al. 2007; He and Häder 2002). UV-A radiation that is not directly absorbed by DNA, can still induce DNA damage either by producing a secondary photoreaction of existing DNA photoproducts or *via* indirect photosensitizing reactions (Hargreaves et al. 2007).

Under enhanced UV-B radiation, light-sensitive matters such as phenols and aromatic amino acids transfer the received radiation energy to neighbouring oxygen molecules, as well as other UV-B-induced NADPH oxidation and/or peroxides, leading to increased ROS free radicals (Zlatev et al. 2012; Brosché and Strid 2003). The DNA is particularly sensitive to UV-B radiation which causes photo transformations resulting to the formation of pyrimidine dimers like cyclobutane. CPDs and pyrimidine (6-4) pyrimidinone dimers (6-4 PPs), thus hindering transcription and duplication of DNA (Kalbin et al. 2001).

Their relative formation frequency the formation of the two major photoproducts CPDs and pyrimidine (6-4) pyrimidone photoproducts [(6-4) PPs] depends on wavelength and dose of light (Varghese 1972; Davies 1995), although, the yield of (6-4) PP is slightly less than CPDs (Mitchell and Nairn 1989). Other minor photoproducts are also generated, such as pyrimidine hydrate, thymine glycols, and dipurine adducts (Demple and Linn 1982; Bose et al. 1983; Kumar et al. 1991). In CPDs, a cyclobutane ring covalently links the two adjacent pyrimidines, whereas in (6-4) PP, the C_6 position of one pyrimidine is covalently linked to C_4 position of the adjacent pyrimidine. Both classes of lesions distort the DNA helix. The 5'-TC (6-4) PP was found to be less mutagenic than the 5'-TT (6-4) product but more mutagenic than a CPD at the sequence 5'-TT. CPDs and 6-4PPs induce a bend or kink of 7-9° and 44°, respectively (Wang and Taylor 1991; Kim et al. 1995).

It was assumed that initial photoproducts are converted into double strand breaks (DSBs) during DNA replication, due to not a distinct process, that is, “collapse of replication forks” (Rastogi and Sinha 2011). Oxidative stress commonly results in single strand breaks (SSBs) or DSBs in the native DNA molecule (He and Häder 2002). Among different types of damages, DNA DSBs are the most deleterious, since they affect both strands of DNA and can lead to the loss of genetic material. CPDs have been reported to inhibit the progress of DNA polymerases. If unrepaired, a single CPD is sufficient to completely eliminate expression of a transcriptional unit. The template activity of DNA was found to be affected mainly under UV-B radiation and an additional band (≤2000 bp) appeared under UV-B radiation (Rastogi et al. 2014a). There is evidence that the stalled mammalian RNA polymerase II remains bound to the site of the obstruction. Persisting lesions may thus not only reduce the overall concentration of free RNA polymerase but also eliminate transcription of the gene in which they are located. Every CPD acts as a block to transcription and replication, and only a small fraction of dimers results in a mutation (Britt 1995; 1996). Therefore, these DNA lesions, if unrepaired, may interfere with DNA transcription and replication and can lead to misreading of the genetic code and cause mutations and death.

UV-induced formation of thymine dimers (T<>T CPD) has been observed in several cyanobacteria such as *Anabaena* sp., *Nostoc* sp., *Scytonema* sp. and *Rivularia* sp. (Sinha et al. 2001). UV-induced formation of CPDs has also been reported in the economically important cyanobacterium *A. platensis* (Gao et al. 2008). UVR-induced delay in chromosome replication was detected in a marine cyanobacterium *Prochlorococcus marinus* PCC 9511 (Kolowrat et al. 2010). Recently, Rastogi et al. (2014a) have reported UV-induced T<>T CPD in *Rivularia* sp. HKAR-4. Of all the essential cellular components that can be impaired by UVR, nuclear chromatin is a critical target for two main reasons: DNA controls ontogenetic development, and genetic damage (mutations) may be transmitted to the subsequent generation (Sinha and Häder 2002).

Lamare et al. (2006) suggested that the rate of repair of CPDs is influenced by life history and environment rather than phylogeny. Gao et al. (2008) have reported temperature and biomass dependent CPDs formation under UV stress in the commercially important cyanobacterium *A. platensis*.

Among them, CPDs and 6-4 PPs are the major DNA lesions formed by UV-B irradiation. Two adjacent cytosines are considered as mutation hotspots of UV-B and UV-C radiations (Ravanat et al. 2001). It has been found that T-T and T-C sequences are more photoreactive than C-T and C-C sequences (Courdavault et al. 2005). The amounts of CPDs and 6-4PPs are about 75 and 25%, respectively, of the total UV-mediated DNA damage products (Sinha and Häder 2002); however, the ratio between the yield of CPDs and 6-4PPs mainly depends upon the two adjacent bases involved in the formation of dimers (Douki and Cadet 2001).

UV-induced formation of thymine dimers (T<>T CPD) has been observed in several cyanobacteria such as *Anabaena* sp., *Nostoc* sp., *Scytonema* sp. and *Rivularia* sp. (Sinha et al. 2001, Rastogi et al. 2011). UV-induced formation of CPDs has also been reported in the economically important cyanobacterium *A. platensis* (Gao et al. 2008). UVR-induced delay in chromosome replication was detected in a marine cyanobacterium *Prochlorococcus marinus* PCC 9511 (Kolowrat et al. 2010). Recently, Rastogi et al. (2014) have reported UV-induced T<>T CPD in *Rivularia* sp. HKAR-4. CPDs correspond to the formation of a four-member ring structure involving C5 and C6 of both neighboring bases whereas 6-4 PPs are formed by a noncyclic bond between C6 (of the 5′-end) and C4 (of the 3′-end) of the involved pyrimidines *via* spontaneous rearrangement of the oxetane (when the 3′-end is thymine) or azetidine (when the 3′-end is cytosine) intermediates (Friedberg et al. 2006; Pattison and Davies 2006).

The 6-4 PPs are converted into their Dewar valence isomers upon exposure to UV-B or UV-A radiation that may further undergo reversion to the 6-4 PPs upon exposure to short wavelength UVR (Taylor et al. 1990, Rastogi et al. 2010a). A few CPD lesions (cis-syn or trans-anti isomers) can also be detected in aqueous solutions by UV-C irradiation (Douki et al. 2003). UV-A induced formation of CPDs has also been observed in bacteria as well as in eukaryotic cells and whole skin (Tyrrell 1973; Kvam and Tyrrell 1997;

Perdiz 2000). UV-C is widely used in laboratory investigations because of its maximal absorption by DNA, producing more photoproducts than the UV-A and UV-B radiations, which are also physiologically relevant UV wavelengths that also cause DNA damage (Kiefer 2007). UV-B for instance causes the formation of pyrimidine dimers, but does so less efficiently than UV-C (You et al. 2000; Rastogi et al. 2010b). In comparison to UV-B, the wavelength of UV-A has poor efficiency in inducing the DNA damage because they are not absorbed by native DNA. However, UV-A or visible light photon (up to 670-700 nm) are still able to induce DNA damage either by producing a secondary photoreaction of existing DNA photoproducts or *via* indirect photosensitization reactions (Hargreaves et al. 2007; Jiang et al. 2009). UV-A damages DNA by inducing DNA adduct formation by photooxidation reactions and by the excitation of endogenous (porphyrins and flavins) and exogenous (psoralens, tetracycline, promazine, and methylene blue) photosensitizers (Epe 1991; Kvam and Tyrrell 1997; Douki et al. 1999). PAR or UV-A is absorbed more weakly by DNA compared with UV-B, and is therefore proportionately less effective in the induction of CPDs. Several workers have reported that UV-A radiation generates CPDs directly (Jiang et al. 2009); however, some other investigators support the view that these CPDs are produced indirectly (Cadet et al. 2005). In addition, UV-A-mediated photosensitization can induce 8-oxo G formation or an excess accumulation of CPD (Epe 1991). UV-B has moderate effects on *16s* rDNA in terms of their band intensity and expression of template activity of certain RAPD gene fragments in *A. variabilis* PCC 7937 (Rastogi et al. 2011, 2014a). Vetter et al. (1999) observed diurnal cycling of DNA damage by both UV-B and UV-A, as measured by CPD formation. Dewar isomers are the main DNA lesions produced by UV-B radiation that can alter the molecular structure of genome leading to chronic mutagenesis and death of the cell (Friedberg et al. 2006). UV sources containing a higher proportion of radiation bordering between UV-B and UV-A, such as solar UVR, should produce a higher proportion of Dewar isomers since the photoisomerization is most efficient around 320 nm, which corresponds to the UV absorption maximum of 6-4 PPs. Consequently it has been suggested that all 6-4 PPs should be converted into Dewar isomers upon exposure to sunlight (Taylor et al. 1990).

Browman et al. (2000), however, showed a significant effect of UV-B, but not of UV-A, on DNA damage. It has been found that T-T and T-C sequences are more photoreactive than C-T and C-C sequences (Courdavault et al. 2005). The diastereoisomers of pyrimidine dimers (Figure 1) can be observed in free solution that differ in the orientation of the two pyrimidine rings relative to the cyclobutane ring, and on the relative orientations of the C5-C6 bonds in each pyrimidine base (Pattison and Davies 2006).

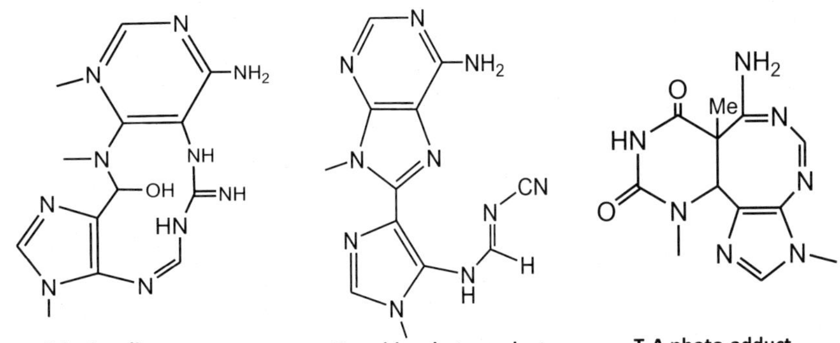

Figure 1. Various photoproducts induced by UVR (A) Adenine dimer (B) Photoadduct (C) T-A Photoadduct.

It has been demonstrated that the main photoproducts are cis-syn configured CPD lesions, while trans-syn-configured CPD lesions are formed in much less quantity (Carell and Epple 1998). In double stranded B-DNA, where the dimer entails two adjoining pyrimidine bases on the same DNA strand, only the syn isomers can be generated, whereas the cis isomer is preferred over the trans isomer to a great extent (Ravanat et al. 2001). The incidence of trans-syn isomer in single-stranded or denatured DNA is more common because of the increased flexibility of the DNA backbone. CPDs are formed at higher quantity by cycloaddition reaction between two pyrimidine bases (Carell and Epple 1998) in single stranded DNA (ss DNA) and at the flexible ends of poly (dA)-(dT) tracts, but not at their rigid centre (Becker and Wang 1989; Lyamichev 1991). Bending of DNA towards the minor groove reduces CPDs formation (Pehrson and Cohen 1992). One of the transcription factors, TATA-box binding protein (TBP), promotes the selective formation of 6-4 PPs in the TATA-box, where the DNA is bent, but CPDs are formed preferentially at the edge of the TATA box and outside where the DNA is not bent (Aboussekhra and Thoma 1999). Mapping of CPDs in the nucleosome core regions of UV-treated chromatin has revealed the formation of CPDs with an average distance of 10.3 bases, away from the surface of histones (Gale et al. 1987).

Rastogi et al. (2011) reported that the formation of thymine dimer is a wavelength dependent process and in *A. variabilis* PCC 7937 thymine dimer formation takes place only under UV-B radiation and not in PAR or UV-A radiations. In *A. variabilis* PCC 7937 low doses of UV-B radiation shows template binding activity of certain DNA gene fragment that may be either functionally harmful or beneficial and therefore needs further investigation. Kumar et al. (2004) observed a hypochromic effect in genomic DNA of the *Anabaena* strain BT2 after UVR exposure and reported damaged template activity of DNA following UVR exposure using a PCR-based assay such as RAPD, rDNA amplification and ARDA (restriction analysis of *16s* rDNA). A fluorimetric analysis of DNA unwinding (FADU) was used by He et al. (2002) to quantify the DNA strand breaks induced by ROS under UV-A and UV-B exposure in the cyanobacterium

Anabaena sp. To counteract these genome integrity threats, all organisms have evolved the DNA damage response (DDR) (Hu et al. 2016; Yoshiyama 2013). The DDR is a pathway that transduces the DNA damage signal into activation of various pathways that leads to DNA repair, cell cycle checkpoint and programmed cell death. Billi (2011) reported the lack of extensive DNA damage, upon desiccation which implies that protection mechanisms were triggered in *Chroococcidiopsis* sp. and must have been effective in preventing DNA damage caused by the lower doses of monochromatic and polychromatic UV radiation.

UV-Induced Purine Photoproducts

Even though dipyrimidine photoproducts are the preferential outcome of UV-B radiation, the biological importance of UVR-induced modifications of DNA purine bases has also been recognized (Duker and Gallagher 1988). These comprise the photoproducts that involve, at least, one adenine (A) residue that undergoes photocycloaddition reactions with contiguous adenine or thymine (T) upon exposure to UVB radiation (Kumar et al. 1987).

It has been established that exposure of DNA to UVR led to the formation of a number of oxidation products of purine such as 8-oxo-7, 8-dihydroguanyl (8-oxoGua), 8-oxo-Ade, 2,6-diamino-4-hydroxy-5-formamidoguanine (FapyGua), FapyAde, and oxazolone (Doetsch et al. 1995, Hall et al. 1996) the photoproducts that involve, at least, one adenine residue that undergoes photocycloaddition reactions with contiguous adenine or thymine upon exposure to UV-B radiation (Kumar et al. 1987; Koning et al. 1990). Cells damaged by UVR and oxidative stress can survive, but often remain in the G1-phase of the cell cycle for longer periods of time than normal cells, and may even remain in a permanent state of cellular growth arrest (Evan and Littlewood 1998). Algal species originally isolated from high-UVR locations (Larson et al. 1990; Rautenberger et al. 2013) that are subject to higher UV-B radiation (Agustí et al. 2015) are particularly resistant to UV-B radiation indicate that the history of UVR experienced by different taxonomic groups may also account for their diverse responses to UV-B radiation (Jin et al. 2017).

The sensitivity of *Anacystis nidulans* to UVR increases when the cells are treated with chloramphenicol prior to irradiation (Bhattacharjee 1977). Exposure of *Synechocystis* sp. strain PCC 6308 to a sublethal dose of UV light results in decreased DNA degradation after a subsequent challenge dose, and this decrease is abolished by protein synthesis inhibitors administered between the exposures to UV light (O'Brien et al. 1982). Low UV-B radiation could produce some accumulated inhibitory effects on PSII activity in both *Microcystis* 912 and *Synechococcus* 8102 under iron-deficient conditions in long-term experiment (Li et al. 2017).

Photolyase

A phenomenon named photoreactivation was independently discovered by Kelner (1949) and by Dulbecco (1949). Bacteria inactivated by UVR could be reactivated when exposed to visible light (Kelner 1949); similarly, bacteriophages that had been UV-inactivated could be reactivated when infected bacteria were exposed to blue light. Rupert (Rupert 1960) demonstrated that photoreactivation is caused by an enzyme. It was later named DNA photolyase (PL). The removal of using the energy of light (Long wavelength UV-A/blue light) is known as photoreactivation. The process of photoreactivation is executed by means of a photoreactivating enzyme known as "photolyase (PHR)," which is well conserved and found throughout the three domains of life like bacteria, fungi, plants, invertebrates and many vertebrates. Since DNA photolyases are found in a number of archaebacteria, they are considered to be ancient repair proteins, which may have helped in the evolution of the earliest organisms on primordial Earth. (Carell and Epple 1998; Sinha and Häder 2002). PHRs are rich in GC content (Rajneesh et al. 2017c). PHRs, can restore two native pyrimidines from a single UV-induced photoproduct in a light dependent manner (Figure 2).

DNA lesions such as CPDs or 6-4 PPs are removed in the presence of DNA PHR enzyme CPD photolyase and 6-4 photolyase, respectively. The enzyme binds specifically to the CPDs (CPD photolyase) or 6-4PPs (6-4 photolyase) and directly monomerizes the cyclobutane ring of the pyr <> pyr, using the energy of visible/blue-light and protects the genome from deleterious effects of UVR (Kim 1992; Essen and Klar 2006). Werbins and coworkers have described the occurrence of photoreactivation in cyanobacteria. The major photoreactivating factor, phrA, in the cyanobacterium, *Synechocystis* sp. strain PCC 6803 was reported by Ng et al. (2000) which codes for a CPD-specific DNA photolyase. It was also confirmed that PhrA contains the same cofactors as in the *Synechococcus* sp. strain 6301 DNA PHR. The over-expressed PhrA protein in *Synechocystis* sp. strain PCC 6803 constituted about 0.6% of the total protein (Ng et al. 2000). The genes for PHR homologs have been identified in *Synechocystis* sp. PCC 6803 and were functionally characterized for their roles in photoreactivation (Hitomi et al. 2000; Ng and Pakrasi 2001). Han et al. (2001) have also reported photoreactivation of UV-B induced inhibition of photosynthesis in *Anabaena* sp. PHRs have been also purified and characterized, from *A. nidulans* (Yasui et al. 1988). It was found that blue light plays the most effective role in photoreactivation against UV-B than any other wavelength band in cyanobacteria. Kumar et al. (2003) has also observed the significant role of photoreactivating light in growth.

The enzyme DNA PHRs are monomeric flavin-dependent repair enzymes having the molecular weight of about 45-66 kDa with 420-616 amino acid residues. Ten to twenty enzyme molecules are believed to scan the genome for UV lesions in every cell nucleus. Each one binds tightly to an encountered CPD with a Kd=108 M (thymine dimer).

It consists of two known cofactors, a catalytic cofactor such as either 5,10-methenyltetrahydrofolate (MTHF) in *Escherichia coli* (Johnson 1988) or 8-hydroxy-5-deaza-riboflavin (8-HDF) in *A. nidulans* (Eker et al. 1990) (absorption maxima of ~380 and ~440 nm, respectively), and a light-harvesting cofactor such as deprotonated reduced flavin adenine dinucleotide (FADH). FAD is not only essential for enzyme catalysis but also for damage recognition, while MTHF or 8-HDF is required to enhance the rate of DNA repair 10 to 100 fold under low light conditions. Unlike the nucleotide-like chromophores (e.g., FAD, FMN, and 8-HDF) deeply buried in the N-terminal α/β domain, the photoantenna MTHF is located in a shallow cleft between the α-helical and α/β domains and partially sticks out at the enzyme's surface (Zhong 2015). Long wavelength UV-A or blue excitation energy is absorbed by either MTHF or 8-HDF (λ_{max} ~380 and ~450 nm) and transferred to the catalytic cofactor ($FADH^-$). The flavin in the excited state transfers an electron to the CPD, splitting the cyclobutane ring and the electron is transferred back to the flavin resulting in the generation of the two canonical bases (Thoma 1999; Sancar 1994). FAD in PLs can be found in three redox states, namely fully oxidized (FAD_{ox}), semi-reduced ($FADH^+$) and fully reduced ($FADH^-$) forms (Maul et al. 2008). Among them, the fully reduced $FADH^-$, which absorbs 350-500 nm light, is the repair-active form. FAD_{ox} and $FADH^*$ can be converted to $FADH^-$ by a separate photoreaction called photoactivation, which typically involves an intraprotein electron transfer (ET) through a chain of three tryptophan residues (Aubert 2000), exceptionally four in animal (6-4) PLs (Müller 2015). A second chromophore (mostly a pterin or a deazaflavin) is present in many PLs and serves as a light-harvesting antenna that transfers excitation energy to the catalytic FAD cofactor, thereby increasing the effective absorption cross section of the enzyme. The crystal structures of CPD PHR of *E. coli* and *A. nidulans* suggest that upon binding to DNA, the enzymes flip the pyrimidine dimer out of the duplex into a hole that contains the catalytic cofactor (Park et al. 1995; Tamada et al. 1997).

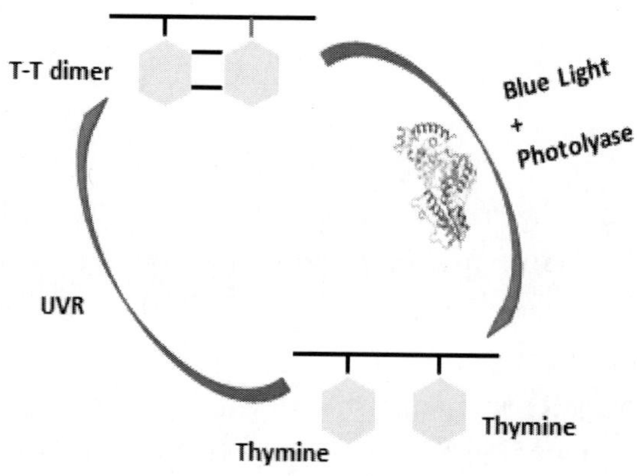

Figure 2. Mechanism of damage reversal by photolyase.

Upon photoexcitation of the fully reduced $FADH^-$, an electron is transferred to the CPD in 30- 250 ps, forming an anion radical of the CPD and the semireduced $FADH^+$ radical (MacFarlane et al. 2003). Kao et al. (2005) captured the catalytic intermediate of flavin radical ($FADH^•$) and directly proved the electron-transfer radical mechanism. The cyclobutane ring then splits almost spontaneously and rearranges into one intact base and the anion radical of the other base. The excess electron of the latter is finally returned to $FADH^*$ in ~1 ns (Kao et al. 2005; Espagne et al. 2009; Thiagarajan et al. 2011), leading to the restoration of the second intact base and the catalytically active $FADH^-$. Second photo-antenna folate (or deazaflavin) is excited by a blue light photon to a high-energy state with a lifetime of 0.5 ns (MTHF) or 2.0 ns (8-HDF). The repaired DNA is released in ~50 üs (Kondoh et al. 2011), and the enzyme is ready to bind and repair another CPD.

The entire catalytic cycle is complete in 1.2 ns, and the enzyme repairs T<>T with a quantum yield of 0.9 (Tan et al. 2014; Tan et al. 2015; Zhong 2015). Rather than depending on the ultrafast photoinduced process, repair efficiency was found to rely on synergistic relationships among all DNA repair steps (i.e., three electron transfer and two bond-breaking steps) (Tan et al. 2015). CPD repair by PHR is a single-photon process, and its reaction quantum yield is as high as 40-100% (Sancar 2003; Hitomi et al. 1997; Thiagarajan et al. 2010). Losses were attributed to (i) decay of the excited state $FADH^{-*}$ to the ground state (intrinsic time constant ~1.3 ns in competition with ET to the CPD and to (ii) back ET from the reduced lesion to $FADH^*$ in competition with cleavage of the two intra dimer bonds. Once the PHR has been bound to a CPD, the efficiency of photoreactivation is extremely high with one dimer split for almost every blue light photon absorbed (Britt 1996). Tan et al. (2014) found that the enzyme exhibits colors ranging from purple to orange depending on the redox state of the flavin cofactor.

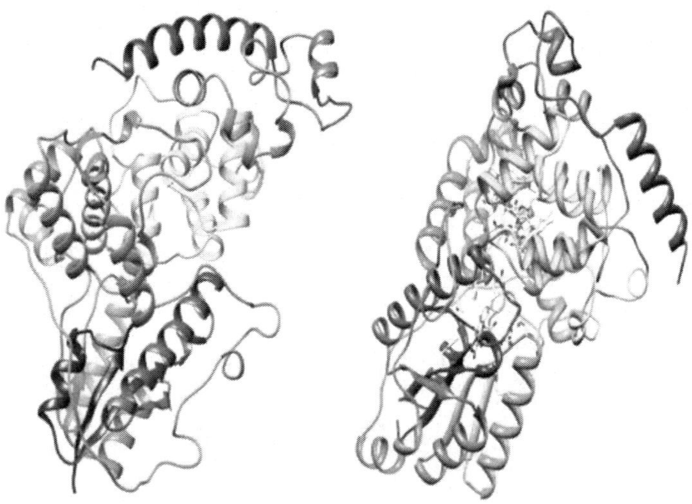

Figure 3. Homology model 3D structures of deoxyribodipyrimidine photolyase.

The functions of CPD and (6-4) PHRs are highly selective; CPD PHR repairs CPD, but not (6-4) PP, and vice-versa. Despite similar structures and repair mechanisms using electron transfer from FAD, their sequential identity is only ~25% (Yasui et al. 1988; Todo et al. 1993; Sancar et al. 1984; Kim et al. 1996). Yamada et al. (2016) reported that the triple mutant of (6-4) PHR is able to repair the CPD photoproduct. In contrast, the (6-4) photoproduct is not repaired by the reverse mutation of CPD PHR, and eight additional mutations (total of 11 mutations) in CPD PHR are not sufficient. Phylogenetic analysis suggests that CPD PHR may be of evolutionary origin. If so, then CPD PHRs as the ancestor have gained the repair ability of (6-4) PP during evolution. The quantum yield (or efficiency) of photorepair of (6-4) PPs was found to be ~4 times (Hitomi 1997) or even ~100 times (Kim 1994) lower than for CPD repair. These findings suggested that the repair reaction of (6-4) PLs could share part of the reaction steps of CPD PLs, but that substantial differences remain. Figure 3 shows the 3D structure of PHR.

In the formation of the (6-4) PPs, a Paternò-Büchi reaction first yields an oxetane-bridged (orazetidine-bridged for cytosine at 30) intermediate (Taylor 1994). This structure is thermodynamically unstable and rearranges (~2 ms for T (ox) T in aqueous solution (Marguet and Markovitsi 2005), to form a (6-4) PP. Based on the homology between photolyases, it was initially proposed that (6-4) PLs first catalyze the formation of a four membered ring, similar to the cyclobutane for CPDs, in the dark. This intermediate was proposed to be the very oxetane (or azetidine) bridged compound found in the formation of the lesion. Photorepair would then follow just as for CPD PLs. For the Dewar PP formed at TT sites, T(Dew)T, no significant repair by (6-4) PL was observed (Glas et al. 2009; Yamamoto et al. 2006). The crystal structure of (6-4) PL complexed with DNA carrying an analogue of the T(Dew)C lesion revealed that this Dewar PP was accommodated in the active site in the same way as the corresponding (6-4) PP (Glas et al. 2010). Concerning the reaction mechanism of repair of the T(Dew)C lesion by (6-4) PL, theoretical and experimental studies (Glas et al. 2010; Fingerhut et al. 2012) agree that that the first step is reversion of the Dewar PP to a (6-4) PP upon electron injection by photoexcited $FADH^{-}$. The (6-4) PP would subsequently be repaired as discussed above. The reversion (isomerization) of the Dewar PP would be the rate-limiting step of overall repair.

Photoreactivation of damaged DNA have been observed in several species of cyanobacteria such as *Agmenellum quadruplicatum* (Van Baalen 1968), *A. nidulans* (Tang and Asato 1978), *Gleocapsa alpicola* (*Synechocystis* sp. strain PCC 6308) (O'Brien and Houghton 1982a, 1982b), *Plectonema boryanum* (Werbin and Rupert 1968; Saito and Werbin 1970), *Anabaena doliolum* (Srivastava et al. 1971) and *Anabaena* sp. (Levine and Thiel 1987). The major photoreactivating factor, phrA, in the cyanobacterium *Synechocystis* sp. strain PCC 6803 codes for a CPD-specific DNA PHR (Ng et al. 2000). In *E. coli,* the phr gene codes for deoxyribodipyrimidine photolyase that, with the cofactor folic acid, binds in the dark to the thymine dimer. When the cell is then exposed to light, folic acid absorbs a photon and uses the energy to break the cyclobutane ring of the thymine dimer; the PHR then leaves the DNA. However, how DNA photolyases find lesions in the DNA molecule is still not clear (Essen and Klar 2006). It has been observed that about 240 KJ/mol of energy is captured upon absorption, out of which about 125 KJ/mol energy is consumed during the initial electron transfer from the excited FADH to CPD lesions (Heelis et al. 1995). The splitting of CPD lesion proceeds rapidly within 0.6 ns (MacFarlane and Stanley 2003; Kao et al. 2005). The back-transfer of electrons from the CPD lesion to the FADH radical is efficiently avoided by the enzyme before completion of cleavage of the cyclobutane ring (Kim et al. 1992). With the help of ultrafast femtosecond laser spectroscopy, MacFarlane and Stanley (MacFarlane and Stanley 2003) have suggested that PHR enzyme is indeed left in the semiquinonid state after accomplishment of repair of the CPD lesion. However, Kavakli and Sancar (2004) have analyzed the role of intraprotein electron transfer in photoreactivation by DNA PHR and found that photoreduction process is not a regular part of the PHR photocycle under physiological conditions, because the enzyme may undergo at least 25 repair cycles before losing its activity.

Two articles showed that the apoproteins of the blue-light photoreceptors (cryptochrome) of *Arabidopsis thaliana* (Ahmad and Cashmore 1993) and mustard (*Sinapis alba*) (Batschauer 1993) are very similar to the microbial CPD PHR and are structurally related, although the photoreceptors function in signal transduction and not in DNA repair. The purified photoreceptors contained the folate and flavin chromophores characteristic of the folate class of PHRs. (Malhotra et al. 1995) These findings suggested a previously unknown mechanism of signal transduction, i.e., photo-induced electron transfer that does not involve a net redox reaction.

Todo et al. (1996) have shown that the three systems, the CPD PHR, the 6-4 photolyase and the plant blue-light photoreceptors, are evolutionary and mechanistically related. PHRs are classified based on different criteria:

1) According the photoproduct they recognize, they are classified as CPD-PHRs (Phr of *E. coli*; Brash et al. 1985) or 6,4- PHRs (PhrB of *Agrobacterium fabrum*; Graf et al. 2015).
2) Based on sequence analysis of the CPD-PHRs, they are divided into three classes: I, II and III.

CPD-PHR class I are found in microorganisms (*Anacystis nidulans* and *E. coli*), class II are mainly present in higher eukaryotic organisms, but also in prokaryotes and viruses, and CPD- PHRs type III (PhrA of *Agrobacterium tumefaciens*) are closely related to cryptochromes plant photoreceptors (Wang et al. 2015; Sancar 2016). The effects of temperature were optimized on the photoreversal of thymine dimers in *A. variabilis* PCC 7937. The optimum efficiency for photoreversal of thymine dimers was found at 20^0C, suggesting the role of different temperature for repair competence in *A. variabilis* PCC 7937 (Rastogi 2011). This temperature dependence of photorepair possibly reflects species specific discrepancy in the temperatures experienced in their natural habitat, in relation to the temperatures chosen to observe photorepair. Repair of *synechocystis* sp. strain PCC 6308 occurs when directly placed under blue light after irradiation but photoreactivation ceases if the cells were placed immediately under dark after irradiation (Williams et al. 1979). Dimer excision occurs under both photoreactivating and non-photoreactivating conditions in *Synechocystis* sp. strain PCC 6308 (O'Brien and Houghton 1982a). In the absence of photoreactivating light, the enzyme binds to pyr <> pyr and stimulates the removal of UV damage by stimulating the NER system *in vivo* or *in vitro* and defense against DNA damage even in the absence of light (Sancar and Smith 1989). Incubation experiments indicated that during the day repair rates are low when compared to DNA damage accumulation rates, suggesting that an essential part of the DNA damage is repaired after sunset by excision (dark) repair (Boelen 2001). *dsDNA* content restored from 63.8% to 81.3% in *M. viridis*, and from 67.5% to 75.2% in *Anabaena* which was due to irreversible DNA damage by UV-B radiation in both cyanobacteria (Mitchell and Karentz 1993). Thus, the cyanobacteria repaired UV-B induced oxidative damage through the effective repair ability of DNA and the photosynthetic system (Chen et al. 2012).

Fang et al. (2011) and Tuteja et al. (2001) found that silicon dramatically enhanced the expression of photolyase (PL) which reverted UV-induced dimers. Levine and Thiel (1987) reported that DNA damage induced by UV radiation was repaired with much greater efficiency by photoreactivation than by the other repair systems (including UV-inducible repair) in *Anabaena* sp. Low temperature is known to decrease the efficiency of PHR and generally the photoreactivation of tymidine dimers (Vass et al. 2013), which leads to decreased DNA repair activity.

Excision Repair

UV sensitive mutants that are defective in repair pathways might unravel interesting changes in UV sensitivity rhythm, if indeed there was a link to DNA repair status. Interestingly, a mutant known to be completely deficient in thymine dimer excision repair showed a highly dampened rhythm compared to other UV sensitive mutants (Chaudhari et al. 2015). This observation suggested that the variations could also originate by modulations/regulation of DNA excision repair machinery in the cells. With the help of a repair assay involving cell-free extracts, we showed that the repair efficiency itself was higher in dark versus light phase culture cell-free extracts providing at least a partial mechanistic justification for higher survival of cells against UV in dark incubated cultures.

Base Excision Repair (BER)

The BER pathway has probably been evolved to protect cells from DNA damage induced by hydrolytic deamination, strong alkylating agents, ionizing radiation (IR), or by different intracellular metabolites and, indirectly, also by UVR *via* generation of ROS and proceeds through a series of repair complexes that act at the site of DNA damage. Parsons et al. (2008) have discovered that the formation of DNA repair complexes on damaged DNA stabilizes BER proteins. On the contrary, BER proteins that are not involved in repair are ubiquitylated by carboxyl terminus of Hsc70 interacting protein (CHIP) and subsequently degraded by the proteasome. BER pathway likely evolved to cope with the high level of spontaneous decay products that are formed in DNA, as well as those damages created upon reactions with natural endogenous chemicals, most notably ROS. BER predominantly deals with non-bulky small nucleobase lesions, excising and replacing incorrect (uracil) or damaged (3-methyladenine, 8-oxoG) bases derived from deamination, alkylation or oxidation (Figure 4).

There are several variations in the BER theme, yet each pathway shares a set of common elements and generally invokes the five following steps:

1) Recognition and removal of an incorrect or damaged substrate base by a DNA glycosylase to create an abasic site intermediate
2) Abasic site incision by an apurinic/apyrimidinic (AP) endonuclease or AP lyase
3) Removal of the remaining sugar fragment by a lyase or phosphodiesterase
4) Gap filling by a DNA polymerase
5) Sealing of the nick by a DNA ligase

The key enzymes involved in BER are DNA glycosylases and AP nucleases. Cadet et al. (2005) measured the kinetics of *Escherichia coli* endonuclease III and formamidopyrimidine, DNA N-glycosylase (Fpg.) to recognize and excise several oxidized pyrimidine and purine bases. DNA glycosylases remove different types of modified bases by cleavage of the N-glycosidic bond between the base and the 2-deoxyribose moiety of the nucleotide residues. DNA glycosylases remove different types of modified or damaged bases by cleavage of the N-glycosidic bond between the base and the 2-deoxyribose moieties of the nucleotide residues. Depending on the biochemical modification there are different, highly specific DNA glycosylases have been reported such as:

1) Uracil glycosylase,
2) methyladenine glycosylase,
3) UV endonucleases,
4) endonuclease III/thymine glycol DNA glycosylase,

5) endonuclease VIII,
6) A-G mismatch DNA glycosylase,
7) 5-hydroxymethyl and 5-formyluracil DNA glycosylases
8) fapy/8-oxoguanine DNA glycosylase

Different DNA glycosylases remove different kinds of damage, and the specificity of the repair pathway is determined by the type of glycosylase involve.

BER can be subdivided in to two forms i.e., shortpatch BER (SP-BER) and long-patch BER (LPBER). DNA having one nucleotide lesion is removed by short-patch BER (SP-BER) whereas two/more nucleotide lesion is repaired by longpatch BER (LP-BER) pathway (Fortini and Dogliotti 2007; Almeida and Sobol 2007). It is assumed that majority of repair takes place through SP-BER, initiated either by monofunctional or by bifunctional glycosylase (Almeida and Sobol 2007). After base removal by DNA glycosylase, the apurinic/apyrimidinic (AP) site is excised by an AP endonuclease or an AP lyase, which nicks the DNA strand 5' or 3' to the AP site, respectively, followed by the excision of the remaining desoxyribose phosphate residue by a phosphodiesterase. Subsequently, the gap is filled by a repair DNA polymerase and the strand is linked by a DNA ligase (Sakumi and Sekiguchi 1990; Seeberg et al. 1995).

DNA Gycolylase

There are several DNA glycosylases reported in eukarotic system (Lucas-Lled′o et al. 2011) and all possess binding pockets that selectively accommodate a set of target base modifications (Hegde et al. 2008; Robertson et al. 2009; Dalhus et al. 2009; Friedman and Stivers 2010). The high-resolution structures of several DNA glycosylases, together with mutational analysis and emerging data on three-dimensional structures of enzyme-substrate complexes, provide detailed mechanistic information on this class of repair enzymes. An excellent study in this mode (Thayer 1995) establishes a novel helix-hairpin-helix motif and a [4Fe-4S] cluster loop motif as DNA-binding structures in *E. coli* endonuclease III, the enzyme that excises thymine glycol, cytosine hydrates, and ring-fragmented pyrimidines from DNA.

DNA glycosylases gently pinch the DNA while scanning, ultimately bending DNA at the position of the damaged base to create a widened and flattened minor groove. This localized DNA distortion promotes the damaged base to flip out of the double helix and enter the binding site of the enzyme for surveying and protein-substrate complex formation. This event, known as 'base flipping,' is an integral step in the enzymatic process of all DNA glycosylases. DNA glycosylases are frequently classified as either mono-functional or bi-functional based on their catalytic mechanism and ability to execute AP lyase strand cleavage activity (Krokan and Standal 1997; Mitra 2001).

Mono-functional glycosylases use a water molecule as a nucleophile to attack the aromatic carbon of the damaged base to promote base release, generating an abasic site product that is identical to that formed upon spontaneous DNA depurination or depyrimidination. This abasic site is a substrate for an AP endonuclease.

On contrary, the bi-functional glycosylases use an active site amine moiety as a nucleophile to excise the damaged base, and generate a covalent Schiff base protein-DNA intermediate during the catalytic process. This class of DNA glycosylases not only exhibits base excision activity, but also harbors the capacity to incise at the abasic site to create a SSB with a non-conventional 3'-terminus. Cleavage of the phosphodiester backbone of DNA by a bi-functional glycosylase occurs within the phosphodiester linkage 3' to the AP site either by β -elimination or by two consecutive elimination steps. Either of these incision events creates a DNA nick with a 3'-blocking residue, a phosphor-α,β-unsaturated aldehyde (PUA) or phosphate (PO_4), respectively, which requires further processing by a specific enzyme to provide a suitable substrate for a DNA polymerase. The gene for the DNA repair enzyme Fpg (formamydopyrimidine-DNA glycosylase) has been reported from *Synechococcus elongatus* and was suggested to be involved in photoprotection against oxidative damage (Mühlenhoff 2000).

AP Endonuclease

After base removal by a DNA glycosylase, an AP endonuclease incises the DNA backbone immediately 5' to the abasic site to generate a strand break with a priming 3'-OH group and a non-conventional 5'-deoxyribose phosphate (dRP). *E. coli* has two such enzymes, exonuclease III and endonuclease IV, which have overlapping specificities and are known to be involved in the repair of abasic sites (Cunningham et al. 1986). (In spite of its name, the main function and activity of exonuclease III is as an AP endonuclease) Both enzymes cut DNA by a hydrolytic reaction on the 5' side of the base-free residue (Warner 1980). The major AP endonuclease in mammals is APE1, which accounts for more than 95% of the total cellular AP site incision activity. APE1 executes an acid-base catalytic reaction to incise the phosphodiester bond (P-O) of DNA, and this hydrolytic reaction mechanism is promoted by the metal ion Mg^{2+}. Mg^{2+} is thought to stabilize the developing charge on the phosphate oxygen atoms, strategically positioning them for the subsequent nucleophilic attack by an active site-generated hydroxyl radical.

There are several amino acid residues that are required for full APE1 nuclease activity. For instance, the catalytic reaction of APE1 is mediated through a triad structure involving Asp283-His309-H_2O (Mol et al. 2000). Asp283 and His309 are active site residues that form a hydrogen bond that has been postulated to allow the histidine residue to function as the general base in the reaction, accepting a proton from water to generate the reactive hydroxyl nucleophile (Gorman et al.1997).

In addition, a 3' to 5' exonuclease activity of APE1 may serve as an autonomous proofreading function to remove mismatched or damaged nucleotides during the synthesis step of BER (Chou and Cheng 2002; Hadi 2002). Minor changes are repaired by the BER pathway that cleaves the glycosidic bond, releasing the damaged base. Further processing is required to remove the deoxyribose so that the appropriate nucleotide can be reinserted with a polymerase, and the ends are then sealed with a ligase. The BER pathway is particularly relevant to oxidative damage, because most reactions involve the addition of only one or two new oxygen atoms to the purine or pyrimidine base (Burrows 2009). For successively bulkier and more complex damages such as cross-links to other DNA strands or to proteins, more complicated pathways are involved. Dianov and Lindhal (1994) reconstituted BER with purified enzymes and found the results which are similar to those obtained with crude cell extracts, indicating that no additional enzymes or accessory protein factors are required for the reaction. However, a minor alternative pathway also exists and generates longer repair patches. This route is initiated in the same way by a DNA glycosylase and an AP endonuclease, but subsequent DNA strand displacement and resynthesis would be expected to occur in a way closely related to the later stages of NER or lagging-strand DNA replication, and to require similar activities. The arrangement provides increased versatility for this major repair process, which is used to counteract spontaneous cellular DNA damage.

Nucleotide Excision Repair (NER)

NER is one of the most versatile and flexible repair systems that recover a wide range of DNA lesions, including CPDs and 6-4 PPs caused by UVR, bulky chemical adducts, DNA-intrastrand crosslinks, and some forms of oxidative damage (Yang 2011). This mechanism is present in most organisms and well documented in eukaryotes (Rouillon and White, 2011). Discovery of NER was first described in *E. Coli* (Pettijohn and Hanawalt 1964) where about six proteins such as UvrA, B, and C (known as ABC-complex, which shows excinuclease activity), UvrD (helicase II), DNA polymerase I (pol. I), and DNA ligase are recruited to complete the repair (Lin and Sancar 1990; Orren et al.1992; Petit and sancar 1999, Truglio et al. 2006). The excision reaction is genetically controlled by the uvr genes in *E. coli* (Howard-Flanders 1966) and XP genes in humans (Cleaver 1968; Cleaver and Bootsma 1975). Six genes, uvrA, uvrB, uvrC, uvrD, polA, and lig, are required for this process in *E. coli*. Figure 4 gives an outline of NER pathway.

UvrA, UvrB, and UvrC proteins, the gene products of uvrA, uvrB, and uvrC, respectively, initiate nucleotide excision repair by incising DNA on both sides of the damaged nucleotide (Sancar and Rupp 1983; Sancar 1996). This damage-specific and dual incision nuclease activity is termed (A)BC excinuclease. (A)BC excinuclease incises the DNA at the 8th phosphodiester bond 5', and at the 4th or '5th phosphodiester bond 3', to the damaged nucleotide. UvrA dimer and UvrB form the damage recognition complexv $(UvrA)_2(UvrB)$, in an ATP-hydrolysis-dependent reaction (Orren and Sancar 1989; 1990). This complex is guided by the DNA-binding affinity of UvrA to track along the DNA and bind to the damage site (Oh and Grossman 1987). Upon binding of the UvrAB complex at the site of damage, the DNA is kinked (Shi et al. 1992) and locally (1-3 bp) denatured (Lin et al. 1992). UvrB has the six motifs of the D-E-A-D family of RNA helicases (Schmid and Linder 1992) and under special experimental conditions such tracking can displace oligomers of <50 nucleotides (Oh and Grossman 1987), which has led to the suggestion that the UvrAB complex is a helicase. UvrA dimer then dissociates from the UvrB-damaged DNA complex, which allows recognition by UvrC (Orren and Sancar 1989; Bertrand- Burggraf et al. 1991). UvrC, which has no affinity for UvrB and no selectivity for damaged DNA, binds specifically to the UvrB-damaged DNA complex. In this complex, first, the 3' incision is made by UvrB, then the DNA is straightened and renatured, which allows UvrC to make the 5' incision (Lin et al. 1992; Lin and Sancar 1992). UvrC and the damage-containing 12- or 13-mer is released by helicase II (Orren et al. 1992). DNA polymerase I releases UvrB and fills in the gap without nick translation. UvrB is released from the gapped DNA.

Genes Associated with (A)BC Exonuclease

The genes of the (A)BC excinuclease subunits, uvrA, uvrB and uvrC, have been cloned, sequenced, and the gene products over produced in large quantities for *in vitro* characterization (Thomas et al. 1985). The gene encoding UvrA protein is 2820 bp long and its expression is under SOS regulation (Husain et al. 1986). Upon SOS induction, the cellular level of UvrA protein increases about 10-folds from 25 molecules in an uninduced cell to 250 molecules in an induced cell. UvrA protein is an ATPase and a DNA-binding protein (Seeberg and Steinum 1982). UvrA is a dimer under physiological conditions and binds one ATP/ADP per dimer (Orren and Sancar 1989; Myles et al. 1991). Consistent with nudeotide binding activity and ATPase activity, sequence analysis of the uvrA gene reveals two Walker A- and B-type sequences that are found in most ATPases (Husain et al. 1986). In addition UvrA contains two zinc fingers (Navaratnam et al. 1989) which are apparently involved in DNA binding (Myles and Sancar 1991).

The *uvrB* gene is 2019 bp long and is under the control of SOS regulation (Arikan et al. 1986). Upon SOS induction, the cellular level of UvrB protetn increases approximately four folds from 500 to 2000 molecules per cell. The uvrB gene encodes a monomeric protein. The amino acid sequence reveals Walker A and B sequences as well as sequences homologous to the seven sequence motifs found in more than 30 (mostly RNA) helicases (Gorbalenya et al. 1990; Arenas and Abelson 1991; Schmid and Linder 1992). UvrB is a UvrA and damaged-DNA-dependent ATPase and this activity is required for loading of UvrB onto the damaged DNA (Lin et al. 1992). UvrB by itself is not a DNA-blnding protein, but it can be loaded onto damaged DNA by UvrA to form a very stable UvrB-DNA complex (Orren and Sancar 1989). The UvrA-mediated binding of UvrB to DNA is by hydrophobic interactions, as evidenced by the extreme stability of the complex to high Ionic strength. The uvrC gene is 1830 bp long and encodes a protein (Sancar et al. 1984; Lin and Sancar 1990). UvrC protein binds to single-stranded and double-stranded DNA non-specifically. UvrC does not associate with either ot the Uvr proteins but binds to the UvrB-DNA complex with high specificity. The C-terminal 314 amino acids of UvrC contain all that is required for the repair activity of the protein (Lin and Sancar 1991). An alternative nuclease, Cho, can substitute for UvrC at the 30 incision site (Moolenaar et al. 2002).

Even though UvrA is required for the nucleotide excision repair activity of (A)BC excinuclease, it is not involved directly in the incision reaction (Orren and Sancar 1989, 1990). UvrA catalytically delivers UvrB to damaged DNA and makes UvrB bind to damaged DNA, and then dissociates from the UvrB-damaged DNA complex, this activity of UvrA has been called molecular matchmaking' (Orren et al. 1992). During the matchmaking step, the structures of both UvrB and DNA are changed. UvrB, which has no affinity to DNA, is changed to a DNA-binding protein by exposing aliphatic and aromatic residues on UvrB for binding to DNA by hydrophobic interactions. The structure of DNA is also changed upon binding, it is both kinked (Shi et al. 1992; Lin et al. 1992) and locally denatured (Lin et al. 1992), which might create the proper structure for binding of UvrB. (A)BC excinuclease utilizes a well-organized sequential mechanism accomplishing the ultimate goal. This sequential mechanism provides substrate selectivity during each step of the reaction and provides (A)BC excinuclease with great flexibility for identifying a specific DNA structure as a substrate or rejecting it as a non-substrate (Lin and Sancar 1992).

Dual Incision

It is the characteristic property of nucleotide excision repair by (A)BC excinuclease that two incisions are made on both sides of the damaged base (Sancar and Rupp 1983). Extensive *in vivo/vitro* site-specific mutagenesis studies on the uvrB and uvrC genes indicate that UvrB contains the 3' active site (Lin et al. 1992) and UvrC contains the 5' active site (Lin and Sancar 1992) for the incisions. Furthermore, the 3' incision by UvrB precedes the 5' incision by UvrC. This finding has led to several revisions of the earlier model for the reaction mechanism of (A)BC excinuclease.

First, UvrB also contains an active site for the incision and thus UvrC is not the sole nuclease of (A)BC excinuclease.

Second, the stepwise incision mechanism indicates that the dual incisions are not concerted and explains the 3' uncoupled incisions observed with certain DNA substrates and using aged UvrC protein (Selby and Sancar 1988). Third, the finding that UvrC mutants abolish only the 5' incision explains the observation that DNA purified from ultraviolet irradiated *E. coli* uvrC$^-$ strains contains nicks whereas that isolated from uvrA$^-$ or uvrB$^-$ strains do not (Sancar and Sancar 1988). The nicks in the *E. coli* uvrC$^-$ strain after UVR apparently originate from the 3' incision activity of UvrB bound to mutant UvrC. Fourth, the stoichiometry of the incision complex is unambiguously determined as one UvrB and one UvrC molecule per damage site (Lin and Sancar 1992).

Post Incision

Post-incision events include enzymatic turnover of UvrB and UvrC, removal of the damage-containing oligonucleotide, and repair synthesis. Helicase II, DNA polymerase I, and DNA ligase are required for the post-incision events (Sancar and Sancar 1988). After the dual incisions, UvrB and UvrC remain bound to DNA; helicase II turns over UvrC and releases the damaged oligonucleotide (Orren et al. 1992). The released oligonucleotide is in protein-free form, which means that if is not bound to either UvrC or helicase II. As nicks on DNA are substrates for helicase II (Runyon and Lohman 1989), it is possible that helicase II recognizes the 5' nick of the DNA and uses its 3'→5' helicase activity to remove the damaged oligonucleotide and force UvrC off the gapped DNA-UvrB complex. An alternative possibility is that there is a specific interaction between the post-incision complex and helicase II. UvrB is not turned over by helicase II. Turnover of UvrB requires DNA polymerase I (Orren et al. 1992). DNA polymerase I recognizes the 3'-0H of the 5' incision and fills in the gap with its polymerase activity and simultaneously dislodges UvrB from the DNA. The remaining nick after repair synthesis is sealed by DNA ligase.

NER can be subdivided into differentially regulated subpathways such as global genome NER (GG-NER) and transcription-coupled NER (TC-NER). Repair of DNA lesions over the entire genome, referred to as global genome repair (GGR), and repair of transcription-blocking lesions present in transcribed DNA strands, referred to as transcription coupled repair (TCR). Both repair systems removed a wide range of UV-induced DNA lesions in a sequential way that includes damage recognition, opening of DNA double helix at damage site, and dual incisions on both sides of the lesion followed by resynthesis and ligation (Nouspikel 2009). NER activity can be actively reduced by the application of nodularin, as reported by Lankoff et al. (2008).

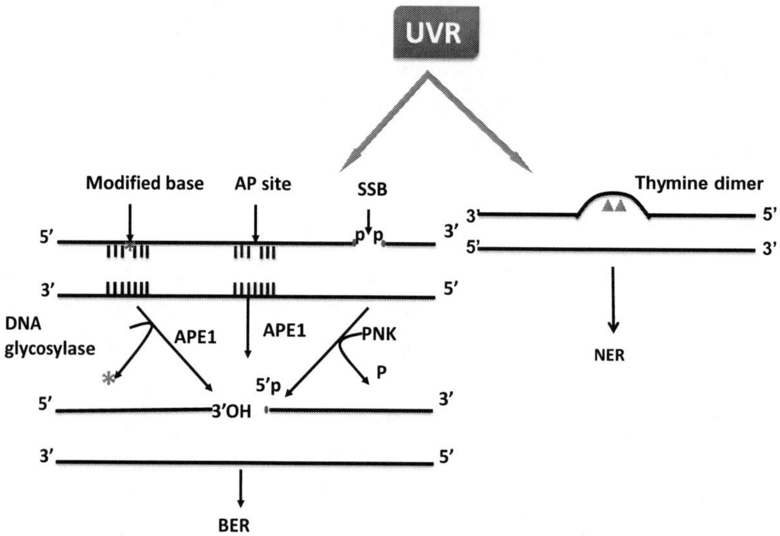

Figure 4. Diagramme showing steps involved in excision repair pathway.

Recombinational Repair

Recombination is one of the most important processes involved in DNA repair, ensuring the transmission of correct genetic information from parents to offspring. Recombination repair efficiently repairs DSBs as well as SSBs in damaged DNA by a series of complex biochemical reactions. Recombinational repair fills the daughter strand gap by transferring a preexisting complementary strand from a homologous region of DNA to the site opposite the damage (Rastogi et al. 2015).

The lesion is left unrepaired, but the cell manages to get through another round of replication, and the damaged base is now available as a substrate for excision repair. When the complementary strand is obtained from the newly replicated sister chromatid, the resulting repair is error free (Britt 1996). Recombination is a series of complex biochemical reactions involving at least 20 gene products in *E. coli*.

The RecA protein forms a right-handed helical nucleoprotein filament on the DNA and carries out a homology search followed by a strand-exchange reaction that is essential for recombination (Friedberg 1995; Kowalczykowski 1994). The lethal effects of DSBs can be conquered by the existence of two independent pathway, such as homologous recombination (HR) and nonhomologous end joining (NHEJ).

Homologous recombination is an error free pathway, since it requires an extensive region of sequence homology between the damaged and template strands. *E. coli* recA gene and its eukaryotic homologs RAD51s are the best recombination genes (Thacker 2005). The recA gene encodes a DNA-dependent ATPase that binds to ssDNA and promotes strand invasion and exchange between homologous DNA molecules (McEntee et al. 1980). An increase in the transcript level of recA and the concomitant increase in the abundance of the corresponding 37-38 kDa polypeptide was reported in *A. variabilis* after UV exposure (Owttrim and Coleman 1989). Murphy et al. (1987) characterized the recA gene from *Synechococcus* sp. strain PCC 7002. The recA mutant in *Synechocystis* sp. PCC 6803 was highly sensitive to UV-light but displayed a similar rate of growth as the wild-type strain under low-light conditions. It was concluded that RecA was essential under conditions where DNA was damaged (Minda et al. 2005). RecA protein binds to duplex DNA carrying a single-stranded gap, and forms a long helical complex that pairs with intact homologous duplex (Cox and Lehman 1987). This protein then promotes a strand exchange, which proceeds 5 ' to 3 ' with respect to the single strand in the gap and continues with very little delay when there are pyrimidine dimers in the exchanged strands (Livneh and Lehman 1982).

Recombination repair occurs in the following way:

1) RecA helical filaments are formed at the postreplication gap.
2) The RecA protein promotes homologous pairing with the intact sister duplex, which may or may not be nicked by the "cutting in trans" activity (Ross and Howard-Flanders 1977).
3) Reciprocal strand exchange takes place, partly facilitated by polymerization from the free 3' terminus of the damaged duplex and using the complementary strand of the intact duplex as a template (copy choice). The latter reaction results in two paired but, because of topological constraints, not fully hydrogen-bonded molecules.
4) Ligation and spontaneous isomerization produce the recombinational intermediate with two Holliday crossovers.
5) Resolution of the crossovers in step E yields two uninterrupted duplexes. Depending on which way the Holliday crossovers are resolved the exchange may transfer only a limited length of a single strand, or it may involve both parental and daughter strands, in which case damaged bases such as pyrimidine dimers in parental strands will be distributed among all four strands (Ganesan 1974).

The structure described at step 4 can be generated *in vitro* with, in addition to RecA, the appropriate constellation of enzymes and artificial substrates. For complete reconstitution of recombinational repair *in vitro*, RecA, PolI, "nicking in trans" activity, ligase, and Holliday resolvase are needed. Even though both the "cutting in trans" and resolvase activities have been demonstrated *in vivo* in *E. coli*, efforts to purify them have failed. It was assumed by Sancar et al. (1988) that the recombinational intermediate (Holliday) can be generated without cutting the sister duplex early in the reaction, and therefore it is quite likely that the "cutting in trans" activity and resolvase may be one and the same enzyme. Recombinational repair contributes Ito cellular survival in two ways, by bypassing adducts that block replication and by making the adduct-containing region double-stranded so that the modified base may be removed by ABC excinuclease. It is also needed for repair of DNA that contains interstrand crosslinks.

DSB repair *via* HR is the resection of 5' ends to produce a 3' ssDNA overhang by means of an exonuclease (such as RecBCD in *E. coli*). Rad51 (a functional homolog of the E. coli RecA) (Ira et al. 2004) is the central protein in HR, binds the exposed single-stranded tails forming a nucleoprotein filament and this early step is promoted by a Rad55/Rad57 protein heterodimer (Sung 1997) by overcoming the inhibitory effects of the heterotrimeric single-stranded DNA binding protein RPA (Wolner et al. 2003). The Rad51 nucleoprotein filament in association with other repair protein searches the genome for an intact copy of the broken DNA on the sister chromatid to form a heteroduplex joint molecules or D-loop that is matured in to a Holliday junctions (HJs). HJ is then resolved to give crossover products. In *E. coli*, this HJ is resolved by the positioning of RuvABC resolvasome.

When HR is inactivated, an alternate pathway, that is, nonhomologous end joining becomes operative for the repair of DSBs (Critchlow and Jackson 1998) that also involves a multiprotein complex and has been found in organisms ranging from a few prokaryotes to mammals. Nonhomologous end-joining applies major DNA doublestrand break repair (DSBR) mechanism that was first described in mammalian cells (Bassing and Alt 2004; Lieber et al. 2004). NHEJ is an error prone, alternate pathway that essentially joins broken chromosomal ends independent of sequence homology.

NHEJ involves the religation of two broken ends of a chromosome in the absence of long sequence homology. NHEJ can function either with only a few bases homology between the repaired chromosome ends, known as microhomology, or without microhomology in a blunt-directed repair. During religation, exogenous DNA may be captured in the chromosome, leading to insertion of DNA into the genome (Popa 2011). Prokaryotic NHEJ pathway was predicted from whole-genome analyses, and later shown to be functional in *B. subtillis* (Aravind and Koonin 2001). The prokaryotic NHEJ is similar to the eukaryotic system in its reliance on a DNA end-binding Ku protein and a dedicated ATP-dependent DNA ligase (Lig4 in eukaryotes and LigD in prokaryotes). Contrary to the eukaryotic system that includes various factors promoting the end processing and ligation stages, in the prokaryotic system the ATP-dependent ligase includes an additional nuclease domain that enables interaction between the Ku and the LigD proteins, thus forming a two-component NHEJ system. In *Bacillus subtilis*, the gene ykou/v has been found that encodes a polypeptide with ligase, primase and nuclease domains. (Shuman and Glickman 2007).

SOS Repair

The accumulation of massive amount of DNA lesions within the cells under different specific physiological responses (Majchrzak et al. 2006) may lead to the occurrence of SOS repair system which was well described in *E. coli*, where the involvement of more than 40 genes have been found (Courcelle et al. 2001). The term "SOS repair" was given by Radman (1974). It has been found that the bacterial NER is linked with all DNA damage response through a network of reactions, known as SOS response (Figure 5) (Michel 2005).

The SOS response begins with activation of the RecA protein to a protease that cleaves the LexA repressor, which normally prevents transcription of the many genes under the control of the SOS pathway. The SOS pathway is induced in response to various DNA-damaging agents, including UV light and mitomycin C, and is also induced by inhibition of DNA replication by agents such as nalidixic acid (Walker 1985). Some of the responses of the cell to induction of the SOS pathway include reactivation of UV irradiated bacteriophage (Weigle reactivation) (Levine 1987), mutagenesis of UV-irradiated cellular or phage DNA, repair of double strand breaks in the DNA, alleviation of restriction, and induction of repair pathways that allow long patch excision repair (Walker 1984; Witkin 1976; Walker 1985).

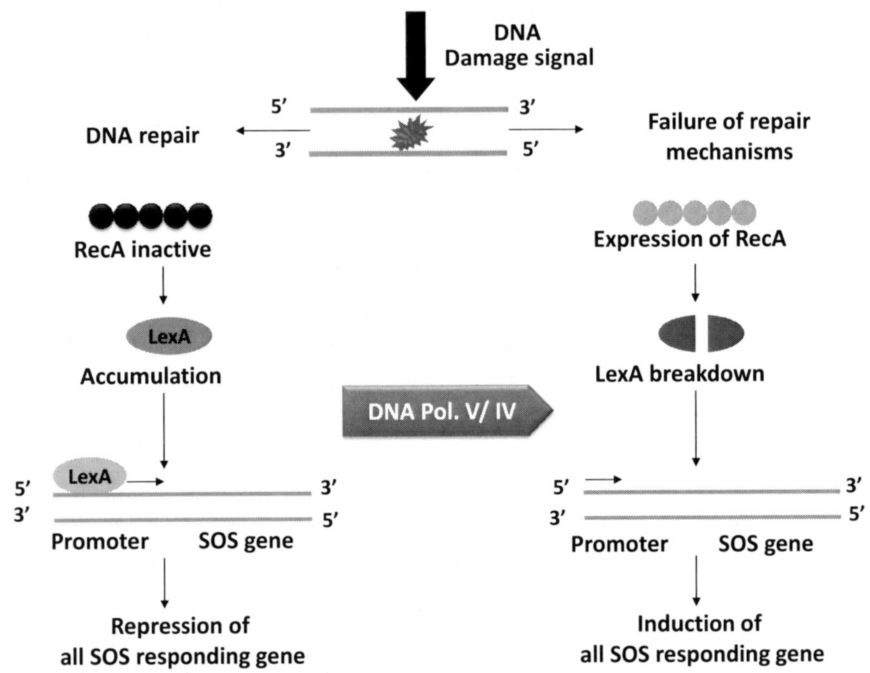

Figure 5. Simplified outline of regulation of SOS response by RecA and LexA genes.

As large genomic damage or failure of all specific repair systems occurs, the SOS repair system is started by interaction of two crucial proteins such as the RecA (36 kDa) and the LexA (27 kDa) repressor which controls the expression of SOS genes by binding to their promoters (Courcelle et al. 2001). The occurrence of SOS system together with its complete mechanisms remain obscure in cyanobacteria, however, it has been assumed that LexA appears to function as the transcriptional regulator of the key SOS response genes in most cyanobacterial genomes (Li et al. 2010).

The *recA* gene encodes a DNA-dependent ATPase that binds to single-stranded DNA and promotes strand invasion and exchange between homologous DNA molecules. As a result of massive DNA damage and failure of all possible repair mechanisms, RecA proteins is expressed, which activate the auto breakdown of LexA proteins, allowing the induction of all SOS responding genes (Rastogi et al. 2010a; Martins-Pinheiro et al. 2007).

LexA is a typical transcriptional repressor, consisting of two domains connected by a flexible hinge. A palindromic DNA sequence, termed the SOS box, is recognized by the dimeric repressor and a consensus sequence has been determined. The SOS boxes are found upstream of the subordinate genes and, for different genes, the number of operator boxes can vary between one and three. The second protein involved in the regulation of the SOS response is RecA, although this protein serves several other functions in the cell (important for homologous recombination). Within the frame of the SOS response, however, RecA mediates the induction of the genes coordinately repressed by LexA. In the presence of DNA damaging agents or UVR, single-stranded DNA fragments appear in the cell. The activity of RecA is changed upon binding to such single-stranded DNA sites. The activated RecA form further facilitates a latent capacity of LexA to autodigest. Upon the LexA proteolytic cleavage, the repression imposed by LexA is released, resulting in the expression of the genes involved in the DNA repair.

A number of genes (or operon) collectively known as din (damage inducible) gene such as *uvrA*, *uvrB*, *cho* (uvrC homolog) and *uvrD* of *E. coli* NER take part in SOS response (Courcelle et al. 2001; Van Houten et al. 2002). Each of the SOS-induced damage-inducible (din) or sos genes has near its promoter/operator site a specific 20-nucleotide-long "SOS-box" (also named, LexA-box) (Janion 2008). The SOS response is induced with damage signal but it is highly mutagenic due to engagement of error-prone DNA polymeraseV (UmuC/$UmuD_2$ complex) (Fuchs et al. 2004) and DNA polymerase IV (Goodman 2000) in *E. coli*.

The pathway of SOS response is reversed when damages are repaired through the damage specific mechanisms. SOS response is highly mutagenic due to involvement of DNA polymerase V/IV. Li et al. (2010) has analysed the LexA regulons in cyanobacteria and found that in most cyanobacterial genomes, LexA appears to function as the transcriptional regulator of the key SOS response genes. An increase in the transcript levels of recA and a concomitant increase in the abundance of the corresponding

37-38 kDa polypeptide have been reported in *A. variabilis* following UVR (Owttrim and Coleman 1989). Activation of RecA is reversible and prevails as long as the single-stranded DNA is abundant. Once DNA lesions have been repaired, RecA ceases to be activated and non-cleaved LexA protein returns to its normal levels, repressing again the transcription of SOS genes.

However, Domain et al. (2004) and Patterson-Fortin et al. (2006) independently showed strong evidence that cyanobacteria do not have an *E. coli* type SOS regulon. Using DNA microarrays, Domain et al. (2004) identified 57 genes with expression that is altered in response to CyLexA depletion in *Synechocystis* sp. PCC 6803, none of them being predicted to operate in DNA metabolism. On the other hand, Patterson-Fortin et al. (2006) observed that lexA is not induced upon UV-irradiation, suggesting that CyLexA is not required for survival following DNA damage. These facts argue against the involvement of CyLexA in the regulation of DNA repair. Instead, most of the CyLexA-responsive genes were known to be involved in carbon assimilation or controlled by carbon availability. Therefore the term suggests CyLexA, referring to the alternative function of this transcription factor in *Synechocystis* sp. PCC 6803, and possibly in cyanobacteria (Oliviera and Lindbad 2009). A computational analysis across a large number of cyanobacterial genome sequences suggests that LexA has a similar role to that in *E. coli* in most cyanobacteria, though not in *Synechocystis* sp. PCC 6803 (Li et al. 2010), the most common model species used for cyanobacteria research.

Apoptosis

In case of cell damage beyond the execution of effective repair, organisms can save their life by selective cell death, also known as programmed cell death (PCD). Cyanobacteria may also undergo apoptosis or PCD when a cell is damaged beyond repair. DNA damage and production of the predominant lesions such as CPDs, 6-4 PPs and certain other lesions (Sinha and Häder 2002) as a result of UVR, IR, oxidative stress, replication or recombination errors as well as from environmental and therapeutic genotoxins is one of the principal triggers of apoptosis (Nagata 1997; Norbury and Zhivotovsky 2004; Batista et al. 2009).

An autocatalytic PCD has been shown to operate in the nitrogen-fixing cyanobacterium *Trichodesmium* sp. and *Anabaena flos-aquae* was found to be induced by high irradiance, iron starvation and oxidative stress (Ning et al. 2002; Berman-Frank et al. 2004). Similarly, recent data on *Microcystis aeruginosa* showed that H_2O_2 treatment induces PCD in this organism, and catalase was capable of inhibiting PCD, implicating the role of PCD under oxidative stress (Ross et al. 2006). Several genes encoding for caspases, enzymes involved in PCD in eukaryotes, have also been recently reported in the sequenced genome of *M. aeruginosa* (Frangeul et al. 2008). The caspase activity as well as proteins reacting to human caspase-3 antibodies was reported in *Trichodesmium* sp. (Berman-Frank et al. 2004). The freshwater cyanobacterium *Anabaena* has also been found to activate PCD and increases general protease activity after exposure to univalent-cation salts (Ning et al. 2002). Thus, it is clear that the process of PCD plays an important role under oxidative stress in cyanobacteria. Marine cyanobacteria were found to produce a wide range of compounds that revealed apoptotic properties (Costa et al. 2012). Apoptosis can be induced by both intrinsic and extrinsic signals, by multiple agents, as the natural flavonoid quercetin, the representative reactive oxygen species H_2O_2 or even the UVR (Lytvyn et al. 2010). However, the exact mechanism of UV-induced PCD in cyanobacteria is still not very clear.

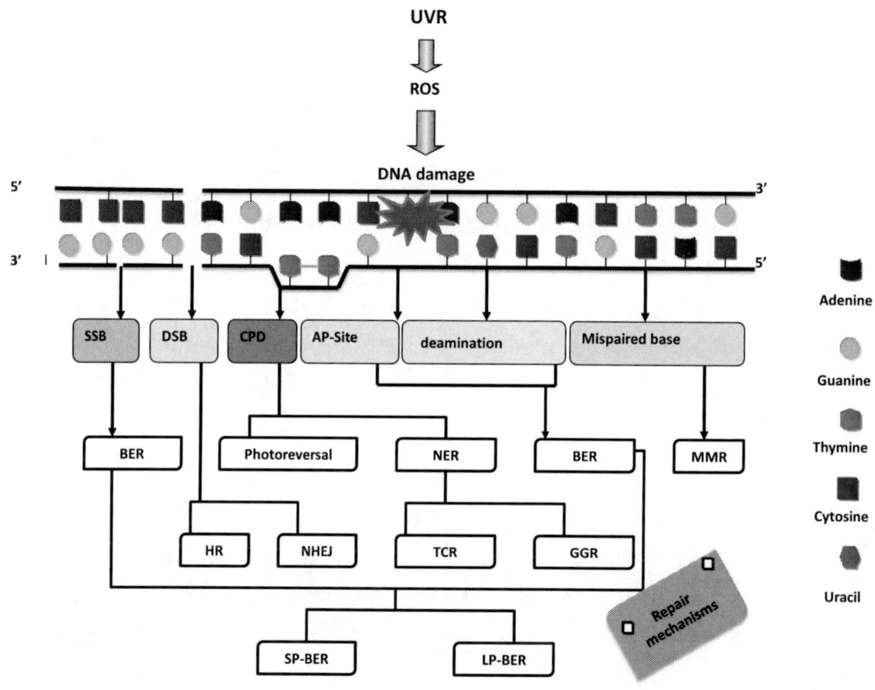

Figure 6. DNA damages caused by UV radiation and the associated repair pathways (Adapted from Rastogi et al., 2010).

Conclusion

The reduced water content influences photoproduct formation without causing a change in the conformational state of DNA (Lindsay and Murrell 1983), i.e., the reduced water content has a protective effect on the DNA. The presence of multiple genome copies in some bacteria, including cyanobacteria (Simon 1980), is often cited as a possible means to circumvent damage to DNA. A more likely reason for multiple genome copies is to force mandatory condensation of the nucleoplasm, providing some structural protection to nucleoids in the dried state. In this regard, it may be of utmost importance to study the role of histone-like DNA binding proteins in desiccation tolerance because these proteins play an important role in maintaining the nucleoid organization and regulate DNA repair following UV irradiation (Palchaudhuri et al. 1998). Figure 6 shows the overview of various types of damages caused by UVR and several repair strategies against such damages.

ACKNOWLEDGMENTS

Rajneesh is thankful to DBT, Govt. of India for the research grant in form of SRF. Jainendra Pathak, Vidya singh, Deepak Kumar Singh and Abha Pandey are thankful to CSIR, New Delhi, India for the financial support in the form of SRF. Haseen Ahmed and SP Singh acknowledge UGC, New Delhi, India for financial support in the form of SRF and Early Career Research Award respectively.

REFERENCES

Aboussekhra A and Thoma F (1999) TATA-binding protein promotes the selective formation of UV-induced (6-4) photoproducts and modulates DNA repair in the TATA box. *EMBO J*, 18: 433-443.

Agustí S, Llabrés M, Carreja B, Fernández M and Duarte CM (2015) Contrasting sensitivity of marine biota to UV-B radiation between Southern and Northern hemispheres. *Estuar Coast*, 38: 1126-1133.

Ahmad M and Cashmore AR (1993) The HY4 gene of *Arabidopsis thaliana* encodes a protein with characteristics of a blue-light receptor. *Nature*, 366: 162-166.

Almeida KH and Sobol RW (2007) A unified view of base excision repair: lesion-dependent protein complexes regulated by post-translational modification. *DNA repair*, 6: 695-711.

Apel K and Hirt H (2004) Reactive oxygen species: metabolism, oxidative and signal transduction. *Annu Rev Plant Biol*, 55: 373-399.

Aravind L and Koonin EV (2001) Prokaryotic homologs of the eukaryotic DNA end-binding protein Ku, novel domains in the Ku protein and prediction of a prokaryotic double-strand break repair system. *Genome Res*, 11: 1365-1374.

Arenas J and Abelson J (1991) Requirement of the RNA helicase-like protein PRP22 for release of messenger RNA from spliceosomes. *Nature*, 349: 487.

Arikan E, Kulkami MS, Thomas DC and Sancar A (1986) Sequences of the *E. coli* uvrB gene and protein. *Nucleic Acids Res*, 14: 2637-2650.

Aubert C, Vos P MH, Mathis AP, Eker M and Brettel K (2000) Intraprotein radical transfer during photoactivation of DNA photolyase. *Nature*, 405: 586-590.

Bassing CH and Alt FW (2004) The cellular response to general and programmed DNA double strand breaks. *DNA Repair*, 3: 781-796.

Batista LFZ, Kaina B, Meneghini R and Menck CFM (2009) How DNA lesions are turned into powerful killing structures: Insights from UV-induced apoptosis. *Mutat Res Rev*, 681: 197-208.

Batschauer A (1993) A plant gene for photolyase: an enzyme catalyzing the repair of UV-light-induced DNA damage. *Plant J*, 4: 705-709.

Becker MM and Wang Z (1989) Origin of ultraviolet damage in DNA. *J Mol Biol*, 210: 429-438.

Berman-Frank I, Bidle KD, Haramaty L and Falkowski PG (2004) The demise of the marine cyanobacterium, *Trichodesmium* sp., *via* an autocatalyzed cell death pathway. *Limnol Oceanogr*, 49: 997-1005.

Bertrand-Burggraf E, Selby CP, Hearst JE and Sancar A (1991) Identification of the different intermediates in the interaction of ABC excinuclease with its substrates by DNase I footprinting on two uniquely modified oligonucleotides. *J mol Biol*, 219: 27-36.

Bhattacharjee SK (1977) Unstable protein mediated ultraviolet light resistance in *Anacystis nidulans*. *Nature*, 269: 82-83.

Billi D, Viaggiu E, Cockell CS, Rabbow E, Horneck G and Onofri S (2011) Damage escape and repair in dried *Chroococcidiopsis* spp. from hot and cold deserts exposed to simulated space and Martian conditions. *Astrobiol*, 11: 65-73.

Björn LO (2007) Stratospheric ozone, ultraviolet radiation, and cryptogams. *Biol Cons*, 135: 326-333.

Boelen P, de Boer MK, Kraay GW, Veldhuis MJ and Buma AG (2000) UVBR-induced DNA damage in natural marine picoplankton assemblages in the tropical Atlantic Ocean. *Mar Ecol Prog Ser*, 193: 1-9.

Boelen P, Veldhuis MJ and Buma AG (2001) Accumulation and removal of UVBR-induced DNA damage in marine tropical plankton subjected to mixed and simulated non-mixed conditions. *Aquat Microb Ecol*. 24: 265-274.

Bose SN, Davies RJ, Sethi SK and McCloskey JA (1983). Formation of an adenine-thymine photoadduct in the deoxydinucleoside monophosphate d(TpA) and in DNA. *Science*, 220: 723-725.

Brack W, Altenburger R, Kuster E, Meissner B, Wenzel KD and Schuurmann G (2003) Identification of toxic products of Anthracene photomodification in simulated sunlight. *Environ Toxicol Chem*, 22: 2228-2237.

Brash DE, Franklin WA, Sancar GB, Sancar A and Haseltine WA (1985) *Escherichia coli* DNA photolyase reverses cyclobutane pyrimidine dimers but not pyrimidine-pyrimidone (6-4) photoproducts. *J Biol Chem*, 260: 11438-11441.

Bray CM and West CE (2005) DNA repair mechanisms in plants: crucial sensors and effectors for the maintenance of genome integrity. *New Phytol*, 168: 511-528.

Britt AB (1995) Repair of DNA damage induced by ultraviolet radiation. *Plant Physiol*, 108: 891-896.

Britt AB (1996) DNA damage and repair in plants. *Annu Rev Plant Biol*, 47: 75-100.

Brosché M and Strid Å (2003) Molecular events following perception of ultraviolet-Bradiation by plants. *Physiol Plant*, 117: 1-10.

Browman HI, Alonso Rodriguez C, Beland F, Cullen JJ, Davis RJ, Kouwenberg JHM, Kuhn P, McArthur B, Runge JA, St-Pierre JF and Vetter RD (2000) The impact of ultraviolet radiation on marine crustacean zooplankton and ichthyoplankton: a synthesis of results from the estuary and Gulf of St. Lawrence. *Canada Mar Ecol Prog Ser*, 199: 293-311.

Buma AGJ, van Hannen EJ, Roza L, Veldhuis MJW and Gieskes WWC (1995) Monitoring ultraviolet-B-induced DNA damage in individual diatom cells by irnmunofluorescent thymine dimer detection. *J Phycol*, 31: 314-321.

Burrows CJ (2009) Surviving an oxygen atmosphere: DNA damage and repair. In: Zaikowski L, Friedrich JM and Seidel SR, editors. *Chemical Evolution II: From Origins of Life to Modern Society*: ACS Symposium Series, American Chemical Society, 1025: 146-156.

Cadet J, Courdavault S, Ravanat JL and Douki T (2005) UV-B and UV-A radiation-mediated damage to isolated and cellular DNA. *Pure Appl Chem*, 77: 947-961.

Camus L and Gulliksen B (2005) Antioxidant defense properties of Arctic amphipods: comparison between deep, sublittoral and surface-water species. *Mar Biol*, 146: 355-362.

Carell T and Epple R (1998) Repair of UV light induced DNA lesions: a comparative study with model compounds. *Eur J Org Chem*, 7: 1245-1258.

Castenholz RW and Garcia-Pichel F (2000) Cyanobacterial responses to UV-radiation. In: *The Ecology of Cyanobacteria*. Springer Netherlands, pp. 591-611.

Chatterjee N and Walker GC (2017) Mechanisms of DNA damage, repair, and mutagenesis. *Environ Mol Mutagen*, 58: 235-263.

Chaudhari VR, Vyawahare A, Bhattacharjee SK and Rao BJ (2015) Enhanced excision repair and lack of PSII activity contribute to higher UV survival of *Chlamydomonas reinhardtii* cells in dark. *Plant Physiol Biochem*, 88: 60-69.

Chen L, Xie M, Bi Y, Wang G, Deng S and Liu Y (2012) The combined effects of UV-B radiation and herbicides on photosynthesis, antioxidant enzymes and DNA damage in two bloom-forming cyanobacteria. *Ecotoxicol Environ Saf*, 80: 224-230.

Chen Y, Chen T, Shen S, Zheng M, Guo Y, Lin J, Baluska F and Samaj J (2006) Differential display proteomic analysis of *Picea meyeri* pollen germination and pollen tube growth after actin depolymerization by latrunculin B. *Plant J*, 47: 194-195.

Chou KM and Cheng YC (2002) An exonucleolytic activity of human apurinic/apyrimidinic endonuclease on 3' mispaired DNA. *Nature*, 415: 655-659.

Cleaver EJ (1968) Defective Repair Replication of DNA in Xeroderma Pigmentosum. *Nature*, 218: 652-656.

Cleaver JE and Bootsma D (1975) Xeroderma pigmentosum: biochemical and genetic characteristics. *Annu Rev Genet*, 9: 19-38.

Costa M, Costa-Rodrigues J, Fernandes MH, Barros P, Vasconcelos V and Martins R (2012) Marine cyanobacteria compounds with anticancer properties: A review on the implication of apoptosis. *Mar Drugs*, 10: 2181-2207.

Courcelle J, Khodursky A, Peter B, Brown PO and Hanawalt PC (2001) Comparative gene expression profiles following UV exposure in wild-type and SOS deficient *Escherichia coli*. *Genet*, 58: 41-64.

Courdavault S, Baudouin C, Charveron M, Canguilhem B, Favier A, Cadet J and Douki T (2005) Repair of the three main types of bipyrimidine DNA photoproducts in human keratinocytes exposed to UV-B and UV-A radiations. *DNA Repair*, 4: 836-844.

Cox MM and Lehman IR (1987) Enzymes of general recombination. *Annu Rev Biochem*, 56: 229-262.

Critchlow SE and Jackson SP (1998) DNA end-joining: from yeast to man. *Trends Biochem Sci*, 23: 394-398.

Cunningham RP, Saporito SM, Spitzer SG and Weiss B (1986) Endonuclease IV (nfo) mutant of *Escherichia coli*. *Bacteriol*, 168: 1120-1127.

Dalhus B, Laerdahl JK, Backe PH and Bjoras M (2009) DNA base repair-recognition and initiation of catalysis. *FEMS Microbiol Rev*, 33: 1044-1078.

Dany AL, Douki T, Triantaphylides C and Cadet J (2001) Repair of the main UV-induced thymine dimeric lesions within *Arabidopsis thaliana* DNA: evidence for the major involvement of photoreactivation pathways. *J Photochem Photobiol B Biol*, 65: 127-135.

Davies RJ (1995) Royal Irish Academy Medal Lecture. Ultraviolet radiation damage in DNA. *Biochem Soc Trans*, 23: 407-418.

Demple B and Linn S (1982) 5, 6-Saturated thymine lesions in DNA: Production by ultraviolet light or hydrogen peroxide. *Nucleic Acids Res*, 10: 3781-3789.

Dianov G and Lindahl T (1994) Reconstitution of the DNA base excision-repair pathway. *Curr Biol*, 4: 1069-1076.

Doetsch PW, Zastawny TH, Martin AM and Dizdaroglu M (1995) Monomeric base damage products from adenine, guanine and thymine induced by exposure of DNA to ultraviolet radiation. *Biochem*, 34: 737-742.

Domain F, Houot L, Chauvat F and Cassier-Chauvat C (2004) Function and regulation of the cyanobacterial genes lexA, recA and ruvB: LexA is critical to the survival of cells facing inorganic carbon starvation. *Mol Microbiol*, 53: 65-80.

Douki T and Cadet J (2001) Individual determination of the yield of the main UV-induced dimeric pyrimidine photoproducts in DNA suggests a high mutagenicity of CC photolesions. *Biochem*, 40: 2495-2501.

Douki T, Laporte G and Cadet J (2003) Inter-strand photoproducts are produced in high yield within A-DNA exposed to UV-C radiation. *Nucleic Acids Res*, 3: 3134-3142.

Douki T, Perdiz D, Grof P, Kuluncsics Z, Moustacchi E, Cadet J and Sage E (1999) Oxidation of guanine in cellular DNA by solar UV radiation: Biological role. *Photochem Photobiol*, 70: 184-190.

Duker NJ and Gallagher PE (1988) Purine photoproducts. *Photochem Photobiol*, 48: 35-39.

Dulbecco R (1949) Reactivation of ultraviolet-inactivated bacteriophage by visible light. *Nature*, 163: 949-950.

Eker APM, Kooiman P, Hessels JKC and Yasui A (1990) DNA photoreactivating enzyme from the cyanobacterium *Anacystis nidulans*. *J Biol Chem*, 265: 8009-8015.

Epe B (1991) Genotoxicity of singlet oxygen. *Chem Biol Interact*, 80: 239-260.

Espagne A, Byrdin M, Eker APM and Brettel K (2009) Very fast product release and catalytic turnover of DNA photolyase. *Chem Bio Chem*, 10: 1777-1780.

Essen LO and Klar T (2006) Light-driven DNA repair by photolyases. *Cell Mol Life Sci*, 63: 1266-1277.

Evan G and Littlewood T (1998) A matter of life and cell death. *Science*, 281: 1317-1322.

Fang, CX, Wang QS, Yu Y, Huang LK, Wu XC and Lin WX (2011) Silicon and its uptaking gene Lsi 1 in regulation of rice UV-B tolerance. *Acta Agron Sin*, 37: 1005-1011.

Fingerhut BP, Heil K, Kaya E, Oesterling S, de Vivie-Riedle R and Carell T (2012) Mechanism of UV-induced Dewar lesion repair catalysed by DNA (6-4) photolyase. *Chem Sci*, 3: 1794-1797.

Fischer WF (2008) Life before the rise of oxygen. *Nature*, 455: 1051-1052.

Fortini P and Dogliotti E (2007) Base damage and single-strand break repair: mechanisms and functional significance of short- and long-patch repair subpathways. *DNA Repair*, 6: 398-409.

Frangeul L, Quillardet P, Castets A-M, Humbert J-F, Matthijs HCP, Cortez D, Tolonen A, Zhang C-C, Gribaldo S, Kehr J-C, Zilliges Y, Ziemert N, Becker S, Talla E, Latifi A, Billault A, Lepelletier A, Dittmann E, Bouchier C and Tandeau DN (2008) Highly plastic genome of *Microcystis aeruginosa* PCC 7806, a ubiquitous toxic freshwater cyanobacterium. *BMC Genomics*, 9: 274-294.

Friedberg EC, Walker GC and Siede W (1995) *DNA repair and mutagenesis*. ASM Press, Washington, DC, 21: 191-232.

Friedberg EC, Walker GC, Siede W, Wood RD, Schultz RA and Ellenberger T (2006) *DNA Repair and Mutagenesis*, ASM Press, Washington, DC, USA.

Friedman JI and Stivers JT (2010) Detection of damaged DNA bases by DNA glycosylase enzymes. *Biochem*, 49: 4957- 4967.

Fuchs RP, Fujii S and Wagner J (2004) Properties and functions of *Escherichia coli*: Pol IV and Pol V. *Adv Protein Chem*, 69: 229-264.

Gale JM, Nissen KA and Smerdon MJ (1987) UV-induced formation of pyrimidine dimers in nucleosome core DNA is strongly modulated with a period of 10.3 bases. *Proc Natl Acad Sci*, 84: 6644-6648.

Ganesan AK (1974) Persistence of pyrimidine dimers during post-replication repair in ultraviolet light-irradiated *Escherichia coli* K12. *J Mol Biol*, 87: 103-119.

Gao K, Li P, Watanabe T and Helbling EW (2008) Combined effects of ultraviolet radiation and temperature on morphology, photosynthesis, and DNA of *Arthrospira (Spirulina) platensis* (Cyanophyta). *J Phycol*, 44: 777-786.

Gill RT, Katsoulakis E, Schmitt W, Taroncher-Oldenburg G, Misra J and *Stephanopoulos* G (2002) Genome-wide dynamic transcriptional pro fi ling of the light-to-dark transition in *Synechocystis* sp. strain PCC 6803. *J Bacteriol*, 184: 3671-3681.

Glas AF, Kaya E, Schneider S, Heil K, Fazio D, Maul MJ and Carell T (2010) DNA (6-4) photolyases reduce Dewar isomers for isomerization into (6-4) lesions. *J Am Chem Soc*, 132: 3254-3255.

Glas AF, Schneider S, Maul MJ, Hennecke U and Carell T (2009) Crystal structure of the T(6-4)C lesion in complex with a (6-4) DNA photolyase and repair of UV-induced (6-4) and Dewar photolesions. *Chem Eur J*, 15: 10387-10396.

Gorbalenya AE and Koonin EV (1990) Superfamily of UvrA-related NTP-binding proteins: implications for rational classification of recombination/repair systems. *J Mol Biol*. 213: 583-591.

Gorman G, Hanlon D and King W (1997) Some research perspectives on entrepreneurship education, enterprise education and education for small business management: a ten-year literature review. *Int Small Bus J,* 15: 56-77.

Gorman MA, Morera S, Rothwell DG, de La Fortelle E, Mol CD, Tainer JA, Hickson ID and Freemont PS (1997) The crystal structure of the human DNA repair endonuclease HAP1 suggests the recognition of extra-helical deoxyribose at DNA abasic sites. *EMBO J*, 16: 6548-6558.

Graf D, Wesslowski J, Ma H, Scheerer P, Krauß N, Oberpichler I and Lamparter T (2015) Key amino acids in the bacterial (6-4) Photolyase PhrB from *Agrobacterium fabrum*. *PloS one*, 10: e0140955.

Häder D-P and Sinha RP (2005) Solar ultraviolet radiation-induced DNA damage in aquatic organisms: potential environmental impact. *Mutat Res*, 571: 221-233.

Häder D-P, Kumar HD, Smith RC and Worrest RC (2007) Effects of solar UV radiation on aquatic ecosystems and interactions with climate change. *Photochem Photobiol Sci*, 6: 267-285.

Häder D-P, RC Worrest and HD Kumar (1989) Aquatic ecosystems. In: *UNEP Environmental Effects Panel Report*, Nairobi, Kenya, pp. 3948.

Hadi MZ, Ginalski K, Nguyen LH and Wilson DM (2002) Determinants in nuclease specificity of Ape1 and Ape2, human homologues of *Escherichia coli* exonuclease III. *J Mol Biol*, 316: 853-866.

Hall DB, Holmlin RE and Barton JK (1996) Oxidative DNA damage through long range electron transfer. *Nature*, 382: 731-735.

Han T, Sinha RP and Häder DP (2001) UV-A/blue light-induced reactivation of photosynthesis in UV-B irradiated cyanobacterium, *Anabaena* sp. *J Plant Physiol*, 158: 1403-1413.

Hargreaves A, Taiwo FA, Duggan O, Kirk SH and Ahmad SI (2007) Near-ultraviolet photolysis of b-phenylpyruvic acid generates free radicals and results in DNA damage. *J Photochem Photobiol B Biol*, 89: 110-116.

He YY and Häder D-P (2002) UV-B-induced formation of reactive oxygen species and oxidative damage of the cyanobacterium *Anabaena* sp.: protective effects of ascorbic acid and N-acetyl-L-cystein. *J Photochem Photobiol B Biol*, 66: 115-124.

Hegde ML, Hazra TK and Mitra S (2008) Early steps in the DNA base excision/single-strand interruption repair pathway in mammalian cells. *Cell Res*, 18: 27-47.

Hitomi K, Kim ST, Iwai S, Harima N, Otoshi E, Ikenaga M and Todo T (1997) Binding and catalytic properties of Xenopus (6-4) photolyase. *J Biol Chem*, 272: 32591-32598.

Hitomi K, Okamoto K, Daiyasu H, Miyashita H, Iwai S, Toh H, Ishiura M and Todo T (2000) Bacterial cryptochrome and photolyase: characterization of two photolyase-like genes of *Synechocystis* sp. PCC6803. *Nucl Acids Res*, 28: 2353-2362.

Holzinger A and Lütz C (2006) Algae and UV irradiation: effects on ultrastructure and related metabolic functions. *Micron*, 37: 190-207.

Howard FP, Boyce RP and Theriot L (1966) Three Loci in *Escherichia Coli* K-12 That Control Excision of Pyrimidine Dimers and Certain Other Mutagen Products from DNA. *Genetics*, 53: 1119-1136.

Hu Z, Cools T and De Veylder L (2016) Mechanisms used by plants to cope with DNA damage. *Annu Rev Plant Biol*, 67: 439-462.

Husain I, Van Houten B, Thomas DC and Sancar A (1986) Sequences of *Escherichia coli* uvrA gene and protein reveal two potential ATP binding sites. *J Biol Chem*, 26: 4895-4901.

Ira G, Pellicioli A, Balijja A, Wang X, Fiorani S, Carotenuto W, Bressan D, Wan L, Hollingsworth NM and Haber JE (2004) DNA end resection, homologous recombination and DNA damage checkpoint activation require CDK1. *Nature*, 43: 1011-1017.

Jägger J (1985) *Solar-UV action on living cells*. Praeger, New York, USA.

Janion C (2008) Inducible SOS response system of DNA repair and mutagenesis in *Escherichia coli*. *Int J Biol Sci*, 4: 338- 344.

Jiang Y, Rabbi M, Kim M, Ke C, Lee W, Clark RL, Mieczkowski PA and Marszalek PE (2009) UV-A generates pyrimidine dimers in DNA directly. *Biophys J*, 96: 1151-1158.

Jin P, Duarte CM and Agustí S (2017) Contrasting Responses of Marine and Freshwater Photosynthetic Organisms to UVB Radiation: A Meta-Analysis. *Front Mar Sci*, 4: 45-63.

Johnson JL, Hamm-Alvarez S, Payne G, Sancar GB, Rajagopalan KV and Sancar A (1988) Identification of the second chromophore of *Escherichia coli* and yeast DNA photolyases as 5,10- methenyltetrahydrofolate. *Proc Natl Acad Sci USA*, 85: 2046-2050.

Kalbin G, Hidema J, Brosché M, Kumagai T, Bornman JF and Strid Å (2001) UV-B-induced DNA damage and expression of defence genes under UV-B stress: tissue-specific molecular marker analysis in leaves. *Plant Cell Environ*, 24: 983-990.

Kao YT, Saxena C, Wang L, Sancar A and Zhong D (2005) Direct observation of thymine dimer repair in DNA by photolyase. *Proc Natl Acad Sci USA*, 102: 16128-16132.

Karentz D, Cleaver JE and Mitchell DL (1991) Cell survival characteristics and molecular responses of Antarctic phytoplankton to ultraviolet-B radiation. *J Phycol*, 27: 326-341.

Kavakli IH and Sancar A (2004) Analysis of the role of intraprotein electron transfer in photoreactivation by DNA photolyase *in vivo*. *Biochemistry*, 43: 15103-15110.

Kelner A (1949) Effect of visible light on the recovery of *Streptomyces griseus* conidia from ultra-violet irradiation injury. *Proc Natl Acad Sci USA*, 35: 73-79.

Kiefer JR (2007) Effects of Ultraviolet Radiation on DNA. In: *Chromosomal Alterations*. New York: Springer, pp. 39-53.

Kim JK, Patel D and Choi BS (1995) Contrasting structural impacts induced by cis-syn cyclobutane dimer and (6-4) adduct in DNA duplex decamers: implication in mutagenesis and repair activity. *Photochem Photobiol*, 62: 44-50.

Kim ST, Heelis PF and Sancar A (1992) Energy transfer (deazaflavin→$FADH_2$) and electron transfer ($FADH_2$→T<>T) kinetics in *Anacystis nidulans* photolyase. *Biochemistry*, 3: 11244-11248.

Kim ST, Malhotra K, Smith CA, Taylor JS and Sancar A (1994) Characterization of (6-4) photoproduct DNA photolyase. *J Biol Chem*, 269: 8535-8540.

Kim ST, Malhotra K, Taylor JS and Sancar A (1996) Purification and partial characterization of (6-4) photoproduct DNA photolyase from *Xenopus laevis*. *Photochem Photobiol*, 63: 292-295.

Kirk JTO (1994) Optics of UV-B radiation in natural waters. *Arch Hydrobiol Beih*, 43: 1-16.

Kolowrat C, Partensky F, Mella-Flores D, Le Corguillé G, Boutte C, Blot N, Ratin M, Ferréo M, Lecomte X, Gourvil P, Lennon JF, Kehoe DM and Garczarek L (2010) Ultraviolet stress delays chromosome replication in light/dark synchronized cells of the marine cyanobacterium *Prochlorococcus marinus* PCC9511. *BMC Microbiol*, 10: 204.

Kondoh M, Hitomi K, Yamamoto J, Todo T, Iwai S, Getzoff ED and Terazima M (2011) Light-induced conformational change and product release in DNA repair by (6-4) photolyase. *J Am Chem Soc*, 133: 2183-2191.

Koning TMG, Davies RJH and Kaptein R (1990) The solution structure of the intramolecular photoproduct of d(TpA) derived with the use of NMR and a combination of distance geometry and molecular dynamics. *Nucleic Acids Res*, 18: 277-284.

Kowalczykowski SC, Dixon DA, Eggleston AK, Lauder SD and Rehrauer WM (1994) Biochemistry of homologous recombination in *Escherichia coli. Microbiol Rev*, 58: 401-465.

Krokan HE, Standal R and Slupphaug G (1997) DNA glycosylases in the base excision repair of DNA. *Biochem J*, 325: 1-16.

Kumar A, Tyagi MB, Singh N, Tyagi R, Jha PN, Sinha RP and Häder DP (2003) Role of white light in reversing UV-B-mediated effects in the N_2-fixing cyanobacterium *Anabaena* BT2. *J. Photochem Photobiol B Biol*, 71: 35-42.

Kumar S, Joshi PC, Sharma ND, Bose SN, Jeremy R, Davies H, Takeda N and McCloskey JA (1991) Adenine photodimerization in deoxyadenylate sequences: Elucidation of the mechanism through structural studies of a major d(ApA) photoproduct. *Nucleic Acids Res*, 19: 2841-2847.

Kumar S, Sharma ND, Davies RJ, Phillipson DW and McCloskey JA (1987) The isolation and characterisation of a new type of dimeric adenine photoproduct in UV-irradiated deoxyadenylates. *Nucleic Acids Res*, 15: 1199-1216.

Kumar, Sudhir, Koichiro Tamura and Masatoshi N (2004) MEGA3: integrated software for molecular evolutionary genetics analysis and sequence alignment. *Briefings in Bioinformatics*,5: 150-163.

Kunz BA, Anderson HJ, Osmond MJ and Vonarx EJ (2005) Components of nucleotide excision repair and DNA damage tolerance in *Arabidopsis thaliana. Environ Mol Mutagen*, 45: 115-127.

Kvam E and Tyrrell RM (1997) Artificial background and induced levels of oxidative base damage in DNA from human cells. *Carcinogenesis*, 18: 2281-2283.

Lamare MD, Barker MF, Lesser MP and Marshall C (2006) DNA photorepair in echinoid embryos: effects of temperature on repair rate in Antarctic and non-Antarctic species. *J Exp Biol*, 209: 5017-5028.

Lampi MA, Gurska J, McDonald KIC, Fangli X, Huang X-D, Dixon DG and Greenberg BM (2005) Photoinduced toxicity of polycyclic aromatic hydrocarbons to *Daphnia magna*: ultraviolet-mediated effects and the toxicity of polycyclic aromatic hydrocarbon photoproducts. *Environ Toxicol Chem*, 25: 1079-1087.

Larson R, Garrison W and Carlson R (1990) Differential responses of alpine and non-alpine *Aquilegia* species to increased ultraviolet-B radiation. *Plant Cell Physiol*, 13: 983-987.

Latifi A, Ruiz M and Zhang CC (2009) Oxidative stress in cyanobacteria. *FEMS Microbiol Rev*, 33: 258-278.

Levine EL and Thiel TE (1987) UV-inducible DNA repair in the cyanobacteria *Anabaena* sp. *J Bacteriol*, 169: 3988-3993.

Li S, Xu M and Su Z (2010) Computational analysis of LexA regulons in cyanobacteria. *BMC Genomics*, 11: 1-527.

Li ZK, Dai GZ, Juneau P and Qiu BS (2017) Different physiological responses of cyanobacteria to ultraviolet-B radiation under iron-replete and iron-deficient conditions: Implications for underestimating the negative effects of UV-B radiation. *J Phycol*, 53: 425-436.

Lieber MR, Ma Y, Pannicke U and Schwarz K (2004) The mechanism of vertebrate nonhomologous DNA end joining and its role in V(D)J recombination. *DNA Repair*, 3: 817-826.

Lin JJ and Sancar A (1990) Reconstitution of nucleotide excision nuclease with UvrA and UvrB proteins from *Escherichia coli* and UvrC protein from *Bacillus subtilis*. *J Biol Chem*, 265: 21337-21341.

Lin JJ and Sancar A (1991) The C-terminal half of UvrC protein is sufficient to reconstitute (A) BC excinuclease.*Proc Natl Acad Sci USA*, 88: 6824-6828.

Lin JJ and Sancar A (1992) (A)BC excinuclease: the *Escherichia coli* nucleotide excision repair enzyme. *Mol Microbiol*, 6: 2219-2224.

Lin JJ, Phillips AM, Hearst JE and Sancar A (1992) Active site of (A)BC excinuclease. II. Binding, bending, and catalysis mutants of UvrB reveal a direct role in 3'and an indirect role in 5'incision. *J Biol Chem*, 267: 17693-17700.

Lindsay JA and Murrell WG (1983) A comparison of UV-induced DNA photoproducts from isolated and non-isolated developing bacterial forespores. *Biochem Biophys Res Commun*, 113: 618-625.

Livneh Z and Lehman IR (1982) Recombinational bypass of pyrimidine dimers promoted by the recA protein of *Escherichia coli*. *Proc Natl Acad Sci USA*, 79: 3171-3175.

Llabrés M, Agustí S, Alonso-Laita P and Herndl GJ (2010) *Synechococcus* and *Prochlorococcus* cell death induced by UV radiation and the penetration of lethal UVR in the Mediterranean Sea. *Mar Ecol Prog Ser*, 399: 27-37.

Los DA, Suzuki I, Zinchenko VV and Murata N (2008) Stress responses in *Synechocystis*: regulated genes and regulated systems. In: *The cyanobacteria-molecular biology, genomics and evolution*. Caister Academic Press, Norfolk, pp. 117-157.

Lubin D and Jensen EH (1995) Effects of clouds and stratospheric ozone depletion on ultraviolet radiation trends. *Nature*, 377: 710-713.

Lucas-Lledó JI, Maddamsetti R and Lynch M (2011) Phylogenomic analysis of the uracil-DNA glycosylase superfamily. *Mol Biol Evol*. 28: 1307-1317.

Lyamichev V (1991) Unusual conformation of (dA)n.(dT) n-tracts as revealed by cyclobutane thymine-thymine dimer formation. *Nucleic Acids Res*, 19: 4491-4496.

Lytvyn DI, Yemets AI and Blume YB (2010) UV-B overexposure induces programmed cell death in a BY-2 tobacco cell line. *Environ Exp Bot*, 68: 51-57.

MacFarlane AW and Stanley RJ (2003) Cis-Syn thymidine dimer repair by DNA photolyase in real time. *Biochemistry*, 42: 8558-8568.

Majchrzak M, Bowater RP, Staczek P and Parniewski P (2006) SOS repair and supercoiling influence the genetic stability of DNA triplet repeats in *Escherichia coli. J Mol Biol*, 364: 612-624.

Malhotra K, Kim ST, Batschauer A, Dawut L and Sancar A (1995) Putative blue-light photoreceptors from *Arabidopsis thaliana* and *Sinapis alba* with a high degree of sequence homology to DNA photolyase contain the two photolyase cofactors but lack DNA repair activity. *Biochemistry*, 34: 6892-6899.

Marguet S and Markovitsi D (2005) Time-resolved study of thymine dimer formation. *J Am Chem Soc*, 127: 5780-5781.

Martins-Pinheiro M, Marques RCP and Menck CFM (2007) Genome analysis of DNA repair genes in the alpha proteobacterium *Caulobacter crescentus. BMC Microbiol*, 7: 17-31.

Maul MJ, Barends TR, Glas AF, Cryle MJ, Domratcheva T, Schneider S, Schlichting I and Carell T (2008) Crystal structure and mechanism of a DNA (6-4) photolyase. *Angew Chem Int Ed*, 47: 10076-10080.

McEntee K, Weinstock GM and Lehman IR (1980) recA protein-catalyzed strand assimilation: stimulation by *Escherichia coli* single-stranded DNA-binding protein. *Proc Natl Acad Sci USA*, 77: 857-861.

McKenzie RL, Björn LO, Bais A and Ilyasd M (2003) Changes in biologically active ultraviolet radiation reaching the Earth's surface. *Photochem Photobiol Sci*, 2: 5-15.

Meyer-Rochow VB (2000) Risks, especially for the eye, emanating from the rise of solar UV-radiation in the Arctic and Antarctic regions. *Int J Circumpolar Health*, 59: 38-51.

Michel B (2005) After 30 years of study, the bacterial SOS response still surprises us. *PLoS Biol*. 3: e255.

Minda R, Ramchandani J, Joshi VP and Bhattacharjee SK (2005) A homozygous recA mutant of *Synechocystis* PCC6803: construction strategy and characteristics eliciting a novel RecA independent UV-C resistance in dark. *Mol Genet Genomics*, 274: 616-624.

Mitchell DL and Karentz D (1993) The induction and repair of DNA photodamage in the environment. In: *Environmental UV photobiology*. Springer USA, pp. 345-377.

Mitchell DL and Nairn RS (1989) The biology of the (6-4) photoproduct. *Photochem and Photobiol*, 49: 805-819.

Mitra S (2001) DNA glycosylases: specificity and mechanisms. *Prog Nucleic Acid Res Mol Biol*, 68: 189-192.

Mol CD, Izumi T, Mitra S and Tainer JA (2000) DNA-bound structures and mutants reveal abasic DNA binding by APE1 and DNA repair coordination. *Nature*, 403: 451-456.

Moolenaar GF, van Rossum-Fikkert S, van Kesteren M and Goosen N (2002) Cho, a second endonuclease involved in *Escherichia coli* nucleotide excision repair. *Proc Natl Acad Sci USA*, 99: 1467-1472.

Mühlenhoff U (2000) The FAPY-DNA glycosylase (Fpg) is required for survival of the cyanobacterium *Synechococcus elongatus* under high light irradiance. *FEMS Microbiol Lett*, 187: 127-132.

Müller P, Yamamoto J, Martin R, Iwai S and Brettel K (2015) Discovery and functional analysis of a 4th electron-transferring tryptophan conserved exclusively in animal cryptochromes and (6-4) photolyases. *Chem Commun*, 51: 15502-15505.

Murphy RC, Bryant DA, Porter RD and de Marsac NT (1987) Molecular cloning and characterization of the recA gene from the cyanobacterium *Synechococcus* sp. strain PCC 7002. *J Bacteriol*, 169: 2739-2747.

Myles GM, Hearst JE and Sancar A (1991) Site-specific mutagenesis of conserved residues within Walker A and B sequences of *Escherichia coli* UvrA protein. *Biochemistry*, 30: 3824-3834.

Nagata S (1997) Apoptosis by death factor. *Cell*, 88: 355-365.

Navaratnam S, Myles GM, Strange RW and Sancar A (1989) Evidence from extended X-ray absorption fine structure and site-specific mutagenesis for zinc fingers in UvrA protein of *Escherichia coli*. *J Biol Chem*, 264: 16067-16071.

Newman PA, Nash ER, Kawa SR, Montzka SS and Schauffler SM (2006) When will the Antarctic ozone hole recover. *Geophys Res Lett*, 33: LI2814.

Ng WO and Pakrasi HB (2001) DNA photolyase homologs are the major UV resistance factors in the cyanobacterium *Synechocystis* sp. PCC 6803. *Mol Gen Genet*, 264: 924-930.

Ng WO, Zentella R, Wang Y, Taylor JSA and Pakrasi HB (2000) phrA, the major photoreactivating factor in the cyanobacterium *Synechocystis* sp. strain PCC 6803 codes for a cyclobutane-pyrimidine-dimer-specific DNA photolyase. *Arch Microbiol*, 173: 412-417.

Ning SB, Guo HL, Wang L and Song YC (2002) Salt stress induces programmed cell death in prokaryotic organism *Anabaena*. *J Appl Microbiol*, 93: 15-28.

Norbury CJ and Zhivotovsky B (2004) DNA damage-induced apoptosis. *Oncogene*, 23: 2797-2808.

Nouspikel T (2009) Nucleotide excision repair: variations on versatility. *Cell Mol Life Sci*, 66: 994-1009.

O'Brien PA and Houghton IA (1982a) UV induced DNA degradation in the cyanobacterium *Synechocystis* PCC 6803. *Photochem Photobiol*, 36: 417-422.

O'Brien PA and Houghton JA (1982b) Photoreactivation and excision repair on UV induced pyrimidine dimers in the unicellular cyanobacterium *Gleocapsa alpicola* (*Synechocystis* PCC 6803). *Photochem Photobiol*, 35: 359-364.

Oh EY and Grossman L (1987) Helicase properties of the *Escherichia coli* UvrAB protein complex. *Proc Natl Acad Sci USA*, 84: 3638-3642.

Oliveira P and Lindblad P (2009) Transcriptional regulation of the cyanobacterial bidirectional Hox-hydrogenase. *Dalton Trans*, 45: 9990-9996.

Olsson-Francis K, de la Torre R and Cockell CS (2010) Isolation of novel extreme-tolerant cyanobacteria from a rock-dwelling microbial community by using exposure to low Earth orbit. *Appl Environ Microbiol*, 76: 2115-2121.

Orren DK and Sancar A (1989) The (A) BC excinuclease of *Escherichia coli* has only the UvrB and UvrC subunits in the incision complex. *Proc Natl Acad Sci USA*, 86: 5237-5241.

Orren DK and Sancar A (1990) Formation and enzymatic properties of the UvrB. DNA complex. *J Biol Chem*, 265: 15796-15803.

Orren DK, Selby P, Hearst JE and Sancar A (1992) Post incision steps of nucleotide excision repair in *Escherichia coli*. Disassembly of the UvrBC-DNA complex by helicase II and DNA polymerase I. *J Biol Chem*, 267: 780-788.

Owttrim GW and Coleman JR (1989) Regulation of expression and nucleotide sequence of the *Anabaena variabilis* recA gene. *J Bacteriol*, 17: 5713-5719.

Pakker H, Martins RST, Boelen P, Buma AGJ, Nikaido O and Breeman AM (2000) Effects of temperature on the photoreactivation of ultraviolet- B-induced DNA damage in *Palmaria palmata* (Rhodophyta). *J Phycol*, 36: 334-341.

Palchaudhuri S, Tominna B and Leon MA (1998) H-NS Regulates DNA Repair in *Shigella*. *J Bacteriol*, 180: 5260-5262.

Parihar P, Singh S, Singh R, Singh VP and Prasad SM (2015) Changing scenario in plant UV-B research: UV-B from a generic stressor to a specific regulator. *J Photochem Photobiol B Biol*, 153: 334-343.

Park HW, Kim ST, Sancar A and Deisenhofer J (1995) Crystal structure of DNA photolyase from *Escherichia coli*. *Science*, 268: 1866-1872.

Parsons JL, Tait PS, Finch D, Dianova II, Allinson SL and Dianov GL (2008) CHIP-mediated degradation and DNA damage-dependent stabilization regulate base excision repair proteins. *Mol Cell*, 29: 477-487.

Patterson-Fortin LM, Colvin KR and Owttrim GW (2006) A LexA-related protein regulates redox-sensitive expression of the cyanobacterial RNA helicase, crhR. *Nucleic Acids Res*, 34: 3446-3454.

Pattison DI and Davies MJ (2006) Actions of ultraviolet light on cellular structures. In: *Cancer: Cell Structures, Carcinogens and Genomic Instability*. Birkhäuser, Switzerland, pp. 131-157.

Perdiz D, Gróf P, Mezzina M, Nikaido O, Moustacchi E and Sage E (2000) Distribution and repair of bipyrimidine photoproducts in solar UV-irradiated mammalian cells: possible role of dewar photoproducts in solar mutagenesis. *J Biol Chem*, 275: 26732-26742.

Petit C and Sancar A (1999) Nucleotide excision repair: from *E. coli* to man. *Biochimie*, 81: 15-25.

Pettijohn DE and Hanawalt PC (1964) Evidence for repair replication of ultraviolet damaged DNA in bacteria. *J Mol Biol*. 9: 395-410.

Popa RA, Redfield C, Zeldovich N and Balakrishnan H (2011) CryptDB: protecting confidentiality with encrypted query processing. In: *Proceedings of the Twenty-Third ACM Symposium on Operating Systems Principles* pp. 85-100 ACM.

Quesada A and Vincent WF (1997) Strategies of adaptation by Antarctic cyanobacteria to ultraviolet radiation. *Eur J Phycol*, 32: 335-342.

Radman M. (1974) Phenomenology of an inducible mutagenic DNA repair pathway in *Escherichia coli*: SOS repair hypothesis. In: Sherman S, Miller M, Lawrence C and Tabor WH, editors.*Molecular and Environmental aspects of mutagenesis.* Springfield IL: Charles C Thomas publisher, 1974: 128-142.

Rajneesh, Singh SP, Pathak J and Sinha RP (2017a) Cyanobacterial factories for the production of green energy and value-added products: An integrated approach for economic viability. *Renew Sustainable Energy Rev* 69: 578-596.

Rajneesh, Pathak J, Chatterjee A, Singh SP and Sinha RP (2017b) Detection of reactive oxygen species (ROS) in cyanobacteria using the oxidant-sensing probe 2',7'-Dichlorodihydrofluorescein Diacetate (DCFH-DA). *Bio-protocol.* DOI:10.21769/ BioProtoc.xxxx55555111112000.

Rajneesh, Pathak J, Kannaujiya VK, Singh SP and Sinha RP (2017c) Analysis of codon usage patterns of photolyase gene in cyanobacteria of various habitats. 3 Biotech 7:192.Rastogi RP and Sinha RP (2011) Solar ultraviolet radiation-induced DNA damage and protection/repair strategies in cyanobacteria. *Int J Pharma Biosci*, 2: 271-288.

Rastogi RP, Kumar A, Tyagi MB and Sinha RP (2010a) Molecular mechanisms of ultraviolet radiation-induced DNA damage and repair. *J Nucleic Acids*, 2010: 1-32.

Rastogi RP, Singh SP, Häder DP and Sinha RP (2010b) Detection of reactive oxygen species (ROS) by the oxidant sensing probe 2′,7′-dichlorodihydrofluorescein diacetate in the cyanobacterium *Anabaena variabilis* PCC 7937. *Biochem Biophys Res Commun*, 397: 603-607.

Rastogi RP, Singh SP, Häder DP and Sinha RP (2011) Ultraviolet-B-induced DNA damage and photorepair in the cyanobacterium *Anabaena variabilis* PCC 7937. *Environ Exper Bot*, 74: 280-288.

Rastogi RP, Singh SP, Incharoensakdi A, Häder DP and Sinha RP (2014a) Ultraviolet radiation-induced generation of reactive oxygen species, DNA damage and induction of UV-absorbing compounds in the cyanobacterium *Rivularia* sp. HKAR-4. *South Afr J Bot*, 90: 163-169.

Rastogi RP, Sinha RP, Moh SH, Lee TK, Kottuparambil S, Kim YJ, Rhee JS, Choi EM, Brown MT, Häder DP and Han T (2014b) Ultraviolet radiation and cyanobacteria. *J Photochem Photobiol B Biology*, 141: 154-169.

Rautenberger R, Wiencke C and Bischof K (2013) Acclimation to UV radiation and antioxidative defence in the endemic Antarctic brown macroalga *Desmarestia anceps* along a depth gradient. *Polar Biol*, 36: 1779-1789.

Ravanat JL, Douki T and Cadet J (2001) Direct and indirect effects of UV radiation on DNA and its components. *J Photochem Photobiol B Biology*, 63: 88-102.

Robertson AB, Klungland A, Rognes T and Leiros I (2009) DNA repair in mammalian cells: Base excision repair: the long and short of it. *Cell Mol Life Sci*, 66: 981-993.

Roldán-Arjona T and Ariza RR (2009) Repair and tolerance of oxidative DNA damage in plants. *Mut Res*, 681: 169-179.

Ross C, Santiago-Vázquez L and Paul V (2006) Toxin release in response to oxidative stress and programmed cell death in the cyanobacterium *Microcystis aeruginosa. Aquat Toxicol*, 78: 66-73.

Ross P and Howard-Flanders P (1977) Initiation of recA-dependent recombination in *Escherichia coli* (λ): I. Undamaged covalent circular lambda DNA molecules in uvrA+ recA+ lysogenic host cells are cut following superinfection with psoralen-damaged lambda phages. *J Mol Biol*, 117: 137-158.

Rouillon C and White MF (2011) The evolution and mechanisms of nucleotide excision repair proteins. *Res Microbiol,* 162: 19-26.

Runyon GT and Lohman TM (1989) Escherichia coli helicase II (uvrD) protein can completely unwind fully duplex linear and nicked circular DNA. *J Biol Chem*, 264: 17502-17512.

Rupert CS (1960) Photoreactivation of transforming DNA by an enzyme from bakers' yeast. *J Gen Physiol*, 43: 573-595.

Sakumi K and Sekiguchi M (1990) Structures and functions of DNA glycosylases. *Mutation Res*, 236: 161-162.

Sancar A (1994) Mechanisms of DNA excision repair. *Science*, 266:1954-1956.

Sancar A (1994) Structure and function of DNA photolyase. *Biochemistry*, 33: 2-9.

Sancar A (1996) DNA excision repair. *Annu Rev Biochem*, 65: 43-81.

Sancar A (1996) No "end of history" for photolyases. *Science*, 272: 48-49.

Sancar A (2003) Structure and function of DNA photolyase and cryptochrome blue-light photoreceptors. *Chem Rev*, 103: 2203-2237.

Sancar A (2016) Mechanisms of DNA repair by photolyase and excision nuclease (Nobel Lecture). *Angew Chem Int Ed*, 55: 8502-8527.

Sancar A and Rupp WD (1983) A novel repair enzyme: UVR-ABC excision nuclease of *Escherichia coli* cuts a DNA strand on both sides of the damaged region. *Cell*, 33: 249-260.

Sancar A and Sancar GB (1988) DNA repair enzymes.*Annu Rev Biochem*, 57: 29-67.

Sancar GB and Smith FW (1989) Interactions between yeast photolyase and nucleotide excision repair proteins in *Saccharomyces cerevisiae* and *Escherichia coli*. *Mol Cell Biol*, 9: 4767-4776.

Sancar GB, Sancar A and Rupp WD (1984) Sequences of the *E. coli* uvrC gene and protein. *Nucleic Acids Res*, 12: 4593-4608.

Schmid SR and Linder P (1992) D-E-A-D protein family of putative RNA helicases. *Mol Microbiol*, 6: 283-292.

Seeberg E and Steinum AL (1982) Purification and properties of the uvrA protein from *Escherichia coli*. *Proc Natl Acad Sci USA*, 79: 988-992.

Seeberg E, Eide L and Bjørås M (1995) The base excision repair pathway. *Trends Biochem Sci*, 20: 391-397.

Selby CP and Sancar A (1988) ABC excinuclease incises both 5'and 3'to the CC-1065-DNA adduct and its incision activity is stimulated by DNA helicase II and DNA polymerase I. *Biochemistry*. 27: 7184-7188.

Setlow RB, Swenson PA and Carrier WL (1963) Thymine dimers and inhibition of DNA synthesis by ultraviolet radiation of cells. *Science*, 142: 1464-1465.

Shi Q, Thresher R, Sancar A and Griffith J (1992) Electron microscopic study of ABC excinuclease: DNA is sharply bent in the UvrB-DNA complex. *J Mol Biol*, 226: 425-432.

Shuman S and Glickman MS (2007) Bacterial DNA repair by non-homologous end joining. *Nat Rev Microbiol*, 5: 852-861.

Singh SP, Häder DP and Sinha RP (2010) Cyanobacteria and ultraviolet radiation (UVR) stress: mitigation strategies. *Ageing Res Rev*, 9: 79-90.

Singh VP, Srivastava PK and Prasad SM (2012) Impact of low and high fluence rates of UVB radiation on growth and oxidative stress in *Phormidium foveolarum* and *Nostoc muscorum* under copper toxicity: differential display of antioxidants system. *Acta Physiol Plant*, 34: 2225-2239.

Sinha RP and Häder DP (2002) UV-induced DNA damage and repair: a review.*Photochem Photobiol Sci*. 1(4): 225-36.

Sinha RP and Häder DP (2008) UV-protectants in cyanobacteria. *Plant Sci*, 174: 278-289.

Sinha RP, Dautz M and Hader DP (2001) A simple and efficient method for the quantitative analysis of thymine dimers in cyanobacteria, phytoplankton and macroalgae. *Acta Protozool*, 40: 187-196.

Sinha RP, Klisch M, Groniger A and Häder D (1998) Ultroviolet-absorbing/screening in cyanobacteria, phytoplankton and macroalgae. *J Photochem Photobiol B Biol*, 47: 82-94.

Sinha RP, Rastogi RP, Ambasht NK and Häder DP (2008) Life of wetland cyanobacteria under enhancing solar UV-B radiation. *Proc Natl Acad Sci India B*, 78: 53-65.

Smith RC, Prezelin BB, Baker KS, Bidigare RR, Boucher NP, Coley T, Karentz D, MacIntyre S, Matlick HA and Menzies D (1992) Ozone depletion: ultraviolet radiation and phytoplankton biology in Antarctic waters. *Science*, 255: 952-959.

Srivastava BS, Kumar HD and Singh HN (1971) The effect of caffeine and light on killing of the blue-green alga *Anabaena doliolum* by ultraviolet radiation. *Arch Microbiol*, 78: 139-144.

Stanier RY and Cohen-Bazire G (1977) Phototrophic prokaryotes: the cyanobacteria. *Annu Rev Microbiol*, 31: 225-274.

Sung P (1997) Yeast Rad55 and Rad57 proteins form a heterodimer that functions with replication protein A to promote DNA strand exchange by Rad51 recombinase. *Genes Dev*, 11: 1111-1121.

Swenson PA and Setlow RB (1966) Effects of ultraviolet radiation on macromolecular synthesis in *Escherichia coli*. *J Mol Biol*, 15: 201-219.

Tamada T, Kengo K, Yoshiki H, Koji I, Akira Y, Petra E. de R, Eker AP and Kunio M (1997) Crystal structure of DNA photolyase from *Anacystis nidulans*. *Nat Struct Mol Biol*, 4: 887-891.

Tan C, Guo L, Ai Y, Li J, Wang L, Sancar A, Luo Y and Zhong D (2014) Direct determination of resonance energy transfer in photolyase: structural alignment for the functional state. *J Phys Chem A*, 118: 10522-10530.

Tan C, Liu ZY, Li J, Guo XM, Wang LJ, Sancar A and Zhong DP (2015) The molecular origin of high DNA-repair efficiency by photolyase. *Nat Commun*, 6: 7302.

Tang T and Asato Y (1978) Ultraviolet light induction and photoreactivation of thymine dimers in the cyanobacterium *Anacystis nidulans*. *Arch Microbiol*, 118: 193-197.

Taylor JS (1994) Unraveling the molecular pathway from sunlight to skin cancer. *Acc Chem Res*, 27: 76-82.

Taylor JS, Lu HF and Kotyk JJ (1990) Quantitative conversion of the (6-4) photoproduct of TpdC to its Dewar valence isomer upon exposure to simulated sunlight. *Photochem Photobiol*, 51: 161-167.

Thayer MM, Ahem H, Xing DX, Cunningham RP and Tainer JA (1995) Novel DNA-binding motifs in the DNA repair enzyme endonuclease III crystal structure. *EMBO J*, 14: 4108-4120.

Thiagarajan V, Byrdin M, Eker APM, Muller P and Brettel K (2011) Kinetics of cyclobutane thymine dimer splitting by DNA photolyase directly monitored in the UV. *Proc Natl Acad Sci USA*, 108: 9402-9407.

Thiagarajan V, Villette S, Espagne A, Eker APM, Brettel K and Byrdin M (2010) DNA repair by photolyase: a novel substrate with low background absorption around 265 nm for transient absorption studies in the UV. *Biochemistry*, 49: 297-303.

Thoma F (1999) Light, dark in chromatin repair: repair of UV-induced DNA lesions by photolyase and nucleotide excision repair. *EMBO J*, 18: 6585-6598.

Todo T, Ryo H, Yamamoto K, Toh H, Inui T, Ayaki H, Nomura T and Ikenaga M (1996) Similarity among the *Drosophila* (6-4) photolyase, a human photolyase homology, and the DNA photolyase-blue-light photoreceptor family. *Science*, 272: 109-112.

Todo T, Takemori H, Ryo H, Ihara M, Matsunaga T, Nikaido O, Sato K and Nomura T (1993) A new photoreactivating enzyme that specifically repairs ultraviolet light-induced (6-4) photoproducts. *Nature*, 361: 371-374.

Truglio JJ, Croteau DL, Van Houten B and Kisker C (2006) Prokaryotic nucleotide excision repair: the UvrABC system. *Chem Rev*, 106: 233-252.

Tuteja N, Singh MB, Misra MK, Bhalla PL and Tuteja R (2001) Molecular mechanisms of DNA damage and repair: progress in plants. *Crit Rev Biochem Mol Biol*, 36: 337-397.

Tyrrell RM (1973). Induction of pyrimidine dimers in bacterial DNA by 365 nm radiation. *J Photochem Photobiol*, 17: 69-73.

Tyrrell RM (1986) Repalr of genetic damage induced by UV-B (290-320 nm) radiation. In: *Stratospherlc ozone reduction, solar ultraviolet radiation and plant life*. Springer-Verlag, Berlin, pp. 139-149.

Vaishampayan A, Sinha RP, Häder DP, Dey T, Gupta AK, Bhan U and Rao AL (2001) Cyanobacterial biofertilizers in rice agriculture. *Bot Rev*, 67: 453-516.

Van Baalen C (1968) The effects of ultraviolet irradiation on a coccoid bluegreen alga: survival, photosynthesis and photoreactivation. *Plant Physiol*, 43: 1689-1695.

Van Baalen C and O'Donnell K (1972) Action spectra for ultraviolet killing and photoreactivation in the blue-green alga *Agmenellum quadruplicatum*. *Photochem Photobiol*, 15: 269-274.

Van Houten B, Eisen JA and Hanawalt PC (2002) A cut above: discovery of an alternative excision repair pathway in bacteria. *Proc Natl Acad Sci USA*, 99: 2581-2583.

Varghese AJ (1972) Photochemistry of nucleic acids and their constituents. *Photophysiology*, 7: 207-274.

Vass IZ, Kos PB, Sass L, Nagy CI and Vass I (2013) The Ability of Cyanobacterial Cells to Restore UV-B Radiation Induced Damage to Photosystem II is Influenced by Photolyase Dependent DNA Repair. *Photochem Photobiol*. 89: 384-390.

Vetter RD, Kurtzman A and Mori T (1999) Diel cycles of DNA damage and repair in eggs and larvae of Northern Anchovy, Engraulis mordax, exposed to solar ultraviolet radiation. *Photochem Photobiol*, 69: 27-33.

Walker GC (1985) Inducible DNA repair systems. *Annu Rev Biochem*, 54: 425-457.

Wang CI and Taylor JS (1991) Site-specific effect of thymine dimer formation on dAn.dTn tract bending and its biological implications. *Proc Natl Acad Sci USA*, 88: 9072-9076.

Wang J, Du X, Pan W, Wang X and Wu W (2015) Photoactivation of the cryptochrome/photolyase superfamily. *J Photochem Photobiol C Photochem Rev*, 22: 84-102.

Wang JG and Zhang CL (2001) DNA damage and repair of two ecotypes of *Phragmites communis* subjected to water stress. *Acta Bot Sin*, 43: 490-494.

Warner HR, Demple BF, Deutsch WA, Kane CM and Linn S (1980) Apurinic/apyrimidinic endonucleases in repair of pyrimidine dimers and other lesions in DNA. *Proc Natl Acad Sci USA*, 77: 4602-4606.

Williams E, Lambert J, O'Brien P and Houghton JA (1979) Evidence for dark repair of ultraviolet light damage on the blue-green alga *Gleocapsa alpicola*. *Photochem Photobiol*, 29: 543-547.

Witkin EM (1976) Ultraviolet mutagenesis and inducible DNA repair in *Escherichia coli*. *Bacteriol Rev*, 40: 869-907.

Wolner B, van Komen S, Sung P and Peterson CL (2003) Recruitment of the recombinational repair machinery to a DNA double-strand break in yeast. *Mol Cell*, 12(1): 221-232.

Yamada D, Dokainish HM, Iwata T, Yamamoto J, Ishikawa T, Todo T, Iwai S, Getzoff ED, Kitao A and Kandori H (2016) Functional Conversion of CPD and (6-4) Photolyases by Mutation. *Biochemistry*, 55: 4173-4183.

Yamamoto J, Hitomi K, Todo T and Iwai S (2006) Chemical synthesis of oligodeoxyribonucleotides containing the Dewar valence isomer of the (6-4) photoproduct and their use in (6-4) photolyase studies. *Nucleic Acids Res*, 34: 4406-4415.

Yang W (2011) Surviving the sun: repair and bypass of DNA UV lesions. *Protein Sci*, 20: 1781-1789.

Yasui A, Takao M, Oikawa A, Kiener A, Walsh CT and Eker AP (1988) Cloning and characterization of a photolyase gene from the cyanobacterium.*Anacystis nidulans*. *Nucleic Acids Res*, 16: 4447-4463.

Yoshiyama K, Sakaguchi K and Kimura S (2013) DNA damage response in plants: conserved and variable response compared to animals. *Biology*, 2: 1338-1356.

You YH, Szabo PE and Pfeifer GP (2000) Cyclobutane pyrimidine dimers form preferentially at the major p53 mutational hotspot in UVB induced mouse skin tumors. *Carcinogenesis*, 21: 2113-2117.

Zalar A, Tepfer D, Hoffmann SV, Kenney JM and Leach S (2007a) Directed exospermia: I. biological modes of resistance to UV light are implied through absorption spectroscopy of DNA and potential UV screens. *Int J Astrobiol*, 6: 229-240.

Zalar A, Tepfer D, Hoffmann SV, Kollmann A and Leach S (2007b) Directed exospermia: II. VUV-UV spectroscopy of specialized UV screens, including plant flavonoids, suggests using metabolic engineering to improve survival in space. *Int J Astrobiol*, 7: 1-11.

Zeeshan M and Prasad SM (2009) Differential response of growth, photosynthesis, antioxidant enzymes and lipid peroxidation to UV-B radiation in three cyanobacteria. *S Afr J Bot*, 75: 466-474.

Zhong PD (2015) Electron transfer mechanisms of DNA repair by photolyase. *Annu Rev Phys Chem*, 66: 691-715.

Zlatev ZS, Lidon FJC and Kaimakanova M (2012) Plant physiological responses to UV-B radiation. *Emirates J Food Agric*, 24: 481-501.

In: Trends in Life Science Research
Editors: R. P. Sinha and U. P. Shrivastava
ISBN: 978-1-53613-241-0

Chapter 8

SUSTAINABILITY IN AGROECOSYSTEMS: MANAGEMENT STRATEGIES INVOLVING HERBICIDES AND ORGANIC INPUTS

Mahesh Kumar Singh, Biswadip Mukherjee, Chandra Mohan Kumar, Alka Singh and Nandita Ghoshal*
Centre of Advanced Study in Botany, Department of Botany,
Banaras Hindu University, Varanasi, India

ABSTRACT

Long-term sustainable management of agroecosystems has now become a global challenge, along with the major objective of maximizing the crop yield. This has led to the switching of research focus, from 'high input and maximum yields' to 'low input and high efficiency sustainable agriculture'. Such approaches have generated renewed and wide interest in ecological aspects of crop production and sustainable agriculture. Organic inputs, rather than the chemical fertilizers are now considered to have the potential to conserve the soil fertility, in terms of a greater soil C storage, the soil microbial biomass, and the crop productivity in the tropical dryland rainfed agroecosystems, especially in the dry tropics of the Indian subcontinent. Weed infestation is a severe problem in agroecosystems, especially in rainfed croplands, as compared to the irrigated systems. The application of herbicide is a common method to control the weeds. Since the continuous and extensive use of herbicides may differently affect the soil organic matter dynamics, therefore, concerns are raised for the long-term sustainability of agroecosystems. The application of high-quality organic resources along with herbicide, not only negates the harmful side effects of these farm-chemicals, but also benefitted the first crop, in terms of improving the soil microbial biomass, the soil carbon sequestration potential, the soil N balance, and the total crop productivity, whereas these

* Corresponding Author: n_ghoshal @yahoo.co.in, 91-9415987328.

benefits were observed in the subsequent crop, when low quality inputs were added along with the herbicides. The combined application of herbicides with the organic inputs of contrasting resource qualities may thus help in, not only sustaining the long-term soil fertility and productivity, but also, in mitigating the increased atmospheric carbon dioxide concentration through its sequestration in agroecosystem soils.

Keywords: Agroecosystem, carbon dynamics, crop productivity, herbicide, nitrogen dynamics, organic inputs, soil management practices, soil microbial biomass, sustainable development, weed control

INTRODUCTION

There is a global challenge to enhance crop productivity, as it is projected that the world human population will reach 9 billion by 2050. Since the middle of the last century, among other agronomic management practices, farm-chemicals, especially in the form of chemical fertilizers and herbicides have been extensively used, which in turn, had revolutionized the agriculture, by greatly enhancing the yield of various crops. But during the last couple of decades, the agricultural production has come to a standstill, with no further scope of increase in productivity. In a few cases, even decline in the production of certain agricultural crops have been reported (Ramesh et al. 2005), besides causing environment and/or health related problems. The factors like irrigation and farm-chemicals, responsible for ushering in the "Green Revolution" are now being blamed for the slowdown. Further, critics have raised doubts over the feasibility of sustainable agriculture by high input farming. This has resulted in finding ways and means to address this stagnation in productivity and looking for options, where the soil fertility is maintained over a long period of time. The need of the hour is to attain a steady rise in the agricultural yield, without permanent damage to the environment. The judicious use of land by the farmers and the development of new strategies of farming are essential to reduce the environmental degradation. Due to an increasing rate of degradation and contamination of the agricultural environment, many contemporary studies have primarily focussed on enhancing the crop yield with low input and high-efficiency sustainable agriculture. Such approaches are generating renewed interest in the ecological aspects of agriculture, in which, the crop production and related activities are considered as an ecosystem, with a view to optimize the crop yield while conserving the soil, water, energy and other environmental resources (Ghoshal and Singh 1995; Zaller and Kopke 2004).

In several International Summits and Conferences, such as the Rio Summit of 1992, Agenda 21 (UNCED 1992), Article 3.3 and 3.4 of the Kyoto Protocol (UNFCCC 1998), World Summit on Sustainable Development (2002), Eye on Earth Summit of 2010, Rio summit of 2012, concepts, goals, methods and challenges of sustainable agriculture have been discussed. With these conventions, the world community has recognized the strong link between the agriculture and threats to food security and increased poverty on one hand, and the loss of biodiversity, environmental degradation (including soil, water and air), risks of the accelerated greenhouse effect and climate change on the other (IPCC 2014).

AGROECOSYSTEM

An agroecosystem is a spatially and functionally coherent unit of agricultural activity, which includes the living and the nonliving components involved in that unit as well as their interactions. Agroecosystems are man-made ecosystems, designed for a particular goal, and in general, derived from the natural ecosystems e.g., forests and grasslands. These ecosystems, by their nature are perturbed systems, containing a significant component of anthropogenic inputs and outputs necessary for its regulation, with some fraction of the system harvested periodically (Odum 1984). The removal of harvest leads to the lack of effective nutrient conservation. Therefore, the constant augmentation of critical nutrients and energy subsidies of the system are necessary for their maintenance. Agroecosystems are manipulated by frequent, marked anthropogenic modifications of the biotic and the abiotic environments (Coleman and Hendrix 1988). These ecosystems are facing the anthropogenic replacement stage, necessarily unstable, and sustained by high energy and material subsidies (Ghoshal 2003). Therefore the agroecosystems are considered as an open ecosystem, in comparison to the natural, closed ecosystem (Figure 1).

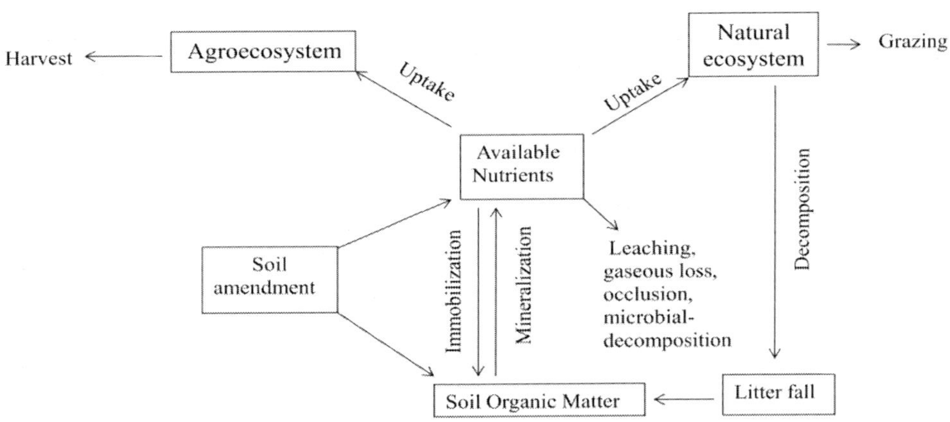

Figure 1. Characteristics of nutrient cycling in natural ecosystem and agroecosystems.

Odum (1984) differentiated the agroecosystems from their natural counterpart in four major aspects: (1) in agroecosystem, wherein, the energy supply is in the form of solar energy, human and animal labour, fertilizers, pesticides, irrigation water and fuel-powered machinery, the natural one depends only on solar power (2), in agroecosystem, the diversity of biota is reduced in comparison to the natural ecosystem, to maximize yield of the desired product, (3) in agroecosystem, where artificial selection produces dominant flora and fauna rather than natural selection, as in natural ecosystem (4) agroecosystems are under external control, whereas the natural ecosystems are under the feedback oriented internal control. Natural ecosystem retains its major portion of litter within the system as input, which, after decomposition, gets added in the system as soil organic matter and therefore, maintains the biogeochemical cycle.

The main emphasis of agroecosystem research is to maximize the crop productivity in order to meet the required foodstuff for the growing human population. But during the past decades, focus of research has shifted from maximizing the crop yield, to retaining the productivity along with sustainability of the agroecosystems. Ecological processes and management practices, responsible for the maintenance of soil fertility in natural ecosystems should be implemented in the agroecosystems for more agricultural production and agroecosystem sustainability (Swift 1987), although the factors of sustainability for the natural and agroecosystem, are different. In natural ecosystem, the stability and maintenance of ecosystem productivity and nutrient cycling are considered as the major factors for sustainability, whereas in the agroecosystem, maintenance of the desired level of crop productivity, over a long period of time is considered as the major factor for sustainability (Ghoshal 2003).

The long-term sustainability of agroecosystem has now become a global issue. About 80% of the total cropland area of the globe is estimated to be under the rainfed farming conditions (FAOSTAT 2005). In India about 66% of the total arable land (100 million ha) is rainfed (Lal 2009). The ratio of precipitation to potential evapotranspiration in dryland agroecosystems varies from 0.05 to 0.65. Due to limited soil moisture and nutrient availability, the rainfed agroecosystems in general, have a low crop productivity, yet immense possibilities exist in these extensive agroecosystems for increasing the crop production, as compared to the irrigated croplands. where crop production has reached a plateau, subsequent to the 'Green Revolution'. Reductions in the crop yield and agricultural productivity have occurred in several dryland countries, with subsequent threats to the global food security. In the dryland soils, the addition of chemical fertilizers is not very effective because the soil is not only poor in available nutrients or carbon but also limited by soil moisture. Organic farming helps to increase the productivity by sustaining the soil fertility and conserving the soil moisture (Carter 1980). Weed infestation may result in the loss of crop yield, ranging from 18- 30% (Anonymous, 2000), to complete failure of crop depending on the extent of weed growth (Gill and Wallia 1979).

Agroecosystem: Potential Carbon Sinks

About 75% of the soil carbon has been lost in the tropical agroecosystems (Lal 2004). The depletion of soil carbon in the agroecosystems occur due to harvesting, that removes a major fraction of the photosynthetically fixed C, and lowers the amount of plant litter returning to the soil (Singh et al. 2009b). Agricultural practices disrupt the soil aggregate structure by making the organic matter more accessible to biological decay, mixing the fresh litter into the soil, and increasing the erosion of C-rich surface layer. It is believed that agroecosystems offer immense opportunities for the soil C sequestration and serve as a large C sink in the context of global climate change. Therefore, nowadays, researchers are looking towards agroecosystem for dual purpose; sustainability and sequestration of carbon in the soil. Carbon sequestration is one of the effective mechanisms for the mitigation of the loss of soil carbon, by slowing or reversing the trend of increase in concentration of carbon dioxide in the atmosphere (Asante et al. 2011). Carbon sequestration in soil is also referred to as the accumulation of carbon in soil, or capturing carbon dioxide from the atmosphere through plants and storing it in the soil in the form of soil organic matter (Singh and Ghoshal 2011). In other words, carbon sequestration is the transfer of atmospheric carbon dioxide into resistant pools in soil with a slow turnover, and storing it firmly, so as to protect it from immediate release (Lal 2004). This includes humification, aggregation, deep incorporation of C in the subsoil and ultimately calcification. The potential to sequester C in the tropical agroecosystem differs from those in the temperate regions; the sequestration rates are generally lower in the tropics with respect to temperate regions due to high temperatures and high microbial activity throughout the year (Lal 2004). Soil is referred to as 'sink' when it stores C and 'source' when it releases C. The changes in the role of soil from 'source' to 'sink' or vice-versa, mainly depends on the land-use as well as the management practices of the agroecosystem (Singh et al. 2015). Any management strategy that enables the soil to act as a 'sink' may eventually help in the mitigation of global climate change. Therefore, such type of management practices should be adopted in agroecosystem which can 'switch-on' the tendency of 'sink' and 'switch-off' or slow down the 'source' tendency of soil.

Sustainable Management Practices

A global problem has emerged to feed the billions of people worldwide, especially when, the rates of malnutrition and hunger are quite high. Appropriate management practices for sustainability and productivity of the agroecosystems have become a global challenge. The major issues in agricultural management are, irrigation, weed and insect control. Exogenous amendments are added to the soil, which is a prerequisite to maintain soil fertility. The addition of organic residues is more effective for the amelioration of soil fertility on one hand and enhancement of the soil moisture content on the other (Singh et al. 2009a). In rainfed dryland agroecosystems, the problem of weed management is more challenging compared to irrigated agroecosystems (Hyvonen and Salonen 2002).

SOIL AMENDMENTS OF VARYING RESOURCE QUALITY

The organic resources play an important role in soil fertility management in the tropics, through their short-term effects on nutrient supply and longer-term contribution to the soil organic matter formation (Palm et al. 2001). These resources can be broadly divided into two groups from the viewpoint of their decomposition patterns. Low quality organic resource (e.g., crop residues), having a higher C:N, lignin:N and polyphenol+lignin:N ratio, releases nutrient slowly or immobilizes nutrients during the early stage of decomposition. On the other hand, high quality organic resources (e.g., animal manure, green manure) having lower C:N, lignin:N, polyphenol+lignin:N ratios, release nutrients rapidly (Singh et al. 2006).

In drylands, the application of soil organic amendments (e.g., crop residues, green manures), among other measures, is generally recommended for the efficient management of soil organic matter and moisture availability (Singh et al. 2007a). It is important to emphasize, not only the quantity of soil organic amendments, but their resource quality as well. In natural ecosystems, governing influence of resource quality on decomposition rate and nutrient release was recognized by Swift et al. (1979), Schlesinger and Hasey (1981) and Melillo et al. (1982) A considerable amount of information has been generated on nutrient contents, resource quality, decomposition, and nutrient release patterns for a variety of organic amendments used in tropical agroecosystems (Palm 1995; Cadisch and Giller 1997; Heal et al. 1997; Mafongoya et al. 1998; Singh et al. 2007a). The potential of these organic inputs to supply nutrients immediately or through residual effects to the subsequent crops depends on their resource quality. In the last decade, a lot of knowledge is acquired about the amendment of organic inputs of varying resource quality in crop productivity and soil fertility, which has emerged as an approach towards sustainable agriculture. The combination of a high quality resource with a low quality resource, alters the rate of decomposition, resulting in the prolonged release of nutrients, which in turn, supports enhanced crop productivity (Singh et al. 2007c), sustained buildup of soil organic matter (Singh et al. 2007a), and minimize the loss of nitrogen from soil (Singh et al. 2007b). Reports are also available that the addition of combined (high and low) resource quality, contribute towards the improvement in crop productivity and soil fertility, in terms of soil C sequestration (Singh et al. 2009b).

WEED CONTROL

The relationship of weeds and man is very old, perhaps since human beings were hunters and gatherers, and use to collect plants of their interest to fulfill the requirements of food and fiber. They rejected other unwanted plants, the weeds. Since then, the struggle has continued between the human beings and the weeds. Weed control is easy in the early or pre-sowing stage, rather than after their establishment. Some common methods used to prevent the germination and/or introduction of weeds in the croplands are: (1) Use of seeds of the desired crop, free from weed seeds, (2) Nursery of the crops, free from weeds and not transplanted with the desired crop seedlings, (3) Irrigation water should be free from weeds/weed seeds and (4) Transportation of weeds/ weed seeds to the field at the time of plowing and harrowing should be avoided.

Some Selected Methods to Control Weeds

Direct method and indirect method are mainly the two principal approaches for controlling weeds in agroecosystems. The indirect methods include plowing, harrowing and water management. Generally, the farmers use intensive tillage practices at the time of seed bed to prevent germination and growth of weeds. Plowing generally mixes the weeds into soil to decompose. With the help of harrow, previously germinated weeds can be destroyed. For a better response, a greater number of harrowing is required. The practice of plowing and harrowing to control weeds is beneficial, but it would be harmful to the soil structure, the microbial biomass or the soil fertility as a whole. This may or may not be compensated with the weed biomass incorporated in the soil that depends on the density and growth of the weeds.

The direct method includes physical and chemical methods of weed control. Weeding is the traditional physical method to control weeds in the agricultural field which includes preparatory land tillage, hand weeding by hoe and hand pulling. The success of manual weeding generally depends on the nature of weeds, the intensity of their infestation, and the crop grown. For example, generally two to three hand weeding is performed for growing the rice crop. Weeding seems to be very expensive and may even lower the crop yield due to late weeding, or in case there is a shortage of laborer. Hand weeding is very difficult during the early stages of growth. It is delayed until the weeds are large enough to be grasped. Therefore, in contrast to physical method, chemical weed control methods like the use of herbicides, is gaining popularity in agroecosystems due to their rapid action and lower cost.

The use of herbicide is an important chemical method for the control of weeds in the cropland. The herbicide may either delay the weed growth, or completely check their growth during the cropping season. The modern agricultural practices involve the intensive use of herbicide due to the lower cost; time bound application, and responsiveness in controlling the weeds. The use of herbicide is not only effective under conditions, when laborers are scarce and expensive but also, where laborers are cheap and easily available, because it eliminates weed from the field in spite of the morphological similarity between the crop and the weed. Thus damage to the crop due to mechanical weeding is taken care of. Further, herbicides also remove those weeds, having an extensive root system. Research must be undertaken to find out the appropriate herbicide, which would be effective at their lower doses and also, economically beneficial to the farmers.

Generally, various types of herbicides are used to overcome the problem of weeds, which is a global issue in the dryland agroecosystems (Singh and Ghoshal 2012) as well as in the irrigated croplands. As farmers continue to realize the usefulness of herbicides, larger quantities are applied to the soil. The knowledge about the fate of these chemical compounds in soil is very important, since they can be leached to contaminate the ground water, or may persist on the top soil. These herbicides are harmful for plants, wild life, microorganisms and even the human beings. These may target and/or reduce the performance of important soil processes like organic matter decomposition, and carbon and nitrogen cycle. The development of precise herbicide application strategies is urgently required to understand the effects of varying environmental conditions on herbicide efficacy. Field trials have shown that the soil organic matter content strongly affects the amount of pesticide like herbicide, fungicide, insecticide, and nematocide, deployed to control specific pests (Jenkinson 1988). Therefore, concerns have been raised over the long-term sustainability of crop production, due to the long-term use of herbicides.

Herbicides are variously classified on the basis of their mode of action, time of application or the chemical composition. According to the mode of action, herbicides are: (1) when the chemicals kill the plants coming in contact, called contact herbicides, e.g., propanil or (2) systemic or translocated herbicides kill plants after their absorption by accelerating or retarding the plant metabolic activities, e.g., 2, 4-D and cyhalofop butyl. According to their relative time of application, these herbicides can be (a) Pre-plant herbicides, applied on the field before the planting of crop. Glyphosate is an example of this class of herbicide. (b) Pre-plant emergence herbicides are applied to the soil surface before the emergence of weeds. This is abbreviated as PRE, e.g., butachlor and (c) Post-emergence herbicides, applied after the emergence of weeds. This class of herbicide is abbreviated as POST, example 2, 4-D. Chemical classification of herbicides refer to the grouping of various herbicides according to their chemical composition. According to the herbicide selectivity, these are categorized as: (i) selective; which kill or retard the growth of some plant species with little or no injury to others, especially the crop, e.g., butachlor, thiobencarb, propanil, and 2, 4-D, (ii) non-selective; herbicide that kills all plant species without selection of the plants and/or crops e.g., glyphosate. However, if the herbicide is applied in the croplands unmindfully, any herbicide may harm or kill the crop. The structural details of some important herbicides are as under:

2, 4-D

2, 4 -D is composed of water-soluble salts, emulsifiable concentrates or esters, and granules. This is a post-emergence, translocated and selective herbicide. It controls sedges and broadleaf weeds rather than the grasses. The application rates vary from 0.4 kg a.i.ha^{-1} to 0.8 kg a.i.ha^{-1}. Chemical name of 2, 4-D is 2,4-dichlorophenoxyacetic acid with chemical formula $C_8H_6Cl_2O_3$ (Figure 2).

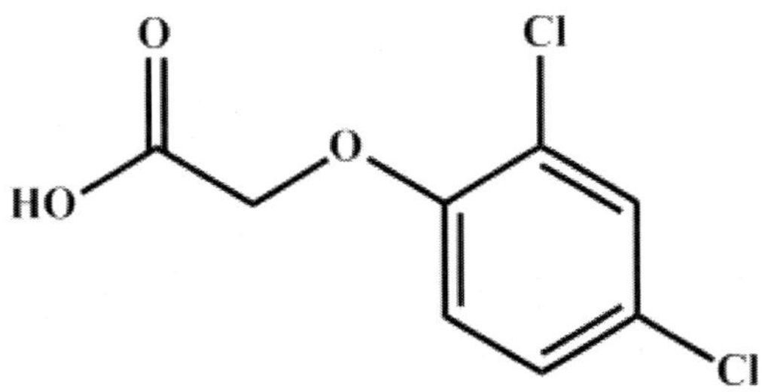

Figure 2. Chemical Structure of 2,4-dichlorophenoxyacetic acid.

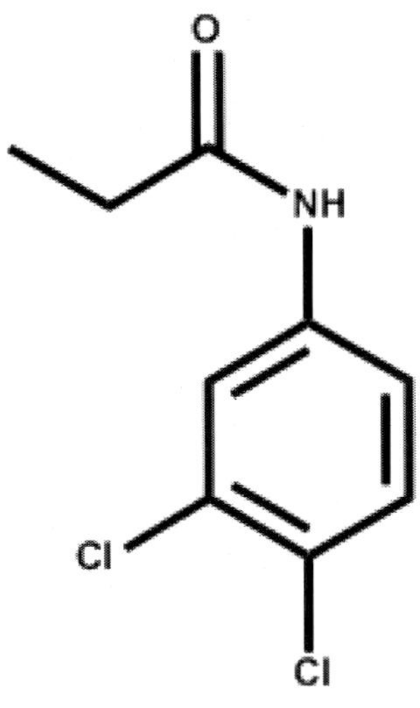

Figure 3. Chemical structure of N-(3,4-Dichlorophenyl) propanamide.

Propanil

It is commonly used in the rice fields to control the grasses. This herbicide is postemergence, contact and selective in nature. At the time of its application it should be ensured that weeds must be exposed to come in contact with the herbicide and the fields must be drained before their use. The dose of propanil ranges from 3 kg a.i. ha^{-1} to 4 kg a.i. ha^{-1} and applied 10 - 20 DAT, or 10 to 20 days after sowing the pre germinated seeds. It would be injurious to crops, if propanil is applied less than 15 days of the application of carbamates and organophosphate insecticides. Chemical name of Propanil is N-(3,4-Dichlorophenyl) propanamide and the chemical formula is $C_9H_9Cl_2NO$ (Figure 3).

Butachlor

Butachlor is the selective herbicide acting before the sprouting. It can be applied either as pre-emergence or early post-emergence herbicide to control a wide range of annual grasses and some broadleaf weeds. The suitable dose of butachlor as pre-emergence herbicide is recommended as 2 kg ha^{-1}. Commercial name of Butachlor is Machete, chemical name is N-(butoxymethyl)-2-chloro-N-(2,6-diethylphenyl)acetamide and the chemical formula is $C_{17}H_{26}ClNO_2$ (Figure 4).

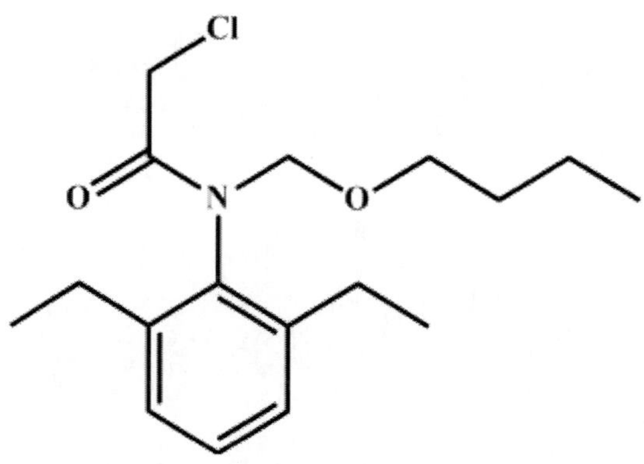

Figure 4. Chemical structure of *N*-(butoxymethyl)-2-chloro-N-(2,6-diethylphenyl)acetamide.

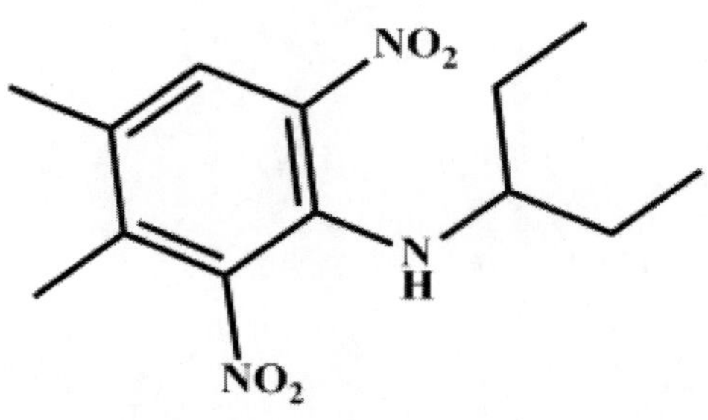

Figure 5. 3,4-dimethyl-2,6-dinitro-N-(pentan-3-yl)aniline.

Pendimethalin

Pendimethalin is widely used against weeds like barnyard grass (*Echinochloa crusgalli*), jimson weed (*Datura stramonium*) and velvet leaf (*Abutilon theophrasti*) but the control of some weeds like *Ipomea* spp., *Amaranthus* spp. and *Sida spinosa* seems difficult. Pre-sowing application of pendimethalin controls efficiently, the weeds like *Amaranthus retroflexus*, *Chenopodium album*, *Polygonum lapathifolium* and *Echinochloa crusgalli* in soybeans. As far as its dose is concerned, the use of pendimethalin at 1.12 kg + metribuzin 0.56 kg/ha is effective and controls nearly 96 per cent of the barnyard grass and significantly enhances the crop yield. It is also useful in the control of weeds, in a stand of soybean (cv. Bragg), as well as the seed yield when applied at 0.75 kg pendimethalin ha^{-1}. Pendimethalin is the only herbicide that controls grass weed as well as the broad leaved weed. As far as its persistence nature is concerned, it remains biologically active up to 25-26 days in a sandy loam soil when applied (1.0 and 1.5 kg/ha), while about 75% pendimethalin is lost in about 45 days. Chemical name of Pendimethalin is 3,4-dimethyl-2,6-dinitro-N-(pentan-3-yl)aniline and the chemical formula is $C_{13}H_{19}N_3O_4$ (Figure 5).

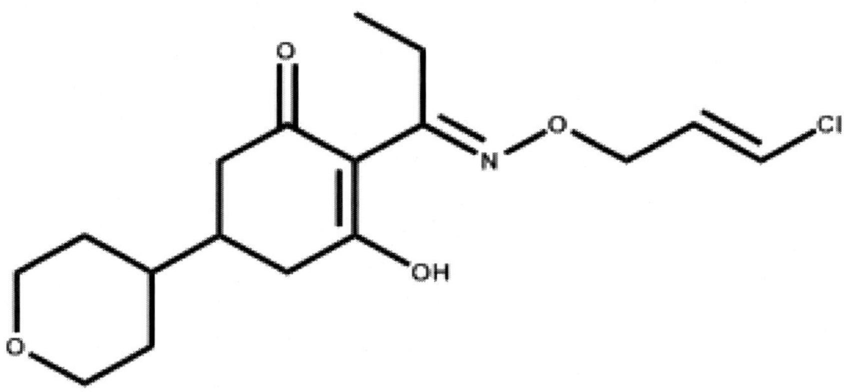

Figure 6. 2-(E)-1-((((*E*)-3-chloroallyl)imino)propyl)-3-hydroxy-5-(tetrahydro-2*H*-pyran-4-yl) cyclohexen-2-enone.

Tepraloxydim

Tepraloxydim is a new generation herbicide and the member of cyclohexanedione group. It acts by the inhibition of the enzyme acetyl CoA carboxylase. It is widely used for the vegetable crops, canola, flax, cotton, soybeans and potato. It is also useful in the control of grass-weeds such as annual meadow-grass, wild-oats, volunteer cereals or black-grass in peas and beans. It can control about 50% of Parthenium hysterophorus, but was ineffective against Celosia argentea, Trianthema monogyna and Cyperus species. When combined with bentazon and ammonium nitrate, it caused phytotoxicity of about 20%, the crop remained stunted throughout the season, and the extent of crop growth was

50%. Chemical name of Tepraloxydim is 2-(E)-1-((((E)-3-chloroallyl)imino) propyl)-3-hydroxy-5-(tetrahydro-2H-pyran-4-yl) cyclohexen-2-enone and the chemical formula is $C_{17}H_{24}ClNO_4$ (Figure 6).

Imazethapyr

Imazethapyr is the group of imidazolinone herbicide, and is used as a broad spectrum herbicide for weed control. It kills weeds by inhibiting the enzyme acetohydroxy acid synthese (AHAS), involved in the synthesis of, the three branched chain aliphatic amino acids, leucine, isoleucine and valine. This inhibition disrupts the protein synthesis, and leads to interference in DNA synthesis, and finally the cell growth. The plants rapidly metabolize the herbicide, making it suitable for the control of grasses and broadleaf weeds, when applied either as a pre-plant incorporation, pre-emergent or post-emergent herbicide. In addition, imazethapyr enhanced the giant foxtail, jimson weed and the velvet leaf control. Chemical name of Imazethapyr is 5-ethyl-2-(4-isopropyl-4-methyl-5-oxo-4,5-dihydro-1H-imidazol-2-yl)-nicotinic acid and the chemical formula is $C_{15}H_{19}N_3O_3$ (Figure 7).

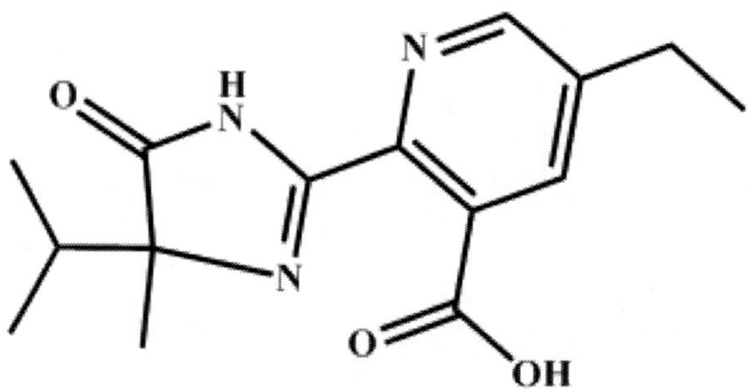

Figure 7. 5-ethyl-2-(4-isopropyl-4-methyl-5-oxo-4,5-dihydro-1H-imidazol-2-yl)-nicotinic acid.

Interaction of Herbicide with Plant and Soil

Herbicides when applied to the soil pass through chemical or microbial transformations, which may lead to the disappearance of their measurable quantity from the soil. This process is commonly known as herbicide degradation, which essentially involves a number of processes, such as herbicide mobility as a leachate or surface water runoffs, and uptake of herbicide by the plant roots. The complete degradation of herbicide depends on the contribution of each of the above processes. The proportion of herbicide left in the soil actually determines the persistence or longevity of the herbicides in a certain soil systems (Locke and Bryson 1997). Soil characteristics such as the pH and organic matter content, significantly affects the herbicide degradation. The degradation and biochemical changes of the herbicides are potentially facilitated by the increased soil organic carbon, microbial and enzyme activity of the soil (Locke et al. 1995). Soils previously amended by herbicides may, also lead to the development of microbial populations, with an increased potential of herbicide degradation. These microbial populations may be sustained by reduced tillage (Wagner et al. 1996). Herbicide sorptive affinity and its solubility in water, temperature, and the moisture content of the soil, finally determine the fate of that particular herbicide in the soil. The rate of application and formulation type of the herbicide is crucial for the effective management. Therefore, systematic knowledge of the process of herbicide degradation in soil, its interaction with water and crops, is urgently required for an effective and sustainable management strategy of the agroecosystem.

Impact of Application of Herbicide along with Organic Resources on Dynamics of Carbon and Nitrogen and Crop Productivity

It has been observed that the difference in resource quality of soil amendments may influence the availability of soil nutrients (Singh et al. 2007a). The application of soil amendments with the herbicide, leads to deviations in the herbicide effectiveness, which results in alterations of factors responsible for the growth of crops as well as the weeds (Jenkinsen 1988), and ultimately affecting the crop yield. However, crop yield is only a small fraction of the total net productivity (TNP) (Ghoshal and Singh 1995). In general, weeds are not removed and remain in the field following the crop harvest, which later on, acts as an important organic input to the soil. The TNP of weeds and the crop makes the total biological productivity of a system. Nitrogen is also an important nutrient required for the productivity of the agroecosystems. To replenish the nitrogen loss through harvest (in the form of aboveground crop biomass), nitrogen input is generally required through fertilizers or organic resources (green manure, animal manure, crop residues). Out of the total nitrogen supplied by the fertilizer, only 40-60% is taken up by the crops, and the rest remains in the soil, vulnerable to loss. Reports are available that when herbicides are applied to the field in combination with organic amendments, the availability of inorganic nitrogen, activities of soil enzymes (particularly β glucosidase, acid and alkaline phosphatase and urease) remained higher throughout the annual cycle (Singh et al. 2012 Singh and Ghoshal 2013). Along with this, the use of a mix combination, tends to increase the crop yield and the soil microbial biomass; the sign of sustainability of such agroecosystems particularly in the drylands (Singh and Ghoshal 2010a; Singh et al. 2016).

Soil Microbial Biomass

Soil microbial biomass comprises the living part of the soil organic matter, excluding the live root fractions and soil organisms larger than 5×10^{-15} m^3 (Jenkinson and Ladd 1981). Soil microbes such as bacteria, fungi, actinomycetes, rotifers and protozoa are the major group constituting soil microbial biomass. Soil microbial biomass is considered as the 'source' (through mineralization) as well as the 'sink' (through immobilization) for various nutrients, due to its rapid turnover rate. Although the soil microbial biomass comprises only a small fraction of the total soil organic matter pool, its rapid turnover rate and quick release of nutrients contribute more to the nutrient cycling as compared to its size (Jenkinson and Ladd 1981). The soil microbial biomass is a large reservoir of critical plant nutrients such as N, P and S, and comprises 73, 372 and 130% of total annual requirements of such nutrients to plants (Smith and Paul 1990). Regular death of soil microorganisms occurs due to changes in the environment, and the dead cells after decomposition and mineralization by the surviving microorganisms, become the available nutrients for plants (Anderson and Dosmch 1980; Jenkinson and Ladd 1981). Soil microbial biomass may directly affect the nutrient cycling of the ecosystem, as it plays a critical role in mineralization and immobilization of plant nutrients. In addition, the soil microbial biomass also plays a significant role in changes in soil aggregation, followed by different management practices. Soil microbial biomass induces the binding of smaller soil particles by producing mucilaginous cementing agents and hyphae.

The common methods for measuring the soil microbial biomass includes: (1) soil fumigation method, in which, a flush of decomposition due to fumigation, is measured (Jenkinson and Powlson 1976). Chloroform is generally used as the fumigant in this method, (2) Substrate induced respiration (SIR), is the measurement of respiratory rate in addition of excess substrate (Anderson and Domsch 1980), (3) Soil analysis for a particular biomass constituent e.g., ATP, muramic acid, hexosamine and nucleic acids (Miller and Casida 1970). (4) Analysis of microorganisms by direct microscopy (Hanssen et al. 1974).

Soil microbial biomass is affected by variations in the quality and the quantity of inputs added to the soil, either through crop residue or as exogenous soil inputs (Singh et al. 2007 a), as well as the changes in the land-use pattern (Singh and Ghoshal 2014). Measurement of soil microbial biomass may thus serve as the vital indicator of changes through management practices. Amendment of a low quality as well as a high quality organic resource shows different effects on the soil microbial biomass. Addition of high quality organic resource (farmyard manure, *Sesbania* shoot) induces higher accumulation than the low quality organic resource (wheat straw) or chemical fertilizer (Ghoshal and Singh 1995; Singh et al. 2007a). However, when a mixture of high and low quality organic resources is applied to the fields, their combined effect is different from both the resources, applied in isolation. Generally, a high quality resource, with low C:N ratio decomposed rapidly and released the mineralizable carbon and nitrogen which gets accumulated and increased the microbial biomass in the early phase of the annual cycle. In contrast, a low quality resource, with a high C:N ratio, higher cellulose and hemicellulose content decomposed slowly and showed early immobilization and late release of nutrients, and delayed accumulation of microbial biomass in the later phase of the annual cycle. Therefore, application of a low quality organic resource, in combination of a high quality organic resource is suggested to be the most effective in sustaining the buildup of microbial biomass and C in the soil (Singh et al. 2007a).

The application of herbicide in the agricultural field has the potential to affect soil organic matter dynamics, especially the soil microbial biomass (Perucci et al. 2000; Vischetti et al. 2002). The soil fertility can be easily linked to the level of soil microbial biomass. The latter may decrease (Wardle and Parkinson 1990; Macur et al. 2007), increase (Debnath et al. 2002; Moreno et al. 2007) or remain unchanged (Wardle and Parkinson 1990; Lupwayi et al. 2007) in response to herbicide application in agricultural fields. Herbicides, when applied to the field, may interact with the soil amendments, and may lead to changes in the soil microbial biomass. The impact of herbicide application alone, on the soil microbial biomass, may turn out to be quite different from that of the combined application, depending upon the resource quality. Reports are available that soil microbial biomass increased, when applied in combination with soil amendments in rainfed agroecosystem of the dry tropics. The extent of increase is more, when herbicides interact with the high quality organic resource, than the low quality (Singh and Ghoshal 2010a; Singh et al. 2016). The application of herbicide in combination with organic soil amendments is the efficient strategy to overcome the adverse effect of herbicide application on the soil microbial biomass and soil fertility.

Carbon Sequestration Potential

The pool of soil carbon is determined by the balance between input and output; higher input relative to the output (positive difference between input and output) results in C sequestration in the soil, but higher output causes C loss. In agricultural soils, the organic carbon storage, calculated as the product of soil organic carbon and bulk density, along with soil depth is considered as one of the major attributes of soil fertility. Lal (2002) has reported soil organic carbon to be less than 1% in the dryland soils, which gets further reduced to 0.5% after cultivation. The rate of organic carbon storage is affected, not only by the environmental conditions, mainly the temperature, soil moisture, and the soil microbial community but also due to the various management practices, like the addition of exogenous organic inputs and the quantity and quality of the litter. The amendment of crop residue with higher C:N ratio (low quality) or mulching of soil with wheat straw have been suggested to be beneficial from the C sequestration point of view

(Paustian et al. 1997; Drinkwater et al. 1998; Jacinthe et al. 2002). The addition of organic input of low quality (wheat straw) either singly or in combination with high quality (*Sesbania shoot*) is reported to show a positive difference, between C input and output in soil (Singh et al. 2009b). In a tropical rainfed agroecosystem, C sequestration potential in terms of C storage capacity was found to be higher where herbicide was applied along with organic amendments compared to the chemical fertilizer. Among the organic inputs, the C storage capacity was highest, when herbicide was applied in combination with low quality resources as compared to the high quality resources. The low quality resources decompose slowly, so persists in soil for a longer duration compared to the high quality resources (Singh et al. 2007a), perhaps facilitating long-term C storage in the soil (Singh 2011). Thus for sustainable agriculture, the combined input of herbicide and low quality input of organic resources (wheat straw) will maintain soil fertility in terms of C sequestration and mitigation of the global climate change.

Nitrogen Dynamics in Soil

Globally it is recognized that nitrogen is the most limiting nutrient in soils (Huber and Thompson 2007), especially in the rainfed agroecosystems; where its cycling, plays a crucial role in the management of nutrient-limited cropland soils (Ghoshal 2002). The fate of nitrogen in soil is determined by the nitrogen balance between input and output in the agroecosystem. Therefore, nitrogen dynamics is considered as the key, to design eco-friendly management practices in order to achieve sustained crop productivity. The sources of nitrogen inputs are the exogenous soil amendments, including the chemical fertilizers, manure (farmyard and/or green manure) and crop residues returned to the soil. Whereas, the major output of nitrogen from the agroecosystem occurs in the form of nitrogen uptake by plants and nitrogen losses in the form of leaching, volatilization, denitrification depending on the natural conditions (Yang et al. 2008). There is a large reserve of nitrogen in the soil organic matter, but its availability to plants is affected by several natural processes including nitrification, denitrification, mineralization and immobilization (either by microorganisms and/or plants). Generally, agroecosystems (especially in the tropical drylands) require supplement of nitrogen through exogenous sources. Chemical fertilizers are an alternative to nitrogen supplement. However, with fertilizer application, the rate of nitrogen release in soil, generally does not match the crop requirement, and therefore, concerns have been raised for the long-term sustainability of such nutrient-limited soils of the agroecosystem. In this context, the organic input of various resource qualities is considered as the sustainable and eco-friendly alternative to meet the N supplement for crops. The suitability of crop input as a source of nitrogen depends on the release of nutrient and nitrogen through mineralization, in synchrony with the crop requirement. The increased rate of nitrogen release facilitates nitrogen loss through various pathways (Goulding 2004); and if the rates of release become too slow, the crop yields may be negatively affected. For an increase in the crop productivity, it is essential to manage both nitrogen as well as the weed. Knowledge of the effect of herbicide on nitrogen dynamics is essential, as extensive use of herbicide along with nitrogen fertilizers may adversely affect the soil microbes and the environment (El-Ghamry et al. 2002). The difference between the rate of nitrogen mineralization and nitrogen uptake represents the availability of inorganic nitrogen in soil (Velthof and Oenema 2010). Therefore, any variations in the rate of N mineralization will affect both the inorganic nitrogen pool in the soil, and nitrogen uptake by the plants. In agroecosystems, both the crops as well as the weeds, compete with each other for nitrogen uptake from the soil. The combined use of herbicides and soil amendments of varying resource quality may affect the N cycle differentially, which may result in variations in the rate of nitrogen mineralization, the inorganic nitrogen pool in soil and the nitrogen uptake by plants.

The input of farmyard manure plays a pivotal role in increasing the rate of N supply, as well as the pool size of available N in the drylands. It also sustains the enhanced N pool throughout the annual cycle, unlike the use of chemical fertilizers, which only initially increases the size of the available nitrogen pool, but cannot sustain it. The amendment of farmyard manure provides a better synchronization of N availability and the crop demand for the next cropping season of the annual cycle (Ghoshal 2002). The combined application of a high and low quality organic resource (*Sesbania* shoot+wheat straw) modulates the nitrogen release and minimizing its loss from the agroecosystem due to a higher synchronization of N release and crop demand (i.e., balanced N immobilization/mineralization). The combined input of *Sesbania*+wheat straw may lead to better nutrient conservation and reduced loss from the agroecosystem in terms of consistent and prolonged availability of nitrogen in the soil (Singh et al. 2007b).

Herbicide when applied in combination with the organic inputs, the inorganic N availability remained higher throughout the annual cycle. However, availability of inorganic N, may vary from the first cropping season to the next (like rice to barley or wheat period) depending on the rate of N mineralization and its uptake by plants. For example, when herbicide was applied to the cropland in combination with *Sesbania* shoot (high quality), inorganic N availability was higher during the first cropping season, i.e., rice period; whereas, the application of herbicide with wheat straw (low quality) facilitated higher availability of inorganic N during the second cropping season i.e., barley period as compared to the rice period (Singh et al. 2012).

Crop Productivity

Along with the ecological benefits, economic incentives in terms of productivity are also important for the implementation of any management practice. The resource qualities of the organic inputs play an important role in regulating the crop yield and the biological productivity. Sum of the total net productivity (TNP) for both the crop and the weeds in each crop period is defined as the total biological productivity for that crop period. The sum of total biological productivity for the cropping seasons within a year is the annual total biological productivity. The total net productivity (TNP) is the sum of the aboveground net productivity (ANP) and the belowground net productivity (BNP) of the crops and the weeds. The sum of successive increments in the shoot and the root biomass in a particular crop period, is referred to as the aboveground net productivity (ANP) and the belowground net productivity (BNP), respectively, of the crops or the weeds.

The organic input of a high quality resource (*Sesbania* shoots) when applied once a year, will be more effective on the first cropping season, like rice, whereas, a low quality input (wheat straw) benefited the succeeding crop, like wheat or barley. The combined input of organic resources (*Sesbania* shoots+wheat straw) achieved a relatively higher biological productivity and grain yield through both the cropping seasons. This is due to an early release of nutrients, as well as N, from the rapidly decomposing *Sesbania* shoots (instantaneous decay constant, $k = 0.028$ day^{-1}, $t_{95} = 110$ days), (Singh et al. 2007a) supporting a higher biological productivity of the first crop i.e., rice; whereas, slower decomposing of wheat straw ($k = 0.0026$ day^{-1}, $t_{95} = 1192$) initially immobilized the nutrients, and delayed mineralization to the next cropping season of the annual cycle i.e., wheat or barley, and is reflected from the enhanced biological productivity (Singh et al. 2007a; Singh and Ghoshal 2010b).

The application of herbicide in the croplands resulted in an increase in the crop yield, yet the total biological productivity (crop and weed total net productivity) decreased relative to the control. However, the application of herbicide in combination with the organic inputs resulted in alterations of the herbicide efficacy. The combined input of herbicide with amendments (like chemical fertilizers, animal manure, green manure, and crop residue), resulted in an increase in the total biological productivity, due to an increase in the total net productivity of the crops and the weeds (Singh and Ghoshal 2010a). The suppression and/or total inhibition of the weeds and the lower competition for the various natural resources is also the reason for a higher crop total net productivity and crop yield when herbicides are applied in the croplands. The increase in the total net productivity due to the application of herbicide was not proportional to the decrease in the weed total net productivity (Singh and Ghoshal 2010a). Since the total biological productivity of a system is generally considered as the index of soil fertility, it may be suggested that the long-term use of herbicide only, without other amendments, could reduce soil fertility, despite achievement of a greater crop yield and therefore, for sustainable agriculture, herbicide should be used in combination of organic inputs.

Conclusion

The organic inputs of varying resource quality play an important role in regulating the soil fertility and crop productivity. The annual input of a high quality resource benefitted the first crop more, (e.g., rice) whereas, a low quality input benefitted the succeeding crop like barley or wheat. The combined input of organic resources seems to be useful in developing a low input, environment friendly, soil management practices in agroecosystems, especially in the rainfed croplands. The combined inputs of high and low quality resources also modulate the N release, leading to higher crop productivity throughout the annual cropping cycle. The amendment of combined input facilitates higher biological productivity and grain yield through both the cropping seasons. The combined input of a high and low quality organic resource resulted in a significant buildup of soil carbon, i.e., a higher carbon sequestration capacity, along with the increased crop productivity. The application of herbicide with soil amendments increased the total biological productivity, due to an increase in the total net productivity of the crop and the weeds. This holds promise to negate the adverse effect of herbicide application on the soil microbial biomass and the total biological productivity. Maintenance of a higher total biological productivity may lead to an increase in the carbon sequestration potential of soil and thus can sustain long-term soil fertility.

Acknowledgments

We are thankful to the Head and the Coordinator, Centre of Advanced Study in Botany, Department of Botany, Banaras Hindu University, Varanasi, for providing the laboratory facilities. University Grants Commission, New Delhi, India, provided financial support in the form of a Major Research Project (p-01/581), University Research Fellowship to Dr. Mahesh Kumar Singh, Dr. Alka Singh, Research Fellowship in Science for Meritorious Students (RFSMS) to Mr. Biswadip Mukherjee, and CAS fellowship to Mr. Chandra Mohan Kumar.

REFERENCES

Anderson JPE, Domsch KH (1980) Quantities of plant nutrients in the microbial biomass of selected soils. *Soil Biol Biochem* 10: 207-213.

Anonymous 1999-2000, Agriculture Statistics of Pakistan, Govt. of Pakistan, Ministry of Food, Agriculture and Livestock. Food and Agriculture Division, Economic Wing Islamabad, 2000.

Asante P, Armstrong GW, Adamowicz WL (2011) Carbon sequestration and the optimal forest harvest decision: A dynamic programming approach considering biomass and dead organic matter. *J Forest Econ* 17: 3-17.

Cadisch G, Giller K (1997) (Eds.), *Driven by Nature: Plant Residue Quality and Decomposition*. CAB International, Wallingford, UK, pp. 409.

Carter LJ (1980) Organic farming becomes "legitimate". *Science* 209: 254-256.

Coleman DC, Hendrix PF (1988) Agroecosystem processes. *In: Concepts of ecosystem Ecology* (Eds. Pomeroy LR, Alberts JJ), Ecological Studies 67 Springer-Verlag, New York, pp. 149-170.

Debnath A, Das AC, Mukherjee D (2002) Rhizosphere effect of herbicides on nitrogen fixing bacteria in relation to availability of nitrogen in rice soil. *J Indian Soc Soil Sci* 50: 463-466

Drinkwater LE, Wagoner P, Sarrantonio M (1998) Legume based cropping systems have reduced carbon and nitrogen losses. *Nature* 396: 262-265.

El-Ghamry AM, Xu JM, Huang CY, Gan J (2002) Microbial response to bensulfuron-methyl treatment in soil. *J Agric Food Chem* 50: 136-139.

FAO Database (2005) Affod and Agriculture Organization, Rome. http://faostat.fao.org/.

Ghoshal N (2002) Available pool and mineralization rate of soil N in a dryland agroecosystem: Effect of organic soil amendment and chemical fertilizer. *Trop Ecol* 43(2): 363-366.

Ghoshal N (2003) Soil fertility management: an ecological approach. *Plant Archives* 3(1): 1-7.

Ghoshal N, Singh KP (1995) Effects of farmyard manure and inorganic fertilizer on the dynamics of soil microbial biomass in a tropical dryland agroecosystems. *Biol Fert Soils* 19: 231-238.

Gill H Wallia U (1979) Chemical weed control in wheat with particular reference to *Phalaris minor* Retz and *Avena fatua* L. *Pesticides* 13: 15-20.

Goulding K (2004) Pathways and losses of fertilizer nitrogen at different scales. *In: Agriculture and the Nitrogen Cycle* (Eds. Mosier AR, Syers KJ, Freney JR), The Scientific Committee on Problems of the Environment (SCOPE). Island Press, Covelo, California, USA, pp. 209-219.

Hanssen JF, Thingstad TF, Golksory J (1974) Evaluation of hyphal lengths and fungal biomass in soil by a membrane filter technique. *Oikos* 25: 102-107.

Haynes RJ (2000) Labile organic matter as an indicator of organic matter quality in arable and pastoral soils in New Zealand. *Soil Biol Biochem* 32: 211-219.

Heal OW, Anderson JM, Swift MJ (1997) Plant litter quality and decomposition: an historical overview. *In: Driven by Nature: Plant Litter Quality and Decomposition.* (Eds. Cadisch G, Giller KE), CAB International, Wallingford, UK, pp. 3-3.

Huber DM, Thompson IA (2007) Nitrogen and plant disease. Mineral nutrition in Minnesota. *Soil Till Res* 55: 127-142.

Hyvonen T, Salonen J (2002) Weed species diversity and community composition in cropping practices at two intensity levels-a six-year experiment. *Plant Ecol* 154: 73-81.

IPCC (2014) *Climate Change 2014: Synthesis Report.* Cambridge University Press, United Kingdom and New York, NY, USA.

Jabro JD, Sainju U, Stevens WB, Evans RG (2008) Carbon dioxide flux as affected by tillage and irrigation in soil converted from perennial forages to annual crops. *J Environ Manag* 88: 1478-1484.

Jacinthe PA, Lal R, Kimble JM (2002) Carbon budget and seasonal carbon dioxide emission from a central Ohio Luvisol as influenced by wheat residue amendment. *Soil Till Res* 67: 147-157.

Jenkinsen DS (1988) Determination of microbial biomass carbon and nitrogen in soil. *In: Advances in Nitrogen Cycling in Agricultural Ecosystems* (Ed. Wilson JR), CAB International, Wallingford, UK, pp. 368-386.

Jenkinson DS, Ladd JN (1981) Microbial biomass in soil measurement and turnover. *In: soil biochemistry* (Eds. Paul EA, Ladd JN), Marcel Dekker, New York, USA, pp. 415-471.

Jenkinson DS, Powlson DS (1976) The effects of biocidal treatments on metabolism in soils-V: A method of measuring soil biomass. *Soil Biol Biochem* 8: 208-213.

Lal R (2002) Soil carbon dynamics in cropland and rangeland. *Environ Poll* 116: 353-362.

Lal R (2004) Soil carbon sequestration impacts on global climate change and food security. *Science* 304: 1623-1627.

Lal R (2009) Soils and world food security. *Soil Till Res* 102: 1-4.

Li C, Frolking S, Butterbach-Bahl K (2005) Carbon sequestration in arable soils is likely to increase nitrous oxide emissions, offsetting reductions in climate radiative forcing. *Clim Change* 72: 321-338.

Locke MA, Bryson CT (1997) Herbicide-Soil Interactions in Reduced Tillage and Plant Residue Management Systems. *Weed Sci Soc Am* 45(2): 307-320.

Locke MA, Zablotowicz RM, Gaston LA (1995) Fluometuron interactions in crop residue-managed soils. *In: Conservation Farming: A focus on Water Quality* (Eds. Kingery WL Buehring N), MAFES Special Bulletin 88-7 Mississippi State, MS: Mississippi Agricultural Forest Experiment Station, pp. 55-58.

Lupwayi NZ, Hanson KG, Harker KN, Clayton GW, Blackshaw RE, O'Donovan JT, Johnson EN, Gan Y, Irvine B, Monreal MA (2007) Soil microbial biomass, functional diversity and enzyme activity in glyphosate-resistant wheat-canola rotations under low-disturbance direct seeding and conventional tillage. *Soil Biol Biochem* 39: 1418-1427.

Macur RE, Wheeler JT, Burr MD, Inskeep WP (2007) Impacts of 2,4-D application on soil microbial community structure and on populations associated with 2,4-D degradation. *Microbiol Res* 162: 37-45.

Mafongoya PL, Giller KE, Palm CA (1998) Decomposition and nitrogen release patterns of tree prunings and litter. *Agrofor Syst* 38: 77-97.

Melillo JM, Aber JD, Muratore JF (1982) Nitrogen and lignin control of hardwood leaf litter decomposition dynamics. *Ecology* 63: 621-626.

Miller WN, Casida LE (1970) Evidence for muramic acid in soil. *Can J Microbiol* 16: 299-304.

Monaco S, Hatch DJ, Sacco D, Bertora C, Grignani C (2008) Changes in chemical and biochemical soil properties induced by 11-yr repeated additions of different organic materials in maize-based forage systems. *Soil Biol Biochem* 40: 608-615.

Moreno JL, Aliaga A, Navarro S, Hernandez T, Garcia C (2007) Effects of atrazine on microbial activity in semiarid soil. *Appl Soil Ecol* 35: 120-127.

Nishio T, Komada M, Arao T, Kanamori T (2001) Simultaneous determination of transformation rates of nitrate in soil. *Japan Agric Res Quart* 35: 11-17.

Odum EP (1984) Properties of agroecosystem. *In: Agricultural Ecosystems- Unifying concepts*. (Eds. R Lawrance, BR Sinner, GJ House), John Wiley, New York, pp. 5-11.

Palm CA (1995) Contribution of agroforestry trees to nutrient requirements in intercropped plants. *Agrofor Syst* 30: 105-124.

Palm CA, Gachengo CN, Delve RJ, Cadisch G, Giller KE (2001) Organic inputs for soil fertility management in tropical agroecosystems: application of an organic resource database. *Agr Ecosyst Environ* 83: 27-42.

Paustian K, Andren O, Janzen HH, Lal R, Smith P, Tian G, Tiessen H, Van Noordwijk M, Woomer PL (1997) Agricultural soils as a sink to mitigate CO_2 emissions. *Soil Use Manag* 13: 230-244.

Perucci P, Dumontet S, Bufo SA, Mazaatura A, Casucci C (2000) Effects of organic amendment and herbicide treatment on soil microbial biomass. *Biol Fertil Soils* 32: 17-23.

Ramesh P, Singh M, Rao AS (2005). Organic farming: Its relevance to the Indian context. *Current Science* 88; 561-568.

Sarkar S, Singh SR, Singh RP (2003) The effect of organic and inorganic fertilizer on soil physical condition and the productivity of a rice-lentil cropping sequence in India. *J Agric Sci* 140: 419-425.

Schlesinger WH, Hasey MH (1981) Decomposition of chaparral shrub foliage: losses of organic and inorganic constituents from deciduous and evergreen leaves. *Ecology* 62: 762-774.

Singh A (2011) *Ecological analysis of microbial biomass dynamics, soil enzyme activities and biological productivity in a grassland and a rainfed agroecosystem receiving herbicide and varying soil amendments*. Doctoral thesis, Banaras Hindu University, Varanasi.

Singh A, Ghoshal N (2010) Impact of herbicide and soil inputs on crop yields in tropical dryland agroecosystems. *Plant Archives* 10(2): 753-756.

Singh A, Ghoshal N (2013) Impact of herbicide and various soil amendments on soil enzymes activities in a tropical rainfed agroecosystem. *Eur J Soil Biol* 54: 56-62.

Singh A, Singh MK, Ghoshal N (2016) Microbial biomass dynamics in a tropical agroecosystem: Influence of herbicide and soil amendments. *Pedosphere* 26(2): 257-264.

Singh KP, Ghoshal N, Singh S (2006) Sustainable dryland agroecosystem: microbial management of soil fertility for improved crop productivity. *In: Ecology, Diversity and Conservation of Plants and Ecosystems in India* (Eds. Pandey HN, Barik SK), Regency Publications, New Delhi, pp. 129-140.

Singh KP, Singh S, Ghoshal N (2009) Soil carbon dioxide flux, carbon sequestration and crop productivity in a tropical dryland agroecosystem: Influence of organic inputs of varying resource quality. *App Soil Ecol* 42: 243-253.

Singh MK, Ghoshal N (2011) Impact of land use change on soil organic carbon content in dry tropics. *Plant Archives* 11(2): 903-906.

Singh MK, Ghoshal N (2014) Variation in soil microbial biomass in the dry tropics: impact of land-use change. *Soil Res* 52: 299-306.

Singh MK, Astley H, Smith P, Ghoshal N (2015) Soil CO_2-C flux and carbon storage in the dry tropics: impact of land-use change involving bioenergy crop plantation. *Biomass Bioen* 83: 123-130

Singh P, Ghoshal N (2010) Variation in total biological productivity and soil microbial biomass in rainfed agroecosystems: Impact of application of herbicide and soil amendments. *Agr Ecosyst Environ* 137: 241-250.

Singh P, Singh RP, Ghoshal N (2012) Influence of Herbicide and Soil Amendments on Soil Nitrogen Dynamics, Microbial Biomass, and Crop Yield in Tropical Dryland Agroecosystems. *Soil Sci Soc Am J* 76: 2208-2220.

Singh S, Ghoshal N, Singh KP (2007a) Variations in soil microbial biomass and crop roots due to differing resource quality inputs in a tropical dryland agroecosystems. *Soil Biol Biochem* 39: 76-86,

Singh S, Ghoshal N Singh KP (2007b) Synchronizing nitrogen availability through application of organic inputs of varying resource quality in a tropical dryland agroecosystem. *App Soil Ecol* 36: 164-175.

Singh S, Ghoshal N, Singh KP (2007c) Influence of resource quality of organic inputs on rice-barley dryland agro-ecosystem: Variations in biological productivity, grain yield, and efficiency of nitrogen use. *Exp Agric* 43: 149-161.

Singh S, Mishra R, Singh A, Ghoshal N, Singh KP (2009) Soil physicochemical properties in a grassland and agroecosystem receiving varying organic inputs. *Soil Sci Soc Am J* 73: 1530-1538.

Smith JL, Paul EA (1990) The significance of microbial biomass estimation. *In: Soil Biochemistry* (Eds. Bollag JM, Stotzky G), Marcel Dekker Inc., New York, USA, 6: 357-396.

Swift MJ (1987) Tropical Soil Biology and Fertility: Interregional Research Planning Workshop. *Biology International*, Special Issue 13. IUBS, Paris.

Swift MJ, Heal OW, Anderson JM (1979) *Decomposition in Terrestrial Ecosystems.* Blackwell Scientific Publications, Oxford 372.

Thakur RC, Bindra AD, Sood RD, Bhargava M (1995) Effect of fertilizer application and green-manuring on physico-chemical properties of soil and grain yield in rice-wheat crop sequence. *Indian J Agron* 40: 4-13.

Velthof GL, Oenema O (2010) Estimation of plant available nitrogen in soils using rapid chemical and biological methods. *Commun Soil Sci Plant Anal* 41: 52-71.

Vischetti C, Casaucci C, Perucci P (2002) Relationship between changes of soil microbial biomass content and imazamox and benfluralin degradation. *Biol Fertil Soils* 35: 13-17.

Wagner SC, Zablotowicz RM, Gaston LA, Locke MA, Kinsella J (1996) Bentazon degradation in soil: Influence of tillage and history of bentazon application. *J Agric Food Chem* 44: 1593-1598.

Wardle DA, Parkinson D (1990) Effects of three herbicides on soil microbial biomass and activity. *Plant Soil* 122: 21-28.

Yang LL, Zhang FS, Mao RZ, Ju XT, Cai XB, Lu YH (2008) Conversion of natural ecosystems to cropland increases the soil net nitrogen mineralization and nitrification in Tibet. *Pedosphere* 18: 699-706.

Zaller JG, Kopke U (2004) Effects of traditional and biodynamic farmyard manure amendment on yields, soil chemical, biochemical and biological properties in a long-term experiment, *Biol Fert Soils* 40: 222-229.

In: Trends in Life Science Research
Editors: R. P. Sinha and U. P. Shrivastava

ISBN: 978-1-53613-241-0

Chapter 9

THE IMPACTS OF UV-B RADIATION ON THE ENZYMES OF NITROGEN METABOLISM IN CYANOBACTERIA

Abha Pandey[1], Haseen Ahmed[1], Deepak K. Singh[1], Vidya Singh[1], Deepak Kumar[1], Rajneesh[1], Jainendra Pathak[1], Shailendra P. Singh[2] and Rajeshwar P. Sinha[1,*]

[1]Laboratory of Photobiology and Molecular Microbiology,
Centre of Advanced Study in Botany, Institute of Science,
Banaras Hindu University, Varanasi, India
[2]Centre of Advanced Study in Botany, Institute of Science,
Banaras Hindu University, Varanasi, India

ABSTRACT

Nitrogen-fixing cyanobacteria possess a central position in the nutrient cycling largely due to their inherent capacity to fix atmospheric nitrogen directly into ammonium, with the help of the enzyme "nitrogenase" thus making it available for use by higher plants. The percentage of total annual ecosystem nitrogen input due to biological nitrogen fixation by cyanobacteria might be as high as 80% and the contribution to total annual nitrogen uptake by plants up to 20%. Recent global climate models predict a further significant loss of stratospheric ozone in the next decades, resulting in an increase in ultraviolet-B (UV-B) radiation reaching the surface of the Earth. UV-B radiation has deleterious effects on a number of metabolic processes such as growth, survival, pigmentation, phycobiliprotein composition, pigmentation, motility, heterocyst

[*]Corresponding Author Email: r.p.sinha@gmx.net.

differentiation, $14CO_2$ uptake and enzymes of nitrogen metabolism in cyanobacteria, as well as several other freshwater and marine phytoplanktons. UV-B-induced stress can interfere with the metabolism of inorganic and organic nitrogen compounds. It affects several points of inorganic nitrogen assimilation pathway in cyanobacteria. UV exposure results in complete inhibition of nitrogenase activity and thus nitrogen-fixing activity is usually inhibited or depressed by UV-B-induced stress in most cyanobacteria. UV-B radiation has been reported to inhibit ammonium (NH_4^+) and nitrite (NO_2^-) uptake rates. In contrast, a significant increase in nitrate (NO_3^-) uptake has been observed under UV-B stress. The activities of nitrate assimilating enzyme, nitrite reductase (NiR), and ammonium assimilating enzymes, glutamine synthetase (GS) and glutamate synthase (GOGAT) are also adversely affected on exposure to UV-B irradiation while nitrate reductase (NR) and glutamate dehydrogenase (GDH; aminating) show a stimulatory response. In this chapter we address the impacts of UV-B radiation on the various enzymes associated with nitrogen metabolism in cyanobacteria.

Keywords: cyanobacteria, glutamine synthetase, nitrate reductase, nitrogenase, ultraviolet-B radiation

ABBREVIATIONS

GS: Glutamine synthetase

GOGAT: Glutamine (amide)-2-oxoglutarate aminotransferase (glutamate synthase)

NR: Nitrate reductase

NiR: Nitrite reductase

INTRODUCTION

Cyanobacteria are phylogenetically a primitive and diverse group of Gram-negative prokaryotes possessing oxygenic photosynthesis, which is similar to that in algae and higher plants. We know from fossil evidence that existence of cyanobacteria on the Earth dates back to Precambrian era (between 2.8 to 3.5 x 10^9 years ago). At that time they might have been the main primary producers of organic matter and the first organisms to release oxygen into the then oxygen-free atmosphere. The advent of oxygenic photosynthesis on Earth probably has increased global biological productivity by a factor of 100-1000 and thereby profoundly affected both geochemical and biological evolution (Hoehler et al. 2001). Thus, cyanobacteria were responsible for a major global evolutionary transformation leading to the development of aerobic metabolism and the subsequent rise of higher plant and animal forms (Schopf 1993; Pace 1997). Cyanobacteria are thought to have survived from a wide spectrum of environmental stresses such as extreme temperatures, drought, salinity, nitrogen starvation, photooxidation, anaerobiosis, osmotic and ultraviolet radiation stress (Tandeau and

Houmard 1993; Sinha and Häder 1996). Members of cyanobacteria are also unique in having a cosmopolitan distribution from hot springs to arctic regions and are therefore expected to face various levels of ultraviolet radiation (UVR) in their natural habitats (Stanier and Cohen-Bazire 1977). Cyanobacterial colonization of oceans, lakes, rivers, hot springs and soils, and their presence as symbiotic organisms in fungi and plants demand high variability in adapting to diverse environmental factors.

Cyanobacterial populations occupy an important place in both aquatic as well as terrestrial ecosystems and achieve net photosynthesis even at temperatures as low as -7°C (Davey 1989) and as high as 75°C (Castenholz 1969). Member of cyanobacteria also possess a central position in nutrient cycling largely owing to their inherent capacity to fix atmospheric N_2 with the help of the enzyme "nitrogenase" directly into ammonium (NH_4^+), a form through which nitrogen enters the food chain. The estimates vary as to the contribution of cyanobacterial nitrogen fixation to the global nitrogen cycle; according to one assumption cyanobacteria fix over 35 million tons of nitrogen annually (Häder et al. 1989). Therefore, cyanobacteria form a prominent component of microbial populations in wetland soils and are of economic importance in rice paddy fields in the tropics, where they significantly contribute to fertility as a natural biofertilizers (Venkataraman 1981; Kannaiyan et al. 1982; Sinha and Häder 1996). The agronomic potential of free-living as well as symbiotic cyanobacteria, particularly *Azolla-Anabaena* symbiosis, have been recognized as one of the most promising biofertilizer systems for wetland soils (Sinha and Häder 1996; Sinha et al. 1998; Vaishampayan et al. 1998). Some cyanobacterial strains that thrive in rice fields release small quantities of the major fertilizing product, ammonia and small nitrogenous polypeptides during active growth, while most of the fixed products are made available mainly through autolysis and decomposition (Martinez 1984).In nature, the process of nitrogen fixation is of utmost importance because it counterbalances the losses of nitrogen from the environment by denitrification (Kuhlbusch et al. 1991). In many instances the availability of nitrogen has been found to be a limiting factor for productivity in natural habitats (Parker et al. 1978; Sinha and Häder 1996).

Recent findings suggest that continuous accumulation of anthropogenically released atmospheric pollutants, such as chlorofluorocarbons (CFCs), chlorocarbons (CCs), organobromides (OBs) and reactive nitrogen species (RNS) including nitric oxide (NO), nitrous oxide (N_2O) and peroxynitrite (ONOO) in the stratosphere has resulted in the depletion of ozone layer which provides shield against UVR (100-400 nm) (Crutzen 1992; Lubin and Jensen 1995; Ravishankara et al. 2009). The process of ozone depletion has been reported at mid latitude and especially in the Antarctic where ozone levels declined by more than 70% during late winter and early spring in the polar vortex (Smith et al. 1992). The decreased density of ozone layer results in increased intensity of harmful UV-B(280-315 nm) reaching on the Earth's surface (Hofmann and Deshler 1991; Madronich et al. 1998; Sahoo et al. 2005).UV-B radiation (<1% of total irradiance), being the most detrimental component of the solar spectrum, has the potential to exert adverse effects on photosynthetic terrestrial and aquatic organisms, thereby affecting the productivity of ecosystems (Karentz et al. 1991a; Vincent and Roy 1993; Williamson 1995; Sinha and Häder 1996).

Both photosynthesis and nitrogen fixation are energy-dependent processes driven by solar energy. Harvesting of solar radiation exposes cyanobacteria simultaneously to harmful doses of UV-B (280-315 nm) and UV-A (315-400 nm) radiation in their natural habitats. In higher plants, reduction of leaf area, fresh and dry weight, lipid contents and photosynthetic activities have been reported in a number of UV-B sensitive plant species (Biggs et al. 1981; Caldwell 1981; Tevini and Teramura 1989). The high-energetic UV-B has the greatest potential for cell damage caused by both direct effects on DNA and proteins and indirect effects *via* the production of reactive oxygen species (ROS) (Karentz et al. 1991b; Vincent and Neale 2000). There are several targets for the potentially toxic ROS including lipids, DNA and proteins. Moreover, damage to the photosynthetic apparatus is also partially mediated by ROS resulting in inhibition of photosynthesis (He and Häder 2002a). In contrast, UV-A radiation which is not absorbed directly by the DNA, can still induce DNA damage either by producing a secondary photoreaction of existing DNA photoproducts or *via* indirect photosensitizing reactions (Hargreaves et al. 2007).Investigations have shown that UV-B irradiation adversely affects key physiological and biochemical life processes, such as morphology, cell differentiation, survival, growth, pigmentation, motility and orientation, N_2 metabolism, phycobiliprotein composition, protein profile, DNA, CO_2 uptake and membrane permeability (Gao et al. 2007; Lesser 2008; Sinha et al. 2008a)of various photosynthetic organisms as they are simultaneously exposed to visible and UVR in their natural habitats.

Cyanobacteria which are simultaneously exposed to photosynthetic active radiation (PAR; 400-700 nm) and UVR have evolved a number of mitigation strategies to overcome the damaging effects (direct and indirect) of UVR. These strategies include avoidance, scavenging of UV-induced ROS by enzymatic and non-enzymatic antioxidants (Middleton and Teramura 1994; Singh et al. 2013), and screening by certain UV-absorbing/screening compounds like MAAs and scytonemin that counteract the deleterious effects of UVR (Karentz et al. 1991c; Dunlap and Shick 1998; Sinha et al. 1998; Pathak et al. 2015a, 2015b, 2017a). Besides this, repair of UV-induced damage of DNA by photoreactivation and excision repair and resynthesis of proteins are also important mechanisms to prevent UV-induced photodamage (Britt 1995; Kim and Sancar 1995).

EFFECT OF UV-B RADIATION ON CYANOBACTERIA

Harvesting of solar energy to perform photosynthesis and nitrogen fixation exposes these cyanobacteria to lethal doses of UV-B and UV-A (315-400 nm) radiations in their natural lit habitats. The highly energetic UVR affects the aquatic cyanobacteria as they can penetrate deep into the water column up to 20 m in the ocean and to a few centimeters in lakes and rivers. The fluence rate of UV-B radiation impinging on the natural habitats seems to be of major concern since UV-B radiation has been reported not only to impair motility and photoorientation (Donkor and Häder 1991) but also to affect a number of physiological and biochemical processes in cyanobacteria (Sinha et al. 2008a; Singh et al. 2010). He and Häder (2002b, 2002c) reported UV-B and UV-A-induced ROS production for the first time in cyanobacteria using a ROS-sensitive probe, 2′, 7′-dichlorodihydrofluorescein diacetate (DCFH-DA).

Differentiation of vegetative cells into heterocysts and akinetes in filamentous cyanobacteria has been reported to be adversely affected by UVR. Growth and survival of cyanobacteria have been reported to be severely affected following UV-B irradiation for different durations. Complete killing and loss of survival was reported in several cyanobacteria within 120-180 min of UV-B exposure (Tyagi et al. 1992; Sinha et al. 1995). Inhibition of growth in the toxic bloom-forming cyanobacterium *Microcystis aeruginosa* FACHB 912 and the non-toxic *Microcystis aeruginosa* FACHB 469 following UV-B exposure has been reported by Zhang et al. (2012). Proteins are one of the main targets of UV-B as the total protein profiles of several cyanobacteria showed a linear decrease in protein content with increasing UV-B exposure time. Kumar et al. (1996) reported complete loss of protein bands between 14.2 and 45 kDa after 90 and 120 min of UV-B exposure in *Nostoc calcicola*. Translation activity was also suppressed by oxidative stress in *Synechocystis* sp. PCC 6803 and oxidation of the elongation factor EF-G was found to be responsible for the inhibited translation activity (Kojima et al. 2007). Aromatic amino acids such as tryptophan, tyrosine, phenylalanine and histidine, having absorption maxima in the UV range (270-290 nm) are also damaged by UV-B radiation (Sinha et al. 1996, 1998). UV-B radiation has been reported to induce a variety of DNA damage. The main DNA lesions which develop by direct UV-B absorption are dimeric photoproducts such as *cis-syn* cyclobutane pyrimidine dimers (CPDs) and pyrimidine (6-4) pyrimidone photoproducts (6-4 PPs) and their Dewar isomers (Sinha and Häder 2002; Häder and Sinha 2005; Singh et al. 2010). UVR-induced formation of thymine dimers in three rice-field cyanobacteria, *Anabaena*, *Nostoc* and *Scytonema* sp., was detected using blotting and the chemiluminescence method (Sinha et al. 2001). UVR has been reported to cause single or double-stranded breaks in the native DNA molecules in various cyanobacterial species (He and Häder 2002a). A rapid decrease in the photosynthetic activity of *Scytonema javanicum* occurs after irradiation with UV-B (Chen et al. 2013). Liet al. (2012) showed that UVR, mainly UV-B, has a detrimental effect on the phosphatase activity of the terrestrial cyanobacterium *Nostoc flagelliforme*. Investigations on photobleaching of photosynthetic pigments of cyanobacteria have shown that the accessory light harvesting pigment phycocyanin (λ_{max} 620 nm) was bleached more rapidly and drastically than any other pigment such as Chl *a* (λ_{max} 437 and 672 nm) or the carotenoids (λ_{max} 485 nm) (Sinha et al. 1995). UV-B irradiation has been found to be responsible for a decrease in the phycobiliprotein content and a disassembly of the phycobilisome complex in a number of cyanobacteria (Sinha et al. 1995, 1997). Reports have shown that UV-B radiation also influences the chlorophyll and carotenoids contents in cyanobacteria (He and Häder 2002b; Han et al. 2003; Gao et al. 2007; Lesser 2008). Certain photosynthetic parameters such as $^{14}CO_2$ uptake, O_2 evolution and ribulose-1,5 bisphosphate carboxylase (RuBISCO) activity in cyanobacteria has been shown to be down-regulated by UVR (Sinha et al. 2008b). The D1 and D2 proteins, major constituents of PS-II reaction centers, are also degraded even after exposure to intermediate levels of UV-B radiation (Sass et al. 1997; Campbell et al. 1998). Zhang et al. (2012) observed a decrease in the content of allophycocyanin or phycocyanin following UV-B exposure. Several groups of mRNAs specifying proteins involved in light harvesting and photosynthesis were also reportedly down-regulated by UV-B treatment, suggesting a reduced photosynthetic activity following UV-B exposure (Huang et al. 2002). UV-B radiation also induces the lipid peroxidation of polyunsaturated fatty acids (PUFA) *via* oxidative damage which subsequently affects the integrity of cellular and thylakoid membranes (He and Häder 2002a). The process of nitrogen fixation is also severely inhibited by UV-B radiation due to the extreme sensitivity of the nitrogenase enzyme (Tyagi et al. 1992; Kumar et al. 2003; Lesser 2008). Complete loss of nitrogenase activity within 25-40 min of UV-B exposure was reported in several rice-field cyanobacteria (Kumar et al. 2003).

Nitrogen Metabolism in Cyanobacteria

Nitrogen (N) is a primary nutrient whose availability constrains the productivity of the biosphere. Although, the Earth's atmosphere contains vast quantities (~78% by volume) of molecular nitrogen (N_2), it gets incorporated into biosphere only through assimilation by microorganism and plants. Nitrogen can constitute as much as about 11% of the dry weight of a cyanobacterial cell (Wolk 1973). All the cyanobacterial species utilize inorganic combined nitrogen *viz.*, NO_3^-, N_2 and NH_4^+ to fulfill their N requirements, but organic sources of N, such as urea, amino acids and some N-containing bases, can also be assimilated by some cyanobacteria. Glutamine synthetase/glutamate synthase (GS/GOGAT) cycle played central position in cyanobacterial N metabolism for ammonium assimilation. Glutamate produced in this pathway is not only the major N donor for the biosynthesis of other N-containing metabolites but is itself a precursor for some amino acids and 5-aminolevulinate, the immediate precursor for porphyrin, phycobilin and chlorophyll biosynthesis. Utilization of the different sources of N requires their passage through the cell wall. The cyanobacterial cell wall is like that of the Gram-negative bacteria that bear an outer membrane outside the peptidoglycan layer(s). The outer membrane contains a special type of proteins, known as 'porins,' which allow the passage of small molecules including amino acids and inorganic ions. There is evidence for the presence of such 'porins' in cyanobacteria (Resch and Gibson 1983; Benz and Bohme 1985). Therefore, it is the cytoplasmic membrane that represents the permeability barrier for the uptake of substances that are sources of N, and some of these substances are known to be taken up by the cyanobacterial cell with the aid of specialized transport systems. The convergence at the level of ammonium and the GS/GOGAT cycle of several N assimilation pathways allows the operation of a regulatory network which helps the cyanobacterial cell to economize the process of N acquisition. This system for N-control is currently a major theme of research in cyanobacteria.

The growth of microorganisms in general depends on the availabilities of basic nutrients containing carbon and nitrogen and, to a lesser extent, phosphorus. Although, carbon metabolism plays a vital role in growth and development of most microorganisms, it is also true that nitrogen metabolism is equally important due to the fact that nitrogen is a component of two essential macromolecules, namely, proteins and nucleic acids. Therefore, a modification in nitrogen metabolism could affect cyanobacterial growth, as well as its biochemical composition which would have consequences on the entire food web (Goes et al. 1995). Information on nitrogen metabolism in cyanobacteria in relation to UV-B-induced stress is very limited. In the following sections we shall describe the changes in the enzymes involved in nitrogen metabolism in some cyanobacteria under the influence of UV-B radiation.

Nitrogen Fixation

Nitrogen fixation is a process by which nitrogen in the Earth's atmosphere is converted into ammonia (NH_3) or other molecules available to living organisms. Atmospheric nitrogen or molecular dinitrogen (N_2) is relatively inert. It does not easily react with other chemicals to form new compounds. The fixation process frees nitrogen atoms from their triply bonded diatomic form, N≡N, to be used in other ways. These prokaryotes include aquatic organisms, such as free-living and symbiotically associated cyanobacteria. Morphologically, cyanobacteria can be grouped into various categories (Table 1). In non-heterocystous cyanobacteria, nitrogenase enzyme responsible for catalysing conversion of molecular nitrogen to ammonium is expressed in all the cells while in heterocystous forms, the enzyme is located in heterocysts. Nitrogen fixation occurs in specialized cells called "heterocysts" that differentiate at regular intervals in a filament in response to an environment that is deficient in fixed nitrogen sources such as ammonium, nitrate, nitrite, urea and some amino acids (Wolk et al. 1999; Zhang et al. 2006; Kumar et al. 2010; Herrero et al. 2013). Heterocysts, which comprise 5-10% of the cells in a filament, have a glycolipid layer that may restrict oxygen diffusion into the cell, lack oxygen-evolving photosystem II activity, and have increased respiration, all of which serve to protect nitrogenase from oxygen (Murry et al. 1984; Walsby 1985, 2007; Murry and Wolk 1989; Valladares et al. 2003).

One of the characteristics of heterocystous cyanobacteria is that they enter into symbiotic association with other plants (Table 2). In symbiosis, the demand for fixed nitrogen is far higher than in free-living state as the cyanobiont is almost entirely responsible for provision of the total fixed nitrogen requirement of the whole symbiosis. To meet such a demand, cyanobacteria in symbiosis differentiate into much higher heterocyst frequency (in some cases up to 80%). In free-living conditions, nitrogenase activity of cyanobacterium shows a linear increase with the increase in heterocyst frequency. Increase in heterocyst frequency of cyanobiont depends on its role in symbiosis and on availability of carbon nutrition in the form of fixed carbon while in symbiosis. In those symbioses where the cyanobiont transfer fixed-N to its partners(s) and in return receive fixed-C, the heterocyst frequency is highest. Cyanobionts supplying fixed-N but not receiving any fixed-C in return come next followed by the cyanobionts that supply both fixed-N and fixed-C to its partner(s). In the last case, the heterocyst frequency of the cyanobionts is almost similar to its free-living counterpart. Microscopic studies have revealed that double and multiple heterocysts occur beyond a heterocyst frequency of 30%. Studies have revealed that nitrogenase activity of a cyanobiont shows a linear relationship with increase in single heterocyst (as in the case of free-living cyanobacteria) rather than to the total heterocyst frequency. In the case of multiple heterocysts, the heterocysts in the middle of a row are not in contact with any vegetative cells at all and those at the end of the chain are connected only to one vegetative cell as against single heterocysts that are connected to two vegetative cells directly on both sides. Hence, the availability of proper amount of photosynthate is a predisposed criterion for efficient nitrogenase activity. The whole focus of the cyanobiont in symbiosis is towards higher create of nitrogen fixation which creates a much higher demand for fixed-C as a source of energy and reductant for nitrogenase activity (Rai 1998). The reductant in the heterocyst for N_2-fixation is ferredoxin, although under iron deficient condition flavodoxin is used as a substitute for ferredoxin (Gallon and Chaplin 1987). Heterocyst lack photosystem II activity and hence there is no generation of reductant from water in the non-cyclic electron transport. Generation of reductant by cyclic electron transport involving light and photosystem I is possible, provided that there is a suitable source of electron. H_2, *via* the action of hydrogenase is a possible source of reductant for

N_2-fixation in the heterocyst. Apart from that, by losing photosystem II, heterocysts protect nitrogenase from O_2 exposure that irreversibly inactivates nitrogenase.

Table 1. Nitrogen-fixing cyanobacterial species

Type	Representative species
A. Unicellular species showing budding or binary fission:	
(i) All strains are aerobic N_2-fixers	*Gloeothece, Synechocystis*
(ii) Some strains are aerobic N_2-fixers	*Synechococcus, Cyanothece*
(iii) Microaerobic/anaerobic N_2-fixers	*Gloeocapsa*
B. Unicellular species showing multiple fission: All strains show microaerobic/anaerobic N_2-fixation	*Dermocarpa, Pleurocapsa, Chroococcidiopsis, Xenococcus, Myxosarcina*
C. Nonheterocystous filamentous species: (i) Microaerobic/anaerobic N_2-fixers (ii) Some strains capable of aerobic fixation	*Plectonema, Lyngbya, Phormidium,Pseudoanabaena Oscillatoria, Microcoleus, Trichodesmium*
D. Heterocystous filamentous anaerobic N_2-fixers dividing in one plane	*Anabaena, Nostoc, Scytonema, Cylindrospermum, Nodularia, Calothrix*
E. Heterocystous filamentous anaerobic N_2-fixers dividing in more than one plane	*Chlorogleopsis, Fischerella*

Table 2.Nitrogen-fixing symbioses involving cyanobacterial species

Symbiotic association		Cyanobacteria	Host (plants)
A	Lichen (lichen fungus, heterocystous cyanobacteria (± green alga)	*Nostoc, Anabaena, Calothrix*	*Collema, Lichina, Peltigera, Placopsis, Lobaria, Stereocaulon, Leptogium, Ephebe*
B	Liverwort-cyanobacteria	*Nostoc*	*Blasia, Cavicularia*
C	Fern-cyanobacteria	*Anabaena*	*Azolla*
D	Cycad-cyanobacteria	*Nostoc, Anabaena*	*Macrozamia, Encephalartos Ceratozamia*
E	Angiosperm-cyanobacteria	*Nostoc*	*Gunnera*
F	Sponge-cyanobacteria	*Oscillatoria*	*Lamellodysidea*

Heterocyst Development

The differentiation of heterocysts from vegetative cells is nitrogen regulated process and is accompanied by morphological, biochemical and genetic changes (Wolk et al. 1994).

The presence of externally available fixed nitrogen sources represses heterocyst differentiation. In cyanobacteria, 2-oxoglutarate, an intermediate in the Krebs cycle, constitutes the signal for nitrogen deprivation (Laurent et al. 2005; Zhang et al. 2006). The Krebs cycle in cyanobacteria is incomplete because of the lack of 2-oxoglutarate dehydrogenase. As a result, 2-oxoglutarate's main function is to serve as a precursor in a variety of biosynthetic reactions. It is the primary carbon skeleton for incorporation of ammonium and is considered the metabolic junction between carbon and nitrogen balance in cyanobacteria (Muro-Pastor et al. 2001; Vazquez-Bermu´dez et al. 2003; Muro-Pastor et al. 2005). Nitrogen limiting conditions result in an increase in the levels of 2-oxoglutarate. NtcA, a transcriptional regulator belonging to the CRP (cyclic AMP receptor protein) family of proteins, senses 2-oxoglutarate levels. The protein is conserved in all cyanobacteria and regulates a number of genes involved in carbon and nitrogen metabolism (Herrero et al. 2004; Marcozzi et al. 2009). In *Anabaena* PCC 7120, NtcA is required for the expression of the genes in pathways for ammonium and nitrate transport and assimilation, dinitrogen fixation and heterocyst development (Wei et al. 1993; Ramasubramanian et al. 1994). The DNA binding activity of NtcA is enhanced in the presence of 2-oxoglutarate, and 2-oxoglutarate is necessary for transcriptional activation by NtcA (Tanigawa et al. 2002; Vazquez-Bermu´dez et al. 2002). HetR is a master regulator of heterocyst development and plays a key role in differentiation and pattern formation (Buikema and Haselkorn 1991). *hetR* is one of the earliest genes induced in differentiating cells and is positively autoregulated. HetR protein is a serine type protease that has autoprotease activity and DNA binding activity that requires formation of a HetR homodimer (Zhou et al. 1998; Huang et al. 2004). The *patS* gene encodes a small peptide inhibitor of heterocyst differentiation and plays an important role in the control of pattern formation (Yoon and Golden 1998, 2001).The hetN is expressed primarily in heterocysts and suggesting that it is involved in the maintenance of heterocyst pattern by suppressing the formation of new heterocysts adjacent to previously formed heterocysts (Callahan and Buikema 2001). Expression of *ntcA* and *hetR* show a mutual dependency during heterocyst development (Muro-Pastor et al. 2002). The *nrrA* gene, showed increased expression of *hetR*, and thus, increased heterocyst frequency (Ehira and Ohmori 2006a, 2006b).The *hetF* and *patA* influences heterocyst development by a positive effect on *hetR* expression. The genes, *hepA*, *hepB* and *hepC* are required for the synthesis or stabilisation of the polysaccharide layer (Wolk et al. 1988; Holland and Wolk 1990; Zhu et al. 1998). The *hglB, hglC, hglD,* and *hglE* genes along with a cluster of nearby genes are required for the synthesis of these glycolipids (Campbell et al. 1997; Fan et al. 2005). DevH, a trans-acting regulatory protein, is required for the formation of the glycolipid layer, either by directly regulating the expression of the genes or indirectly through other gene products (Ramirez et al. 2005).

Biochemistry and Genetic Regulation of N_2 Fixing Cyanobacteria

The basic enzyme complex nitrogenase that catalyses N_2 fixation is common to all N_2 fixing organisms. Kulasooriya (2008) explained the biochemistry including the structure of the nitrogenase enzyme complex and proposed the mechanism of nitrogen reduction and its genetic regulation. The enzyme nitrogenase which functions exclusively in heterocysts under oxic conditions is a conventional Mo-dependent nitrogenase (Nif 1) (Elhai and Wolk 1990; Theil et al. 1995) made up of two different proteins (Table 3). The Fe-protein (dinitrogenase reductase), which is a dimer having a molecular weight ranging from 62,000 to 64,000 daltons. The other component of nitrogenase complex, Mo-Fe protein (dinitrogenase) is a tetramer having a molecular weight ranging from 220,000 to 230,000 daltons. The latter also possess the Fe-Mo-co-factor, which has no peptide residues and has inorganic constituents with an approximate stoichiometry of $Fe_6Mo_4S_4$. The N_2 molecule gets attached to the site of the co-factor and undergoes stepwise reduction through the enzyme bound intermediates N_2H_2 and N_2H_4 to finally produce the enzyme free key intermediate NH_3. This enzyme free end product NH_3 is released in cell free extracts and *in vitro* demonstrations using purified nitrogenase, but immediately incorporated in the intact organisms into the amino acid glutamine by the glutamine synthetase and L-glutamine 2-oxo glutarate 2-amino transferase (glutamate synthase) (GS/GOGAT) pathway. These basic biochemical mechanisms of nitrogen fixation were extensively studied and initially demonstrated in the anaerobic bacterium *Clostridium pasteurianum* and the facultative anaerobe *Klebsiella pneumonia,* but have been found to be similar in all N_2 fixing organisms including the cyanobacteria. Among cyanobacteria the common experimental species studied have been *Anabaena variabilis* and *Anabaena* (*Nostoc*) PCC 7220. A generalized scheme has been accepted for all N_2 fixing microorganisms with minute variations between the major groups such as the proteobacteria and the cyanobacteria. Besides the 2-component nitrogenase complex, other ancillary reactions are needed for the complete N_2 fixation process. It requires Mg^{2+} ions to activate the ATP and accept electrons (reducing power) transported by electron carriers ferredoxins and flavodoxins. This MgATP complex binds to the Fe-protein, which makes it highly reductive and energized. This complex then binds to the Fe-Mo-co centre of the Mo-Fe-protein and makes this site highly energized and reductive. Energy and reducing power (electrons) are accepted by the Mo-Fe-protein only from the reduced and energized Fe-protein and no other source. The energized and highly reduced Fe-Mo-co centre is now capable of reducing triple bonded substrates such as N_2, C_2H_2, CN, iso-CN, azides etc. In an extensive review article Bothe et al. (2010) reported more recent information on the mechanisms of N_2 fixation and H_2 metabolism in cyanobacteria including their genetic regulation. According to this article the pattern of pyruvate degradation, ammonium and H_2 formation by nitrogenase, and H_2 uptake by hydrogenases is not only typical of strict and facultative anaerobic bacteria, but also proceeds in cyanobacteria. It has been reported that a transitional 2:1 state is formed between the two nitrogenase components with the larger Mo-Fe component in the centre and one molecule of the smaller component (Fe-protein) attached at each end. It is the prosthetic group (4Fe-4S) cluster that bridges the sub-unit interface and is ligated by 2 cysteinyl residues from each sub-unit. This cluster accepts reducing equivalents from the electron carriers and also binds MgATP/MgADP. The overall reaction of the reduction of one molecule of nitrogen to produce two molecules of ammonia can be depicted by the following equation from Bothe et al. (2010).

$$8H^+ + 8e^- + N_2 + 16MgATP \rightarrow 2NH_3 + H_2 + 16MgADP + 16Pi$$

Table 3. Some molecular and kinetic properties of enzymes involved in N assimilation in cyanobacteria

Enzyme	Molecular weight	Subunit mol wt	Prosthetic groups	Physiological substrates
Nitrogenase:				
Dinitrogenase reductase	66,000	33,000	$[Fe_4S_4]$ cluster	Fd ATP
Dinitrogenase	226,800	55,800 57,600	4 $[Fe_4S_4]$ clusters 2 Mo-Fe cofactors	N_2
Nitrate reductase	75,000	75,000	Mo cofactor [Fe-S] cluster(s)	Nitrate Fd Flavodoxin
Nitrite reductase	56,500	56,500	Siroheme $[Fe_4S_4]$ cluster	Nitrite Fd Flavodoxin
Glutamine synthetase	600,000	50,000		Ammonium Glutamate ATP
Glutamate synthase	156000	156000	FMN [Fe-S] c1uster(s)?	Glutamine 2-oxoglutarate Fd
$NADP^+$-Glutamate dehydrogenase	208,000	46,000		Ammonium 2-oxoglutarate NADPH
NAD^+-Glutamate dehydrogenase	295,000	48,500		Ammonium 2-oxoglutarate NADH

Genetic Regulation

The two identical subunits of Fe-protein (dinitrogenase reductase) are coded by *nifH* gene and α and β subunits of Mo-Fe protein (dinitrogenase) are encoded by *nifD* (Lammers and Haselkorn 1983; Golden et al. 1985) and *nifK* (Mazur and Chui 1982) genes, respectively (Figure 1). In addition to these three structural genes *nifHDK*, nitrogenase expression requires a total of 20 genes in the enterobacterium *Klebsiella pneumoniae* and all of them are contiguously located on the chromosome. Before nitrogenase can be expressed in cyanobacteria a gene rearrangement has to occur within the *nifD* (Bothe et al. 2010). An 11kb DNA element is excised by a specific enzyme XisA and the two fractions of *nifD* are ligated to permit nitrogenase transcript formation. The gene *xisA* is located on the excised DNA element. This gene re-arrangement occurs in heterocystous cyanobacteria such as *Anabaena variabilis* and *Anabaena* (*Nostoc*) PCC 7120, but not in non-heterocystous N_2 fixing forms. Similar re-arrangements have also been reported in the late stages of heterocyst development of certain other cyanobacteria. Under aerobic conditions, *Anabaena variabilis* ATCC 29413 has also been shown to possess another Mo-dependent nitrogenase (Nif 2) functioning in the vegetative cells (Theil et al. 1995; Theil and Pratte 2001). Two other alternative nitrogenases, one vanadium-dependent nitrogenase encoded by *vnfDGK* genes and another Fe-only-dependent-nitrogenase have also been reported in *Anabaena variabilis* (Theil et al. 1993). However, oxygenic photosynthesis and nitrogen fixation are incompatible processes because nitrogenase is inactivated by oxygen. Cyanobacteria mainly use two mechanisms to separate these activities: a biological circadian clock to separate them temporally, and multicellularity and cellular differentiation to separate them spatially. For example, the unicellular *Cyanothece* sp. strain ATCC 51142 stores glycogen during the day and fixes nitrogen at night (Toepel et al. 2008), whereas the filamentous *Trichodesmium erythraeum* IMS101 fixes nitrogen during the day in groups of specialized cells (Sandh et al. 2009). Heterocyst-forming cyanobacteria differentiate highly specialized cells to provide fixed nitrogen to the vegetative cells in a filament. Heterocysts and vegetative cells are mutually interdependent. Because they lack photosystem II and carbon fixation, heterocysts are dependent on vegetative cells for a source of reductant and carbon, which is probably partially supplied as sucrose (Cumino et al. 2007; Marcozzi et al. 2009). In *Anabaena* PCC 7120, vegetative cells must also supply glutamate to heterocysts, which convert it to glutamine and other amino acids (Martin-Figueroa et al. 2000). In return, vegetative cells obtain fixed nitrogen in the form of amino acids from the heterocysts (Meeks and Elhai 2002). The temporal and spatial distributions of fixed carbon and nitrogen were studied using high-resolution nanometer scale secondary ion mass spectrometry (NanoSIMS), in *Anabaena oscillarioides* (Popa et al. 2007). Newly fixed nitrogen is rapidly exported from heterocysts and distributed to nearby vegetative cells.

In a more recent publication Pratte and Thiel (2014) have provided more evidence on the regulation of nitrogenase gene expression by transcript stability in the cyanobacterium *Anabaena variabilis.* They proposed that the regulation of nitrogenase expression in cyanobacteria is different from that in the proteobacteria. An active nitrogenase requires the products of many genes and several of them are needed for the synthesis and insertion of the Fe-Mo-cofactor. In cyanobacteria they include a single cluster of *nifB, fdxN, nifS, nifU, nifH, nifD, nifK, nifE, nifN, nifX* and *nifW* as well as *hesAB, fdxH* and four open reading frames within the cluster. They also re-state the removal of the 11kb excision factor (Bothe et al. 2010) towards the later stages of heterocyst development. Pratte and Thiel (2014) refer to earlier reports of the *nif* cluster in *Anabaena* having several operons with transcription start sites identified upstream of *nifB* and *nifH*. However, these authors were unable to drive a transcription of *lacZ* using a DNA fragment that included the entire *nifU1-nifH1* intergenic region of *A. variabilis*. They showed that the 5' end of the *nifH1* transcript is a processed transcript rather than a transcription initiation site, and there is no promoter upstream of *nifH1*. In contrast *nifB1* has a promoter that initiates transcription and there is a promoter within the coding region of *nifU1*, which has been verified by transcriptome sequencing (RNA sequencing). They show that the *nif1* cluster in *A. variabilis* from *nifB1* through *nifW1* depends primarily on the *nifB1* promoter together with contributions from weak promoters in the *nifU1* and *nifE1* regions.

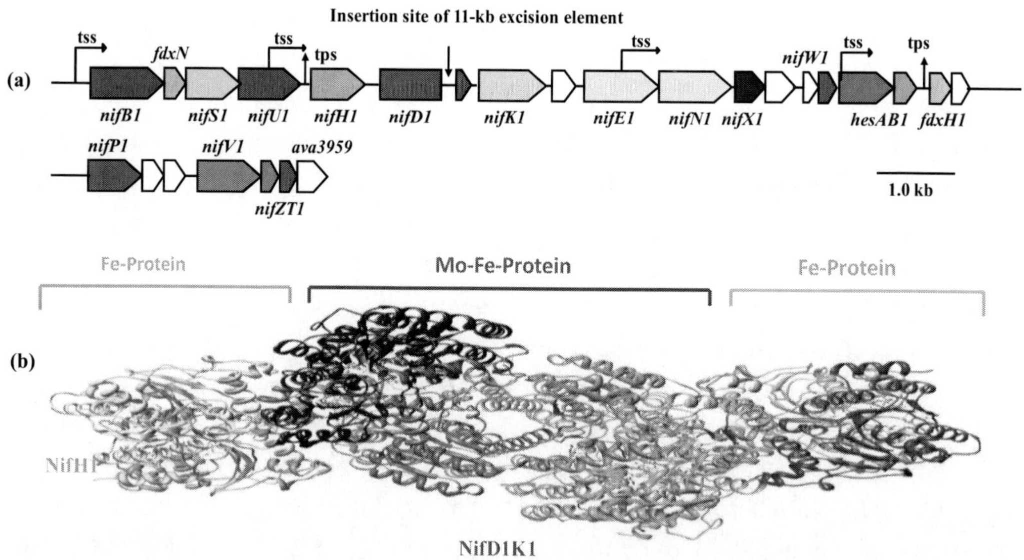

Figure 1. Nitrogenase structure and genes required for its biosynthesis. (a) The nitrogen fixation (nif) gene clusters from *Anabaena variabilis*. The 11-kb excision element in *nifD1* is not shown. tss, transcription start site; tps, transcriptional processing site. White ORFs indicate proteins of unknown function. (b) Structure of the nitrogenase enzyme complex (PDB code 1n2c) showing the MoFe and Fe protein components.

Cyanobacteria do not synthesize nitrogenase when grown in the presence of nitrate or ammonium, and the genetic regulation of nitrogen metabolism in cyanobacteria is different. According to recent findings, expression of nitrogen assimilation genes *hetR* involved in heterocyst differentiation and *glnA* involved in ammonia assimilation, are regulated by the *NtcA* transcriptional regulator. The analysis of *nifH* sequences is widely used as a measure for the potential of a system to fix dinitrogen and for the diversity of diazotrophic organisms in that ecosystem. Until recently *Microcoleus chthonoplastes* has not been assigned as a diazotrophic organism and none of the strains that were correctly identified as *M. chthonoplastes* has been shown to fix dinitrogen. It has repeatedly been reported that these organisms lacked the cyanobacterial *nifH*, the structural gene for dinitrogenase reductase. However, the study by Bolhuis et al. (2010) has shown the presence of the complete nif-gene cluster in the genome of *M. chthonoplastes* PCC 7420 and that the three structural nitrogenase genes *nifHDK*, are present in a collection of axenic strains of *M. chthonoplastes* from distant locations. Researchers have identified the areas in which future studies needs to be concentrated on, such as studying the transcriptional regulators, genes, and signalling pathways i.e., the genetics, their physiology, biochemistry and ecology not only in heterocyst forms, but also in non-heterocystous and in symbiotic cyanobacteria. According to some researchers, one of the largest obstacles in determining the overall importance of symbiotic associations to the global cycling of nitrogen is the lack of consistent rate measurements. Certain studies have combined ^{15}N isotope rate measurements with quantitative PCR approaches, which is a promising path forward. Although molecular and microscopic characterization of nitrogen symbioses has helped to elucidate the diversity and distribution of symbioses, future work should target a consistent approach to relate these to their biogeochemical importance. Another difficulty faced by scientists today is that although it is possible to see the expressions under natural conditions, they have been unable to express these functions in laboratory strains.

Effects of UV-B on Heterocyst Differentiation and Nitrogen-Fixing Enzyme Nitrogenase

Cellular differentiation, i.e., differentiation of vegetative cells into heterocysts and akinetes in filamentous cyanobacteria has been reported to be adversely affected by UVR. Exposure of cells to UV-B radiation has been shown to delay the differentiation of vegetative cells into heterocysts and akinetes in *Anabaena aequalis* (Blakefield and Harris 1994). In *Anabaena* sp. PCC 7120 differentiation of heterocysts has also been reportedly suppressed by solar UVR (Gao et al. 2007) and alteration in the C:N ratio following UV-B exposure was suggested to be responsible for the altered spacing pattern of heterocysts in the filament (Sinha et al. 1996). The spiral filaments of *Arthrospira platensis* have recently been reported to be broken and compressed under solar UVR (Wu et al. 2005; Gao et al. 2008). The filaments of the cyanobacterium *Anabaena flos-aquae* were also broken by UVR while PAR had no effect (Singh et al. 2010). In addition, major heterocyst polypeptides of around 26, 54 and 55 kDa have been shown to be decreased in concentration following UV-B irradiation, suggesting that the multilayered thick wall of heterocysts may be disrupted resulting in the inactivation of the nitrogen fixing enzyme nitrogenase by the penetrating oxygen (Sinha et al. 1996). UV-B-induced membrane disruption, leading to changes in membrane permeability and release of ^{14}C-labelled compounds, has been observed in a number of cyanobacteria (Sinha et al. 1997).

UV-B-induced inactivation of the nitrogen fixing enzyme nitrogenase has been reported in many cyanobacteria (Kumar et al. 1996; Sinha et al. 1996). The process of nitrogen fixation has also been reported to be depressed during midday, when visible and UV irradiance are higher(Peterson et al. 1979). Nitrogenase is a molybdoenzyme and requires ATP and reductants for its activity. There is a possibility that the inhibition of the nitrogenase activity might be due to reduced supply of reductants and ATP following UV treatment. However, the loss of reductants and ATP does not occur immediately in cyanobacteria since many N_2 fixing species are capable to drive nitrogenase activity at the cost of an endogenous pool. It has been suggested that UV irradiation causes inactivation/denaturation of the nitrogenase enzyme and this appears to be a novel phenomenon(Kumar et al. 1996; Sinha et al. 1996).

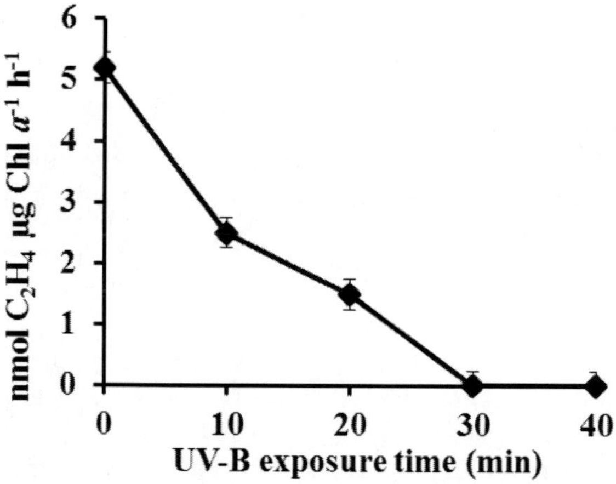

Figure 2. Effect of UV-B radiation on nitrogenase activity of *Nostoc calcicola.* An actively N_2-fixing culture was exposed to UV-B (5 W/m^2) for the indicated time period and thereafter a C_2H_2 reduction assay was performed under fluorescent light. The untreated control cultures showed an activity of 5.2 nmol C_2H_4 µg Chl a^{-1} h^{-1} throughout the period of assay (Adapted from Kumar et al. 1996).

The effects of UV-B irradiation on nitrogenase activity have been studied in the filamentous and heterocystous N_2-fixing cyanobacterium *Nostoc calcicola* (Kumar et al. 1996). Exposure of cultures to UV-B irradiation (5 Wm^{-2}) for as little as 30 min caused complete inactivation of nitrogenase activity (Figure 2). Inhibition of nitrogenase activity by UV-B treatment has also been reported in *Anabaena flos-aquae, Nostoc spongiaeforme, Cylindrospermum* sp. and *Anabaena doliolum* (Newton et al. 1979; Tyagi et al. 1992; Rai et al. 1998; Chris 2012).

Elevated UV-B stress inhibited nitrogenase enzyme activity by 21-30% in roots and by 33-52% in root nodules respectively, was reported in *Vigna mungo* L. var. KM-2 (Sudha and Rajendiran 2014). Similar inhibition of nitrogenase enzyme activity after UV-B exposure was also reported in *Vigna radiata* (L.) Wilczek var. KM-2 (Rajendiran and Ramanujam 2006) and *Sesbania grandiflora* (L.) Pers (Sudaroli Sudha and Rajendiran 2013). The reduction in nitrogenase activity was reported in *Vigna radiate* and *Phaseolus mungo* (Singh 1997). The molecular basis for many of the observed changes, including suppression of nodulation and nitrogenase activity, following UV-B exposure remains unknown (Caldwell et al. 1995).

Perceptible decline in nitrogenase activity was observed for all the three *Anabaena* species, this being maximum in *Anabaena doliolum*, followed by *Anabaena* sp. PCC 7120 and minimum in *Anabaena* L31. (Srivastava et al. 2015). Exposure of cyanobacterial cells to UV-B radiation has been found to delay differentiation of vegetative cells into heterocyst. This is in agreement with the reduction in nitrogenase activity and a severe fall in abundance of proteins required for heterocyst formation which looks apparent in *Anabaena doliolum* which is sensitive towards UV-B radiation.

Exposure of the N_2-fixing cyanobacterium *Anabaena* BT2 to UV-B radiation (2.5 Wm^{-2}) for 30 min resulted in complete loss of nitrogenase activity but 100% cell killing occurred only after a 90 min exposure. Inactivation of nitrogenase activity was not specific to *Anabaena* BT2; other species also showed a similar effect (Table 4). The effects of UV-B radiation on nitrogenase activity indicated an intensity and time-dependent inhibition in UV-B irradiated *S. platensis* 794 (Xue et al. 2011). The time required for 100% killing and inactivation of nitrogenase activity differed in various species, and this difference may be ascribed to the presence of different levels of UV-B protection mechanisms in individual species (Castenholz 1997; Quesada et al. 1999; Sinha et al. 2001). *Scytonema* sp. contains a yellow-brown, lipid-soluble dimeric pigment, scytonemin, located in the extracellular sheath, which has been proposed to serve as a UV-B sun-screen compound (Castenholz 1997; Quesada et al. 1999; Pathak et al. 2017a, 2017b, 2017c). Similarly, many species are known to protect themselves against UV-B damages by synthesizing UV-B-absorbing substances such as MAAs (Sinha et al. 2001; Richa and sinha 2015; Pathak et al. 2017a). It is pertinent to mention here that marine bacterial isolates also display diverse responses to UV-B radiation (Joux et al. 1999). The complete killing of cells after exposure to UV-B radiation might be due to inactivation of a number of processes (Tyagi et al. 1992; Sinha et al. 1996; Sinha et al. 1997; Sinha and Häder 2000). Disintegration of phycobilisomes, followed by release of phycobiliproteins in the medium by exposure of UV-B (2.5 Wm^{-2} or above) for 2-3h to cultures of cyanobacteria, has been reported earlier (Tyagi et al. 1992; Sinha et al. 1996).

Such cultures fail to revive upon transfer to fluorescent light and fresh medium. Inhibition of nitrogenase activity was immediate, since exposure of cultures to UV-B for as little as 5 min elicited some inhibition of activity. The activity of UV-B-inhibited nitrogenase did not recover upon transfer of exposed cells to fluorescent light, suggesting that the inhibition may be due to specific inactivation of the enzyme. By employment of inhibitors of protein synthesis and PS-II activity, it was demonstrated that restoration of nitrogenase activity in a UV-B-treated culture occurred by fresh synthesis of nitrogenase polypeptide. Complete loss of nitrogenase activity seems surprising, because survival of all the test organisms was not affected significantly by 30 min of UV-B treatment.

Table 4. Test for the inactivation of nitrogenase activity in different cyanobacteria following exposure of UV-B radiation (Adopted from Kumar et al. 2003)

Organisms	Time (min) required for the inactivation of nitrogenase activity
Anabaena BT2	30
A. doliolum	30
A. cycadeae	25
Calothrix sp.	40
N. spongiaeforme	30
N. muscorum	25
N. commune	30
Scytonema sp.	40

The virtually instant inhibition induced by UV-B, followed by complete loss of nitrogenase activity, prompted us to seek answers to two questions: a) is the loss of activity due to complete inactivation of nitrogenase polypeptide and/or some conformational changes b)do the reductant and ATP limit the activity? To answer the above hypotheses, the cyanobacterium *Anabaena* BT2 was employed, and UV-B-mediated effects were tested in the presence of inhibitors of protein synthesis and photosystem-II. Decrease in nitrogenase activity may also be due to impact of UV-B in their morphology and heterocyst formation (Gao et al. 2007). That nitrogenase is indeed inactivated/denatured is evident from the fact that restoration of C_2H_2 reduction activity upon transfer of *Anabaena* BT2 to fluorescent light does not occur instantly, but after about 6h. This suggests that cells either do not contain active nitrogenase or, if some enzyme is present, it does not react with its substrate. The activity detectable after 6h appears to be due to *de novo* synthesis of the nitrogenase complex. This is further supported by the data on the effects of chloramphenicol on protein synthesis. Addition of chloramphenicol to UV-B-exposed cells did not restore nitrogenase activity

(Some nitrogenase activity also declined in UV-B unexposed controls treated with chloramphenicol, but complete loss occurred only after 8-9h). Thus, if active nitrogenase was present in UV-B-exposed material, there should have been perceptible but constant activity at least during the initial period of chloramphenicol treatment. Chloramphenicol-treated cultures revived on transfer to inhibitor-free medium, indicating that the concentration of inhibitor used was non-toxic. Delayed appearance of nitrogenase activity also appears to rule out the alternative possibility of conformational change in the enzyme complex. Nitrogenase requires a continued and abundant supply of suitable reductant and ATP (Halbleib and Ludden 2000; Wolk 1992) for conversion of N_2 to NH_3. We presumed that UV-B might impair the synthesis and/or supply of the reductant. To test this hypothesis, we treated the cells with DCMU (3-(3, 4-dichlorophenyl)-1,1-dimethylurea) and a readily utilizable reductant source. Cells treated with DCMU (specific inhibitor of photosystem-II) or incubated in the dark, showed complete loss of nitrogenase activity after 6 and 8 h respectively. In neither case was there an instantaneous loss; rather, the activity declined gradually. When such cultures were transferred to light in DCMU-free medium, the activity started to reappear and by 3h was almost comparable to that in control cultures. In contrast, the UV-B-exposed material failed to restore activity, even with glucose treatment, up to at least 3h. Failure of activity restoration with glucose may be due to the fact that nitrogenase enzyme is probably not present in the cells. The data suggest that DCMU treatment or dark preincubation does not inactivate a presynthesized nitrogenase complex. Furthermore, loss of reductant and ATP supply does not occur immediately under the above conditions, and the time required for the above loss may differ among different species (Kumar et al. 1982; Meeks 1998). These observations suggest that loss of nitrogenase activity after UV-B treatment is not related to the instant impairment of reductant/ATP supply. It is pertinent to mention here that a number of cyanobacteria have been reported to drive nitrogenase activity in the dark or, in the absence of functional PS-II, at the expense of endogenous reductant/ATP for short durations (Meeks 1998). Both transcriptional and post-translational modes of regulation of nitrogenase are known (Halbleib and Ludden 2000). Inhibition of nitrogenase activity by several factors, especially the utilizable inorganic nitrogen sources, has been reported in a number of N_2-fixing microorganisms (Stewart 1980; Halbleib and Ludden 2000). However, these inhibitors do not cause immediate loss of *in vivo* nitrogenase activity, and inhibition does not occur with the purified enzyme *in vitro*. Instant and complete inactivation of nitrogenase both under *in vivo* and *in vitro* conditions has been demonstrated after exposure to O_2, but only at a concentration higher than the ambient level (Fay 1992; Stewart 1980). Inactivation of nitrogenase by ADP-ribosylation of Fe-protein involving the roles of the NAD-dependent enzymes, dinitrogenase reductase ADP-ribosyltransferase (DRAT) and its partner, dinitrogenase reductase-activating glycohydrolase (DRAG) after UV-B treatment does not seem possible, since the inactivation of nitrogenase is not reversible (Halbleib and Ludden

2000). Thus, inactivation of nitrogenase by UV-B appears to be a novel phenomenon. Nitrogenase is very sensitive to UV-B exposure, and inactivation of activity is possibly due to complete damage to the nitrogenase polypeptide. Most probably the amino acids present or the general structure of nitrogenase makes it more susceptible to UV-B radiation. The inhibition of nitrogenase activity by UV-B is by no means specific to the members of cyanobacteria; a similar effect on two other N_2-fixing bacteria has also been observed. Thus, estimation of nitrogenase activity both under laboratory and field conditions may be used as a marker enzyme for assessing the impact of UV-B radiation.

Assimilation of Ammonium

Ammonium assimilation is the process by which ammonium is incorporated into carbon skeletons. Ammonium is the most reduced inorganic form of nitrogen, however, nitrogen is commonly found in nature in less reduced forms as nitrate, nitrite, and dinitrogen. In addition, organic compounds such as urea, amino acids, etc. are also found in certain ecosystems. In general, all these forms of nitrogen require their reduction to ammonium before their incorporation into carbon skeletons, through processes that need reducing equivalents and energy. For this reason, ammonium is the preferred nitrogen source and the presence of this compound plays a crucial role in the control of nitrogen assimilation processes.

Ammonium Uptake

Although ammonia appears to readily permeate biological membranes, ammonium transporters have been characterized in numerous organisms. They are monocomponent permeases which seem to be necessary for uptake of ammonium when it is present at low concentrations (below 1mM) in the extracellular medium or when the organisms grow in a rather acidic medium (which is not the case for cyanobacteria).Three genes encoding ammonium permeases are found in the chromosome of *Synechocystis* sp. strain PCC 6803 (Kaneko et al. 1996; Montesinos et al. 1998). The three genes are expressed at the highest levels when *Synechocystis* cells are incubated in a medium lacking a source of nitrogen, but one of them, *amt1*, is responsible for most of the [^{14}C] methylammonium transport activity when this substrate is provided at micromolar concentrations (Montesinos et al. 1998) or produced intracellularly from the metabolization of others nitrogen sources, is incorporated into carbon skeletons mainly through the sequential action of two enzymes, glutamine synthetase (GS) and glutamate synthase (GOGAT), in a cycle commonly known as the GS-GOGAT pathway.

Ammonium Assimilation Pathways

In general, there are two major pathways by which ammonium can be incorporated into organic constituents of the cell: directly into glutamate by means of glutamate dehydrogenase (GDH; EC 1.4.1.3) or through the sequential operation of two enzymes, glutamine synthetase (GS; EC 6.3.1.2) and glutamate synthase (GOGAT; EC 2.6.2.53), in a cycle commonly known as the GS-GOGAT pathway (Florencio and Reyes 2002; Schubert 1986; Day et al. 2001). This was confirmed by studies using $^{15}N_2$ (Ohyama et al. 1978) and $^{13}N_2$, a short-lived radioactive isotope of N (Meeks et al. 1978), which demonstrated that NH_4^+ was first incorporated into the amide position of glutamine in the reaction catalysed by GS, and that the amide group was subsequently transferred to the 2-carbon of oxoglutarate in the reductive amination reaction carried out by GOGAT. GDH catalyzes the reductive amination of 2-oxoglutarate to glutamate. However, assimilation of ammonium through GDH seems to be quantitatively of low importance under normal growth conditions. GS catalyzes the ATP dependent ligation of glutamate and ammonium rendering glutamine, and GOGAT transfers the amino group of glutamine to 2-oxoglutarate rendering two glutamate molecules. The concerted action of GS and GOGAT yields one molecule of glutamate per molecule of ammonium assimilated.

The Glutamine Synthetase/Glutamate Synthase Pathway

The notion that ammonium assimilation in cyanobacteria takes place primarily *via* the GS/GOGAT cycle originated from the observation that, whereas GDH activity is very low or even undetectable in some cyanobacteria (Hoare et al. 1967; Neilson and Doudoroff 1973), high activity levels of GS could be found in *Anabaena cylindrica* (Dharmawardene et al. 1972). In cyanobacteria, ammonium, either taken up from the outer medium or produced intracellularly, is incorporated into carbon skeletons mainly through the glutamine synthetase-glutamate synthase cycle (Flores and Herrero 1994; Zhou et al. 1998). However, it was the experimental approach set up by Wolk and co-workers which firmly established the operation of the GS/GOGAT cycle as the major pathway for the assimilation of ammonium in cyanobacteria.

Glutamine Synthetase and the *glnA* Gene

GS can account for about 0.5-2% of the total cell protein. The enzyme has been purified to electrophoretic homogeneity from a number of cyanobacteria (Orr et al. 1981;

Florencio and Ramos 1985; Blanco et al. 1989; Merida et al. 1990). The *glnA* gene from *Anabaena* sp. strain PCC 7120 encoding GS was cloned by hybridization using the *glnA* gene from *Escherichia coli* as a probe (Fisher et al. 1981) and sequenced (Tumer et al. 1983). Other cyanobacterial *glnA* genes have been cloned either by hybridization using the *Anabaena* sp. strain PCC 7120 *glnA* gene as a probe (Merida et al.1992; Elmorjani et al. 1992) or by complementation of *E. coli* glutamine auxotrophs (Wagner et al. 1993). The *Calothrix* sp. strain PCC 7601 (Elmorjani et al. 1992), *Synechocystis* sp. strain PCC 6803 and *Synechococcus* sp. strain PCC 7002 (Wagner et al. 1993) *glnA* genes encode polypeptides which are 89, 79 and 75% identical, respectively, to the *Anabaena* sp. strain PCC 7120 GS polypeptide. As expected from the structure of the GS protein, the polypeptide deduced from the cyanobacterial *glnA* gene is strongly similar to other prokaryotic (not only eubacterial) GS polypeptides and shows conserved regions which are found in both prokaryotic and eukaryotic GS polypeptides (Elmorjani et al. 1992). In general, the cyanobacterial *glnA* gene is expressed in *E. coli* and complements *E. coli* glutamine auxotrophs. Expression in *E. coli* of the *Anabaena* sp. strain PCC 7120 *glnA* gene takes place from an *'E. coli-like'* promoter which reads as follows: TTGTGC (-35 box) and TAATAT (-10 box; Turner et al. 1983). In *Anabaena, Calothrix* and *Synechocystis* sp., an identical putative ribosomebinding site (AGGAG) that can be operative in *E.coli* is located five nucleotides upstream from the *glnA* ATG start codon.

In enterobacteria, the GS protein is rapidly inactivated by adenylylation upon addition of ammonium to cultures containing a poor supply of N (Magasanik and Neidhardt 1987), but no evidence for regulation by adenylylation of the cyanobacterial enzyme has been found (Rowell et al. 1977; Stacey et al. 1979; Fisher et al. 1981). On the other hand, all the purified cyanobacterial GS proteins are susceptible to inhibition by some amino acids and nucleotides (Stacey et al. 1979; Florencio and Ramos 1985; Blanco et al. 1989; Merida et al. 1990). Cumulative inhibition has been detected with a mixture of aspartate, serine and AMP, indicating independent binding sites for these three compounds (Stacey et al. 1979) while alanine and glycine share a binding site with serine (Orr and Haselkorn 1981). Feedback inhibition by aspartate and alanine (and serine and glycine), and by ADP and AMP, can be considered a feasible *in vivo* regulatory mechanism for cyanobacterial GS. The GS protein found in ammonium-grown cultures of *Synechocystis* sp. strain PCC 6803 is largely inactive (Merida et al. 1990), and the addition of ammonium to nitrate-grown cultures of this cyanobacterium provokes a short-term inactivation of GS (90-95% inactivation in 25 min) (Merida et al. 1991). Interestingly, inactivation of the GS from *Anabaena* sp. strain PCC 7120 in response to ammonium has been observed in a *Synechocystis* sp. strain PCC 6803 recombinant strain bearing a mutated version of its own *glnA* gene and carrying integrated into its genome the *glnA* gene from *Anabaena* sp. strain PCC 7120 (Merida et al. 1992). *Synechocystis* sp. strain PCC 6803 appears to have a regulatory system for GS which is absent, or much less active, in other commonly studied cyanobacterial strains.

Glutamate Synthase

Cyanobacterial GOGAT uses reduced Fd rather than pyridine nucleotides as an electron donor, thus being more similar to that of the higher plant chloroplasts than to the enterobacterial enzyme (Lea and Miflin 1975). Only one cyanobacterial GOGAT has been purified to electrophoretic homogeneity-that from *Synechococcus* sp. strain PCC 6301. It is a flavoprotein containing one molecule of FMN and consisting of only one polypeptide that may also contain non-heme iron. Feedback inhibition by the reaction product, glutamate, might be a regulatory mechanism for the enzyme.

Glutamate Dehydrogenase

In number of cyanobacterial strains, a direct amination of 2-oxoglutarate catalyzed by GDH can take place. Although no GDH activity is detected in *Synechococcus* sp. strain PCC 6301 (Hoare et al. 1967) or in some other strains (Neilson and Doudoroff 1973), $NADP^+$- and NAD^+-dependent GDH activities have been detected in some cyanobacteria, though generally at very low levels $NADP^+$-GDH has been purified from *Synechocystis* sp. strain PCC 6803 (Florencio et al. 1987) and *Phormidium laminosum* (Martinez-Bilbao et al. 1988), and NAD^+-GDH from *Synechocystis* sp. strain PCC 6803 (Chavez and Candau 1992). Some properties of both enzymes are summarized in Table 3. The two enzymes catalyze the amination of 2-oxoglutarate preferentially over the reverse deaminating reaction, which is consistent with them having a role in ammonium assimilation. The *gdhA* gene encoding $NADP^+$-GDH has recently been cloned from *Synechocystis* sp. strain PCC 6803 by complementation of an *E. coli* glutamate auxotroph. Sequence analysis has shown that the *gdhA*-encoded polypeptide is homologous to GDH from other organisms. *Synechocystis* sp. strain PCC 6803 *gdhA* insertional mutants lacking any $NADP^+$ -GDH activity have been generated that are unaffected in their growth characteristics under standard conditions.

Effect of UV-B on Enzymes of Ammonium Assimilation Pathways

Recently, convincing effects of UV-B radiation on enzyme activities have been described for many metabolic pathways. The elevated level of UV-B negatively affects nitrogen metabolism by affecting activities of nitrate assimilating enzymes, nitrate reductase and nitrite reductase and ammonium assimilating enzymes, glutamine synthetase and glutamate synthase in *Phormidium* as well as in higher plants like *Lablab purpureus* (Arulmozhi and Rajendiran 2014; Srivastava et al. 2014). Srivastava et al. (2014)studied the nitrogen metabolism indices in the two paddy field cyanobacteria i.e.,

Nostoc muscorum and *Phormidium foveolarum* against single and combined doses of low and high UV-B irradiation in combination with the low and high doses of the insecticide chlorpyrifos and their simultaneous impact on the specific growth rates. Moreover, nitrogen metabolism may also be affected by substantial decrease in the activity of ammonia assimilating enzymes i.e., GS and GOGAT, under both the stresses. Decline in GS and GOGAT activities may be attributed to oxidative modification of proteins associated with these enzymes, as reported in *Anabaena doliolum* under enhanced UV-B (Rai et al. 1998).

Rai et al. (1998) reported that the exposure of UV-B (Cu) decreased GS activity in *Anabaena doliolum* which may be attributed to direct action of stress factors on the enzyme complexes. In contrast, the exposure of low dose chlorpyrifos and UV-B alone slightly increased GS and GOGAT activities compared to their respective controls. Besides, the exposure of low UV-B together with any of the chlorpyrifos treatments increased GS and GOGAT activities suggesting improved incorporation of NH_4^+ into glutamate compared to chlorpyrifos treatments alone. Contrary to GS and GOGAT activities, the aminating activity of GDH was greatly increased in both organisms after 24h of high dose of chlorpyrifos and UV-B exposures and after 72h it showed a decreasing trend, however GDH activity was still higher than their respective controls. Under high dose of chlorpyrifos and UV-B exposures, increase in GDH activity suggested that it might have compensated for decreased GS and GOGAT activities. In the literature, contradictory opinions for the existence of GDH activity in cyanobacteria are available. Some workers reported that the enzyme is either absent or present only at very low level (Ramos et al. 1982), however, other groups reported that it was responsible for glutamate synthesis in *Anacystis nidulans* (Meeks et al. 1977). Chávez et al. (1999) showed that the presence of GDH activity is a selective advantage in *Synechocystis* sp. under certain conditions such as during non-exponential growth conditions and could be an alternative pathway to the GS-GOGAT cycle for NH_4^+ assimilation. Skopelitis et al. (2006) have reported that under stress conditions when the GS-GOGAT system is not fully operative, increase in GDH activity may relieve pressure of accumulating toxic amounts of NH_4^+ and provide glutamate for the biosynthesis of protective compounds such as proline. Thus, under high dose of chlorpyrifos and UV-B, increase in GDH activity may be a selective advantage of both organisms to sustain NH_4^+assimilation as evidenced by its higher activity after 24h. However, exposure of both organisms to low UV-B did not significantly influence GDH activity compared to their respective controls. But, exposure of low UV-B together with low dose of chlorpyrifos showed a marked decrease in GDH activity in comparison to chlorpyrifos treatments alone. It suggests that the cells grown under low chlorpyrifos and low UV-B exposures showed a much improved primary route of NH_4^+assimilation i.e., GS-GOGAT pathway in comparison to chlorpyrifos treatments alone.

UV-B may also affect the internal pools of amino acids and the protein contents of phytoplankton cells. Döhler (1995) showed that short-term UV-B exposures can reduce the synthesis of amino acids in batch cultures of *Bellerochea yucatanensis*. However, in another study, Döhler (1984) reported an increase of internal pools of amino acids during 2-day UV-B exposures (5h per day) of batch cultures of marine diatoms from the North Sea. Short-term incubations (7-8h) of natural communities dominated by diatoms from the North Pacific Ocean (Goes et al. 1995) corroborated the results of the latter study, since they showed that internal pools of amino acids increase under natural UVR as compared with excluded UVR conditions. Döhler (1984, 1985) reported that a 2-day exposure (5h per day) decreases protein content in batch cultures of several marine diatoms. Goes et al. (1995) reached the same conclusion with diatoms incubated under natural or excluded UV conditions. However, Buma et al. (1996) exposed cultures of three temperate marine diatoms to different levels of UV-B (on short-time scale) and observed either an increase in cellular protein content under realistic levels of UV-B, or the same protein content as compared with cultures non-exposed under higher levels of UV-B.

NITRATE AND NITRITE ASSIMILATION

Nitrate is probably the most abundant source of combined N for cyanobacterial nutrition. The assimilation of nitrate by cyanobacteria involves nitrate uptake into the cell through an active transport system and its intracellular two-step reduction (*via* nitrite) to ammonium sequentially catalyzed by nitrate reductase (NR) and nitrite reductase (NiR). Nitrite, which can also fulfill the N requirement of cyanobacteria, is taken up into the cell and then reduced to the level of ammonium.

Nitrate and Nitrite Uptake

Because nitrate is frequently found in the environment at relatively low concentrations (in the μM range), specific nitrate uptake systems are required to concentrate this nutrient inside the cells before nitrate reduction can take place. High-affinity nitrate and nitrite uptake takes place in different cyanobacteria through nitrate/nitrite permeases, either an ABC-type transporter or a permease from the major facilitator superfamily (MFS). An ATP-binding cassette (ABC)-type transporter constituted by the products of the *nrtA*, *nrtB*, *nrtC*, and *nrtD* genes is involved in nitrate-nitrite uptake by the freshwater cyanobacteria *Synechococcus* sp. strain PCC 7942 and *Anabaena* sp. strain PCC 7120 (Omata et al. 1993; Luque et al. 1994; Cai and Wolk 1997; Frías et al. 1997; Maeda and Omata 1997), whereas a MFS permease mediates nitrate-nitrite uptake in the marine cyanobacteria *Synechococcus* sp. strain PCC 7002 (Sakamoto et al. 1999) and *Trichodesmium* sp. strain WH9601 (Wang et al. 2000). The gene encoding this permease has been named *nrtP* (Sakamoto et al. 1999) or *napA* (Wang et al. 2000). Whereas the ABC-type nitrate-nitrite transporter likely uses ATP as a source of energy (Flores and Herrero 1994), it is unknown whether the NrtP permeases use a gradient of H^+ or Na^+ (Sakamoto et al. 1999). Like some other MFS permeases, NrtP likely contains 12 transmembrane segments in the form of *a*-helices. ABC-type permeases couple ATP hydrolysis to the transport of solutes across cell membranes. ATP binding and hydrolysis drive conformational changes in the membrane spanning domains of the permease that mediate the transport process (Davidson and Chen 2004).ABC-type uptake permeases are highly efficient transporters. Consistently, Ks values for nitrate of cyanobacterial cells bearing ABC-type nitrate permeases are about 1-4 μM for *Synechococcus* sp. strain PCC 7942 (Tischner and Schmidt 1984; Rodriguez et al.1992) and close to 30 μM for *Anabaena* sp. strain PCC 7120 (Meeks et al. 1983).

Figure 3 presents schemes of genomic regions carrying nitrate assimilation genes in some cyanobacteria with sequenced genomes. It is evident that gene clusters with the structure nitrite reductase-nitrate/nitrite permease-nitrate reductase are found in numerous cyanobacteria. In *Synechococcus* sp. strain PCC 7942 and *Anabaena* sp. strain PCC 7120, the transporter-encoding genes are clustered together with the structural genes for nitrite reductase, *nirA* (Luque et al.1993), and nitrate reductase, *narB* (Rubio et al. 1996), constituting an operon, which we call the *nir* operon, with the structure *nirA-nrtABCD-narB* (Cai and Wolk 1997; Frías et al. 1997). This operon is expressed at high levels only when ammonium is not present in the growth medium, with nitrate or nitrite having some positive effect on the operon mRNA levels (Frías et al. 1997). Additionally, closely linked to the nitrate assimilation (nir) operon, a set of genes is usually found that encode proteins which influence nir operon expression or nitrate assimilation activity. These include the NtcB transcription factor, the NirB and NarM proteins that are required for attaining maximum levels of nitrite reductase and nitrate reductase, respectively. In the genome of *Synechocystis* sp. strain PCC 6803, the *nrtABCD* and *narB* genes are clustered together whereas the *nirA* gene is found in a different location. In *Synechococcus* sp. strain PCC 7002, the *nrtP* gene is clustered with *narB* but they appear to be transcribed independently. The expression of both genes is, however, regulated in a similar way, being high in nitrate-containing medium and low in medium containing ammonium or urea (Sakamoto et al. 1999). In *Trichodesmium* sp. strain WH9601, a gene cluster with the structure *nirA-napA*(*nrtP*)-*narB* is found but its expression has not yet been characterized (Wang et al. 2000).

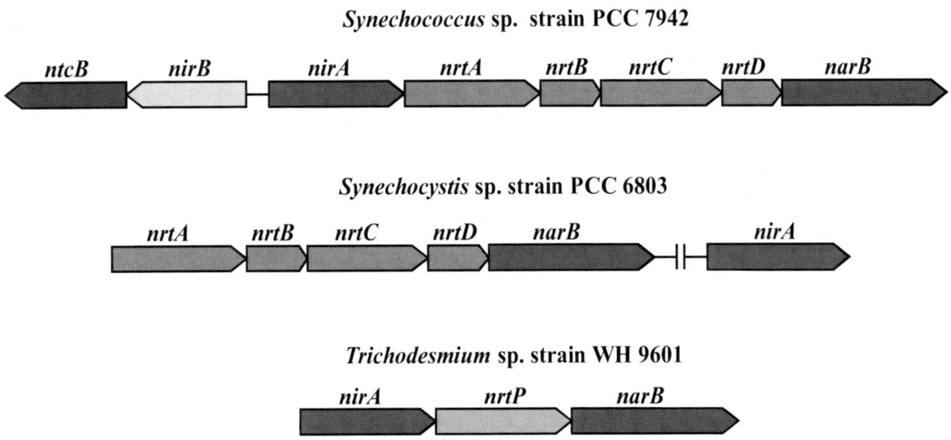

Figure 3. Comparative organization of the nitrate assimilation gene clusters between freshwater (*Synechococcus* sp. strain PCC 7942, *Synechocystis* sp. strain PCC 6803) and marine (*Trichodesmium* sp. strain WH 9601) cyanobacteria.

Nitrate Reduction

Intracellular nitrate is reduced to nitrite in a two electron reaction catalyzed by nitrate reductase (NR), the resulting nitrite being then reduced to ammonium in a six electron reaction catalyzed by nitrite reductase (NiR). Both NR and NiR use reduced Fd as the physiological electron donor (Manzano et al. 1976; Mendez et al. 1981; Arizmendi and Serra 1990). In the reaction catalyzed by NR, two moles of Fd are oxidized per mole of nitrate reduced to nitrite, whereas in that catalyzed by NiR six moles of Fd are oxidized per mole of nitrite reduced to ammonium. Flavodoxin can serve as an electron donor alternative to Fd during growth under iron-limited conditions (Manzano 1977). Both NR and NiR are conveniently assayed *in vitro* by using methyl viologen, chemically reduced with dithionite, as an electron donor (Manzano et al. 1976; Martin-Nieto et al. 1989).

In some cyanobacteria either NR or both NR and NiR have been found firmly bound to thylakoid membranes. This has permitted the demonstration of nitrate or nitrite photoreduction by subcellular preparations in a PSI-dependent reaction (Hattori and Myers 1967; Manzano et al. 1976; Peschek 1979).

Nitrate Reductase

Cyanobacterial nitrate reductases studied to date are proteins of about 75 kDa which appear to be constituted by a single polypeptide (Candau 1979; Mikami and Ida 1984; Martin-Nieto et al. 1992) (Table 3). They contain molybdenum (Peschek 1979; Herrero and Guerrero 1986; Martin-Nieto et al. 1990) which would be integrated into a Mo cofactor (Martin-Nieto et al. 1990), and non-heme iron and acid-labile sulfide (Candau 1979; Mikami and Ida 1984), which in NR from *Piectonema boryanum* might constitute two Fe_2S_2 clusters (Mikami and Ida1984). NR from heterocyst-forming cyanobacteria exhibits a biphasic kinetic behavior with nitrate as the variable substrate that is not generally observed with NR from non-heterocystous strains (MartinNieto et al. 1992). Three genes, namely *narA, narB,* and *narC,* involved in the reduction of nitrate to nitrite were isolated (Kuhlemeier et al. 1984a, 1984b). These three genes are unlinked in the genome of *Synechococcus* sp. The *narB* gene has been sequenced (Andriesse et al. 1990). Based on the similarity of the predicted amino acid sequence of the NarB protein to that of the a-subunit of *E. coli* NR and of other molybdo-proteins, and on the similarity of its predicted amino acid composition to that of the *Piectonema boryanum* NR (Mikami and Ida 1984), it has been suggested that *narB* could represent the structural gene for nitrate reductase (Andriesse et al. 1990). Nothing is currently known about the actual function of the products of the genes *narA* or *narC,* although they could be involved in the biosynthesis of the molybdenum cofactor of NR.

Nitrite Reductase

Cyanobacterial NiR consists of a single polypeptide of 52-68 kDa (Hattori and Uesugi 1968; Manzano 1977; Mendez and Vega 1981; Yabuki et al. 1985; Arizmendi and Serra 1990; Miyaji and Tamura 1992). From spectrophotometric studies of NiR from several cyanobacteria, the presence of siroheme as a prosthetic group has been proposed (Mendez and Vega 1981; Arizmendi and Serra 1990; Miyaji and Tamura 1992). NiR from *Synechococcus* sp. also shows homology to the hemoprotein component of enterobacterial sulfite reductase, and bears the four conserved cysteine residues, as well as other conserved amino acids surrounding these cysteines, that have been attributed structural importance in the configuration of the active site of the sulfite reductase hemoprotein and that are also present in higher-plant nitrite reductase (Ostrowski et al. 1989). *Synechococcus* sp. strain PCC 7942 NiR would therefore bear an active center much alike that present in these other reductases (Luque et al. 1993) in which an Fe_4S_4 cluster is bound to the protein through the S atoms of the above-mentioned four conserved cysteine residues, one of which also binds one molecule of siroheme thus bridging together the two prosthetic groups (Ostrowski et al. 1989). The facts that NiR from a cyanobacterium and higher plants have been characterized as homologous proteins (Luque et al. 1993) and that NiR from higher plants is located within the chloroplast and encoded by a nuclear gene, provide support to the notion that the chloroplast originates from cyanobacterial-like ancestor(s), and that genes from the endosymbiont have been transferred to the nucleus of the host during the course of higher-plant evolution.

Effect of UV-B on Assimilation of Nitrate and Nitrite

It has been previously shown that UV-B affects different steps of the nitrogen metabolism in phytoplankton. Laboratory experiments conducted on marine diatoms grown in batch cultures showed that UV-B depresses NO_3^- uptake during 2-day exposures (5h per day; Döhler and Biermann 1987). In these experiments, Döhler and Biermann (1987) observed that a dark period following the UV-B exposure can allow diatom cells to recover their natural NO_3 uptake capability. Döhler (1992) also observed a decrease in NO_3^- uptake during short-term (6h) laboratory experiments on natural phytoplankton communities dominated by the flagellate *Phaeocystis pouchetii* from the Wadden Sea. Short-term exposure to artificial UV-B (1-5h) of diatom-dominated natural populations from the Weddell Sea also induced a reduction of NO_3^-uptake (Döhler 1997). However, Döhler (1992) showed no variation in NO_3^- uptake in a natural phytoplankton assemblage consisting mainly of the dinoflagellate *Ceratium* spp. after 3h of UV-B exposure. In addition, during a transect from 37^0N to 55^0N in the Pacific Ocean, Behrenfeld et al. (1995) found no correlation between changes in absolute NO_3^- uptake rates and UV-B dose, during short-term (4-8h) surface incubations of natural phytoplankton populations under excluded, natural, and enhanced UV-B.

The study of Srivastava et al. (2014) assessed the responses on NO_3^- and NO_2^- uptake, nitrate assimilating enzymes-nitrate reductase and nitrite reductase in two nitrogen fixing cyanobacteria, *Nostoc muscorum* and *Phormidium foveolarum* exposed to two UV-B doses (low; UV-B_L: 0.5472 kJ m^{-2} and high; UV-B_H: 5.472 kJ m^{-2}) and two doses of the insecticide chlorpyrifos (O,O-diethyl O-3,5,6-trichloro-2-pyridyl phosphorothioate; low i.e., CP_L, 1 μg ml^{-1} and high i.e., CP_H, 2 μg ml^{-1}) singly and in combination. The rate of NO_3^- and NO_2^- uptake and activities of nitrate reductase and nitrite reductase were severely affected when treated either with CP_H or/and UV-B_H. CP_L also reduced all the measured parameters after 24h, however, a stimulatory effect was observed after 72h due to an increase in nitrogen metabolism (and other antioxidant) enzymes during this period. UV-B_L did not cause significant alteration in the studied parameters while in combination with CP doses, it either alleviated the inhibitory effects or further enhanced the CP_L induced activities of these enzymes. Flores and Herrero (1994) showed that the entry of NO_3^- and NO_2^- into cyanobacterial cells is facilitated by ABC-type transporter protein that likely uses ATP as a source of energy. Thus, the decrease in NO_3^- and NO_2^- uptake rates could be due to a decreased ATP pool as a consequence of damaged photosynthetic electron transport chain (Rai et al. 1995).

Further, NiR activity was more sensitive to CP_H and UV-B_H compared to NR activity. It is known that NR and NiR are NO_3^- and NO_2^- inducible, respectively (Ohmori and Hattori 1970). Therefore, under CP_H and UV-B_H exposure, decrease in NR and NiR activities can directly be correlated with decreased NO_3^- and NO_2^- uptake, respectively. It has been shown that sulfhydryl (-SH) groups are required for catalytic activity of NR

(Solomonson and Barber 1990), thus it is possible that CP_H alone and together with UV-B_H might have affected the NR activity by binding to functional -SH groups present in the active sites of this enzyme (Sharma and Dubey 2005). Decline in NiR activity may be associated with the reduced availability of NO_2^- ions, which are considered to be originated primarily from NR-catalyzed NO_3^- reduction. Enhanced UV-B irradiation is also reported to inhibit NR and NiR activities by interfering at transcriptional level in *Spirodela polyrhiza* (Appenroth et al. 1993). In contrast, exposure of UV-B_L marginally stimulates NR and NiR activities compared to the irrespective controls. Moreover, exposure of UV-B_L together with CP increased NR and NiR activities in both organisms suggesting improved NO_3^- and NO_2^- reduction compared to CP treatments alone.

The study of Yadav et al. (2016) has reported a comparative analysis of interactive effect of UV-B and heavy metals (Cd/Zn) on NO_3^- uptake and nitrate reductase in three cyanobacteria (*Anabaena, Microcystis, Nostoc*) revealed additive increase in nitrate uptake upon UV-B+ heavy metal exposure. A significant stimulation in NO_3^- uptake and NR activity following exposure to UV-B in all the three cyanobacteria suggests UV-B-induced structural change(s) in the enzyme/carriers. However, metals alone resulted in inhibition of NO_3^- uptake and NR activity.

Some reports are available on UV-B-induced inhibition of NO_3^- uptake and their assimilatory enzymes (Döhler 1994). However, none of these reports provide an explicit understanding of the NO_3^- uptake and metabolism. The study further demonstrates interactive effects of UV-B and test metals on inhibition kinetics of NO_3^- uptake of *Anabaena, Microcystis, and Nostoc*. Exposure of the test strains to both the doses of UV-B and metals simultaneously showed a decrease in V_{max}. However, no change in apparent Km of NO_3^- uptake was observed in all the three cyanobacteria. This clearly indicates that UV-B in combination with metals produced a noncompetitive type of inhibition of NO_3^- uptake. Likewise metals alone also generated a non-competitive inhibition of NO_3^- uptake. It is particularly interesting to point out that UV-B exposed cells depicted a higher value of V_{max} for NO_3^- uptake as compared to the untreated control.

A noncompetitive inhibition of this nutrient by UV-B suggests for an alteration in the structure of the enzyme(s) responsible for their uptake/transport. Inhibition of inorganic nitrogen uptake by UV-B irradiation has been reported by (Döhler 1986, 1994) and (Behrenfeld et al. 1995) for diatoms and marine planktonic algae. These workers demonstrated that high energetic cost of nitrogen uptake and assimilation results in a tight coupling between nitrogen metabolism and photosynthetic production of ATP and NADPH (Vosjan et al. 1990). This would alter the amount of energy available for incorporation of nitrogen and other nutrients because the transport of nitrogenous compounds takes place through ATP-dependent permease (Valladares et al. 2002). In other words, decrease in uptake of nutrients constitutes the secondary response to the primary UV-B radiation damage of photosynthetic electron transport. In contrast to these, NO_3^- uptake was found to be higher after UV-B exposure. Increase in NO_3^- uptake after

UV-B exposure has also been observed in diatoms *Lithodesmium cariable* and *Thalassiosira rotula* (Döhler 1986). Though, exact mechanism of stimulation of this enzyme under UV-B exposed condition are not known; it seems that an accelerated uptake of NO_3^- due to a changed membrane permeability in UV-B exposed cells or presence of some specific UV-B photoreceptor in the cells can cause subsequent stimulation of nitrate reductase. Further, an increased V_{max} of the enzyme in UV-B exposed cells, demonstrates an increased turnover rate of this enzyme following UV-B exposure. Contrary to this Cd and Zn were found to inactivate NR activity through altering its structure (non-competitive inhibition).

A similar study on responses of nitrate uptake and nitrate reductase following exposure of *Cylindrospermum* sp. to UV-B was also observed (Chris 2012). Nitrate uptake exhibited an increase after the treatment of UV-B. Generally the nutrient uptake is reduced under stress condition and it was earlier reported by some workers also (Fuggi et al.1984; Döhler 1986; Tyagi et al. 1992). The stimulation of nitrate uptake by UVB for short periods and thereafter slow depression is supported well with the findings of Prasad et al. (1998) and Rai et al. (1998) who observed stimulation in NO_3^- uptake in the cyanobacterium *Anabaena doliolum* and in green alga *Chlorella vulgaris*. A possible explanation for the initial stimulation of NO_3^- uptake might be due to the consumption of previously stored ATP and reductant inside the cell. Likewise the inhibition in nitrate uptake at higher doses may be due to depletion in ATP content and carbon assimilation, which are essential for nitrogen metabolism (Oelmueller et al.1988).

A similar trend of nitrate uptake was noticed when UV-B exposed cells were analyzed for nitrate reductase. This could be due to an accelerated uptake of NO_3^- or presence of UV-B photoreceptors in the cells. The inhibition of nitrate reductase in the test alga at higher doses of UV-B may be attributed due to reduced carbon fixation as already reported by Döhler (1985) and Tyagi et al. (1992). The reduction in chlorophyll content was also noticed in this study due to UV-B that might be responsible for inhibition of nitrate reductase activity since nitrate reductase is bound to the chlorophyll-containing membrane fractions.

Rajendiran and Ramanujam (2015) reported enhanced nitrate uptake and NR activity in *Vigna radiata* plants. The effects of UV-B on nitrate reductase (NR) activity have been studied in the filamentous and heterocystous N_2-fixing cyanobacterium *Nostoc calcicola* revealed that nitrate reductase activity was stimulated twofold in comparison to one exposed to fluorescent white light (Kumar et al. 1996). Although, similar to the UV-B effect, stimulation of NR activity has also been reported by exposure of *Oscillataria princeps* to blue light (Kumar et al. 1986). However, stimulation of NR activity following UV-B treatment has been reported in an angiospermic plant, *Crotalaria juncea* (Saralabai et al. 1989). *In vivo* nitrate reductase activity was found to be increased in a hot-spring cyanobacteria *Scytonema* sp. strain HKAR-3 and *Nostoc* sp. strain HKAR-2 following UV-B exposure (Richa et al. 2013a, 2013b).

Conclusion and Future Perspectives

Although we have learned much about nitrogen metabolism under abiotic stresses in many living systems, it is still in its early stage for studies done in cyanobacteria. UV stress can exert, depending on the species, either a stimulatory or an inhibitory effect on most of the inorganic nitrogen pathways. It clearly concludes that nitrogenase is very sensitive to UV-B exposure, and inactivation of activity is possibly due to complete damage to the nitrogenase polypeptide. UV-B radiation causes reduction in NH_4^+ uptake as well as in activities of glutamine synthetase. However, nitrate uptake as well as nitrate reductase activity displays substantial increase in the presence of UV-B stress. Substantial increase in nitrate uptake and nitrate reductase activity under UV-B stress can be attributed to UV-B-induced structural change(s) in the enzyme/carriers and reduced ammonium uptake induces transcription factor NtcA to induce transcription of the nitrate assimilation genes which results in enhanced nitrate uptake and nitrate reductase activity. A recent review on the nitrogen control in cyanobacteria has emphasized the importance of the *ntcA* gene in the regulation of nitrogen assimilation. Much work has been devoted to the effects of nitrogen sources but little study has been done on the effects of abiotic stresses. The effects of UV-B stress on the regulation of this gene at both transcription and translation level might provide further insight into the mechanisms eliciting the signal for UV-B stress adaptation in cyanobacteria which might be amenable for application in higher plants.

References

Arulmozhi D, Rajendiran K (2014) Effect of supplementary ultraviolet-B radiation on nodulation and nitrogen metabolism in *Lablab purpureus* L. var. Goldy. *Int J Geol Earth Environ Sci* 4: 343-346.

Behrenfeld MJ, Lean DRS, Lee H (1995) Ultraviolet-B radiation effects on inorganic nitrogen uptake by natural assemblages of oceanic plankton. *J Phycol* 31: 25-36.

Benz R and Bohme H (1985) Pore formation by an outer membrane protein of the cyanobacterium *Anabaena variabilis*. *Biochim Biophys Acta* 812: 286-292.

Biggs RH, Kossuth SV, Teramura AH (1981) Response of 19 cultivars of soybeans to ultraviolet-B irradiance. *Physiol Plant* 53: 19-26.

Blakefield MK, Harris DO (1994) Delay of cell differentiation in *Anabaena aequalis* caused by UV-B radiation and the role of photoreactivation and excision repair. *Photochem Photobiol* 59: 204-208.

Blanco F, Alana A, Llama MJ and Serra JL (1989) Purification and properties of glutamine synthetase from the non-NT fixing cyanobacterium *Phormidium laminosum. J Bacteriol* 171: 1158-1165.

Bolhuis H, Severin I, Confurius-Guns V, Wollenzien UIA, Stal LJ (2010) Horizontal transfer of the nitrogen fixation gene cluster in the cyanobacterium *Microcoleus chthonoplastes*. *The ISME J* 4: 121-130.

Bothe H, Schmitz O, Yates MO, Newton WE (2010) Nitrogen fixation and hydrogen metabolism in cyanobacteria. *Microbiol Mol Biol Rev*74: 529-551.

Britt AB (1995) Repair of DNA damage induced by ultraviolet radiation. *Plant Physiol* 108 (3): 891-896.

Buikema WJ, Haselkorn R (1991) Isolation and complementation of nitrogen fixation mutants of the cyanobacterium *Anabaena* sp. strain PCC 7120. *J Bacteriol* 173: 1879-1885.

Cai Y, Wolk CP (1997) Nitrogen deprivation of *Anabaena* sp. strain PCC 7120 elicits rapid activation of a gene cluster that is essential for uptake and utilization of nitrate. *J Bacteriol* 179: 258-266.

Caldwell MM (1981) Plant response to solar ultraviolet radiation. *In: Encyclopedia of Plant Physiology*. Vol. 12A, Physiological Plant Ecology I (Eds. Lange OL, Nobel PS, Osmund CB, Ziegler H), Springer-Verlag, Berlin, pp. 169-197.

Caldwell MM, Teramura AH, Tevini M, Bornman JF, Bjorn LO, Kulandaivelu G (1995) Effects of increased solar ultraviolet radiation on terrestrial plants. *Ambio* 24: 166-173.

Callahan SM, Buikema WJ (2001) The role of *HetN* in maintenance of the heterocyst pattern in *Anabaena* sp. PCC 7120. *Mol Microbiol* 40: 941-950.

Campbell D, Eriksson MJ, Öquist G, Gustafsson P, Clarke AK (1998) The cyanobacterium *Synechococcus* resists UV-B by exchanging photosystem II reaction-center D1 proteins. *Proc Natl Acad Sci USA* 95: 364-369.

Campbell EL, Cohen MF, Meeks JC (1997) A polyketidesynthase- like gene is involved in the synthesis of heterocyst glycolipids in *Nostoc punctiforme* strain ATCC 29133. *Arch Microbiol* 167: 251-258.

Castenholz RW (1969) Thermophilic blue-green algae and the thermal environment. *Bacteriol Rev* 33: 476-504.

Castenholz RW (1997) Multiple strategies for UV tolerance in cyanobacteria. *Spectrum* 10: 10-16.

Chávez S, Lucena JM, Reyes JC, Florencio FJ, Candau P (1999) The presence of glutamate dehydrogenase is a selective advantage for the cyanobacterium *Synechocystis* sp. strain PCC 6803 under non exponential growth conditions. *J Bacteriol* 181: 808-813.

Chen L, Deng S, De Philippis R, Tian W, Wu H, Wang J (2013) UV-B resistance as a criterion for the selection of desert microalgae to be utilized for inoculating desert soils. *J Appl Phycol* 25: 1009-1015.

Chris A (2012) Effect of solar ultraviolet-B radiation on growth and enzymes of nitrogen assimilation in *Cylindrospermum* sp. *Asian J Microbiol Biotechnol Environ Sci* 7: 200-203.

Cumino AC, Marcozzi C, Barreiro R, Salerno GL (2007) Carbon cycling in *Anabaena* sp. PCC 7120. Sucrose synthesis in the heterocysts and possible role in nitrogen fixation. *Plant Physiol* 143: 1385-1397.

Davey MC (1989) The effect of freezing and desiccation on photosynthesis and survival of terrestrial Antarctic algae and cyanobacteria. *Polar Biol* 10: 29-36.

Day DA, Poole PS, Tyerman SD, Rosendahl L (2001) Review: ammonia and amino acid transport across symbiotic membranes in nitrogen-fixing legume nodules. *Cell Mol Life Sci* 58: 61-71.

de Marsac NT, Houmard J (1993) Adaptation of cyanobacteria to environmental stimuli: new steps towards molecular mechanisms. *FEMS Microbiol Rev* 104: 119-190.

Dharmawardene MWN, Stewart WDP, Stanley SO (1972) Nitrogenase activity, amino acid pool patterns and amination in blue-green algae. *Planta* 108: 133-145.

Döhler G (1985) Effect of UV-B radiation (290-320 nm) on the nitrogen metabolism of several marine diatoms. *J Plant Physiol* 118: 391-400.

Döhler G (1986) Impact of UV-B radiation on [^{15}N] ammonia and [^{15}N] nitrate uptake of *Dictylum brightweilli. Photobiochem Photobiophys* 11: 115-121.

Döhler G (1994) UV-effects on the nitrogen metabolism of marine phytoplankton and adaptation to UV radiation, in: Biggs, R.H., Joyner, M.E.B. (Eds.), *Stratospheric Ozone Depletion/UV-B Radiationinthe Biosphere*, Springer, Berlin, pp. 163-174.

Donkor V, Häder DP (1991) Effects of solar and ultraviolet radiation on motility, photomovement and pigmentation in filamentous gliding cyanobacteria. *FEMS Microbiol Ecol* 86: 159-168.

Dunlap WC, Shick JM (1998) Ultraviolet radiation-absorbing mycosporine- like amino acids in coral reef organisms: a biochemical and environmental perspective. *J Phycol* 34: 418-430.

Ehira S, Ohmori M (2006a) *NrrA* directly regulates the expression of *hetR* during heterocyst differentiation in the cyanobacterium *Anabaena* sp. strain PCC 7120. *J Bacteriol* 188: 8520-8525.

Ehira S, Ohmori M (2006b) *NrrA*, a nitrogen-responsive response regulator facilitates heterocyst development in the cyanobacterium *Anabaena* sp. strain PCC 7120. *Mol Microbiol* 59: 1692-1703.

Elhai J, Wolk CP (1990) Developmental regulation and spatial pattern of expression of the structural genes for nitrogenase in the cyanobacterium *Anabaena*. *EMBO J* 9: 3379-3388.

Elmorjani K, Liotenberg S, Houmard J, Tandeau de Marsac N (1992) Molecular characterization of the gene encoding glutamine synthetase in the cyanobacterium *Calothrix* sp. PCC 7601. *Biochem Biophys Res Commun* 189: 1296-1302.

Fan Q, Huang G, Lechno-Yossef S, Wolk CP, Kaneko T, Tabata S (2005) Clustered genes required for synthesis and deposition of envelope glycolipids in *Anabaena* sp. strain PCC 7120. *Mol Microbiol* 58: 227-243.

Fay P (1992) Oxygen relations of nitrogen fixation in cyanobacteria. *Microbiol Rev* 56: 340-373.

Fisher R, Tuli R, Haselkorn R (1981) A cloned cyanobacterial gene for glutamine synthetase functions in *Escherichia coli,* but the enzyme is not adenylylated. *Proc Natl Acad Sci USA* 78: 3393-3397.

Florencio FJ, Ramos JL (1985) Purification and characterization of glutamine synthetase from the unicellular cyanobacterium *Anacystis nidulans. Biochim Biophys Acta* 838: 39-48.

Florencio FJ, Reyes JC (2002) *In: Regulation of ammonium assimilation in cyanobacteria* (Eds. Foyer CH, Noctor G) Photosynthetic Nitrogen Assimilation and Associated Carbon and Respiratory Metabolism), Springer Netherlands, pp. 93-113.

Flores E, A Herrero (1994) Assimilatory nitrogen metabolism and its regulation, *In: Molecular Biology of Cyanobacteria* (Ed. Bryant DA), The Kluwer Academic Publishers, Dordrecht, The Netherlands, pp. 487-517.

Frias JE, Flores E, Herrero A (1997) Nitrate assimilation gene cluster from the heterocyst-forming cyanobacterium *Anabaena* sp. strain PCC 7120. *J Bacteriol* 179: 477-486.

Fuggi A, Vona V, Rigano DI, Martino C, Martella A, Rigano C (1984) Evidence for two transport systems for nitrate in the acidophilic thermophilic alga, *Cyanidium caldarium. Arch Microbiol* 137: 281-285.

Gallon JR, Chaplin AE (eds.) (1987) *An introduction to nitrogen fixation*, Cassell, London.

Gao K, Yu H, Brown MT (2007) Solar PAR and UV radiation affects the physiology and morphology of the cyanobacterium *Anabaena* sp. PCC 7120. *J Photochem Photobiol B* 89: 117-124.

Gao K, Li P, Watanabe T, Helbling EW (2008) Combined effects of ultraviolet radiation and temperature on morphology, photosynthesis, and DNA of *Arthrospira* (*Spirulina*) *platensis* (Cyanophyta). *J Phycol* 44:777-786.

Goes JI, Handa N, Taguchi S, Hama T, Saito H (1995) Impact of UV radiation on the production patterns and composition of dissolved free and combined amino acids in marine phytoplankton. *J Plankton Res* 17: 1337-1362.

Golden JW, Robinson SJ, Haselkorn R (1985) Rearrangement of nitrogen fixation genes during heterocyst differentiation in the cyanobacterium *Anabaena. Nature* 314: 419-423.

Häder DP, Worrest RC, Kumar HD (1989) Aquatic ecosystems. *UNEP Environmental Effects Panel Report*, pp. 39-48.

Häder DP, Sinha RP (2005) Solar ultraviolet radiation-induced DNA damage in aquatic organisms: potential environmental imapct. *Mutat Res* 571: 221-233.

Halbleib CM, Ludden PW (2000) Regulation of biological nitrogen fixation. *J Nutr* 130: 1081-1084.

Hargreaves A, Taiwo FA, Duggan O, Kirk SH, Ahmad SI (2007) Near-ultraviolet photolysis of b-phenylpyruvic acid generates free radicals and results in DNA damage. *J Photochem Photobiol B Biol* 89: 110-116.

He YY, Häder DP (2002a) Reactive oxygen species and UV-B: effect on cyanobacteria. *Photochem Photobiol Sci* 1: 729-736.

He YY, Häder DP (2002b) UV-B-induced formation of reactive oxygen species and oxidative damage of the cyanobacterium *Anabaena* sp.: protective effects of ascorbic acid and N-acetyl-L-cysteine. *J Photochem Photobiol B* 66: 115-124.

He YY, Häder DP (2002c) Involvement of reactive oxygen species in the UV-B damage to the cyanobacterium *Anabaena* sp. *J Photochem Photobiol B* 66: 73-80.

Herrero A, Muro-Pastor AM, Valladares A, Flores E (2004) Cellular differentiation and the *NtcA* transcription factor in filamentous cyanobacteria. *FEMS Microbiol Rev* 28: 469-487.

Herrero A, Picossi S, Flores E (2013) Gene expression during heterocyst differentiation. *In: Advances in Botanical Research* (Eds. Franck C, Corinne CC), Academic Press: Waltham, MA, USA, Volume 65, pp. 281-329.

Hoare DS, Hoare SL, Moore RB (1967) The photoassimilation of organic compounds by autotrophic blue-green algae. *J Gen Microbiol* 49: 351-370.

Hoehler TM, Bebout BM, Des Marais DJ (2001) The role of microbial mats in the production of reduced gases on the early earth. *Nature* 412: 324-327.

Hofmann DJ, Deshler T (1991) Evidence from balloon measurements for chemical depletion of stratospheric ozone in the Arctic winter of 1989-90. *Nature* 349: 300-305.

Holland D, Wolk CP (1990) Identification and characterization of *hetA*, a gene that acts early in the process of morphological differentiation of heterocysts. *J Bacteriol* 172: 3131-3137.

Huang L, McCluskey MP, Ni H, LaRossa RA (2002) Global gene expression profiles of the cyanobacterium *Synechocystis* sp. strain PCC 6803 in response to irradiation with UV-B and white light. *J Bacteriol* 184: 6845-6858.

Kaneko T, Sato S, Kotani H, Tanaka A, Asamizu E, Nakamura Y, Miyajima N, Hirosawa M, Sugiura M, Sasamoto S, Kimura T (1996) Sequence analysis of the genome of the unicellular cyanobacterium *Synechocystis* sp. strain PCC6803. II. Sequence determination of the entire genome and assignment of potential protein-coding regions. *DNA Res* 3: 109-136.

Kannaiyan S, Govindarajan K, Lewin HD, Venkataraman GS (1982) Influence of blue green algal application on rice crop. *Madras Agricul J* 69: 1-5.

Karentz D, Cleaver JE, Mitchell DL (1991a) Cell survival characteristics and molecular responses of Antarctic phytoplankton to ultraviolet-B radiation. *J Phycol* 27: 326-341.

Karentz D, Cleaver JE, Mitchell DL (1991b) DNA damage in the Antarctic. *Nature* 350: 28.

Karentz D, McEuen FS, Land MC, Dunlap WC (1991c) A survey of mycosporine-like amino acid compounds in Antarctic marine organisms: potential protection from ultraviolet exposure. *Mar Biol* 108: 157-166.

Kim ST, Sancar (1995) A Photorepair of nonadjacent pyrimidine dimers by DNA photolyase. *Photochem Photobiol* 61: 171-174.

Kojima K, Oshita M, Nanjo Y, Kasai K, Tozawa Y, Hayashi H, Nishiyama Y (2007) Oxidation of elongation factor G inhibits the synthesis of the D1 protein of photosystem II. *Mol Microbiol* 65: 936-947.

Kuhlbusch TA, Lobert JM, Crutzen PJ, Warneck P (1991) Molecular nitrogen emission. Trace nitrification during biomass burning. *Nature* 351: 135-137.

Kuhlemeier CJ, Logtenberg T, Stoorvogel W, Van Heugten HAA, Borrias WE, Van Arkel GA (1984a) Cloning of nitrate reductase genes from the cyanobacterium *Anacystis nidulans*. *J Bacteriol* 159: 36-41.

Kuhlemeier CJ, Teeuwsen VJP,Janssen MJT,van Arkel GA (1984b) Cloning of a third nitrate reductase gene from the cyanobacterium *Anacystis nidulans* R2 using a shuttle cosmid library. *Gene* 31: 109-116.

Kulasooriya SA (2008) Biological Nitrogen Fixation: Fundamentals and Utilization. *Peradeniya Science Publication No: 27*, Science Education Unit, Faculty of Science, University of Peradeniya, Peradeniya, pp. 143.

Kumar A, Tabita FR, Van Baalen C (1982) Isolation and characterization of heterocysts from *Anabaena* sp. strain CA. *Arch Microbiol* 133: 103-109.

Kumar A, Sinha RP, Häder DP (1996) Effect of UV-B on enzymes of nitrogen metabolism in the cyanobacterium *Nostoc calcicola*. *J Plant Physiol* 148: 86-91.

Kumar A, Tyagi MB, Jha PN, Srinivas G, Singh A (2003) Inactivation of cyanobacterial nitrogenase after exposure to ultraviolet-B radiation. *Curr Microbiol* 46: 380-384.

Kumar HD, Jha M, Kumar A (1986) Stimulation of nitrate reductase activiry by blue light in a thermophilic cyanobacterium *Oscillatoria princeps*. *Br Phycol J* 21, 165-168.

Kumar K, Mella-Herrera RA, Golden JW (2010) Cyanobacterial heterocysts. *Cold Spring Harb Perspect Biol*2(4): a000315.

Lammers PJ, Haselkorn R (1983) Sequence of the *nifD* gene coding for α subunit of dinitrogenase from cyanobacterium *Anabaena*. *Proc Natl Acad Sci USA* 80: 4723-4727.

Laurent S, Chen H, Bedu S, Ziarelli F, Peng L, Zhang CC (2005) Nonmetabolizable analogue of 2-oxoglutarate elicits heterocyst differentiation under repressive conditions in *Anabaena* sp. PCC 7120. *Proc Natl Acad Sci* 102: 9907-9912.

Lesser MP (2008) Effects of ultraviolet radiation on productivity and nitrogen fixation in the cyanobacterium, *Anabaena* sp. (Newton's strain). *Hydrobiologia* 598: 1-9.

Li P, Liu W, Gao K (2012) Effects of temperature, pH, and UV radiation on alkaline phosphatase activity in the terrestrial cyanobacterium *Nostoc flagelliforme*. *J Appl Phycol* 25: 1031-1038.

Lubin D, Jensen EH (1995) Effects of clouds and stratospheric ozone depletion on ultraviolet radiation trends. *Nature* 377: 710-713.

Luque I, Flores E, Herrero A (1993) Nitrite reductase gene from *Synechococcus* sp. PCC 7942: homology between cyanobacterial and higher plant nitrite reductases. *Plant Mol Biol* 21: 1201-1205.

Luque I, Flores E, Herrero A (1994) Nitrate and nitrite transport in the cyanobacterium *Synechococcus* sp. PCC 7942 are mediated by the same permease. *Biochim Biophys Acta* 1184: 296-298.

Madronich S, McKenzie R, Bjorn R, Caldwell M (1998) Changes in biologically active ultraviolet radiation reaching the Earth's surface. *J Photochem Photobiol B Biol* 46: 5-19.

Maeda SI, Omata T (1997) Substrate-binding lipoprotein of the cyanobacterium *Synechococcus* sp. strain PCC 7942 involved in the transport of nitrate and nitrite. *J Biol Chem*272: 3036-3041.

Magasanik B, Neidhardt FC (1987) *In: Regulation of carbon and nitrogen utilization.* (Eds. Neidhardt FC, Ingraham JL, Low KB, Magasanik B, Schaechter M, Umbarger HE) *Escherichia coli* and *Salmonella typhimurium.* Cellular and Molecular Biology, American Society for Microbiology, Washington, DC, pp. 1318-1325.

Marcozzi C, Cumino AC, Salerno GL (2009) Role of NtcA, a cyanobacterial global nitrogen regulator, in the regulation of sucrose metabolism gene expression in *Anabaena* sp. PCC 7120. *Arch Microbiol* 191: 255-263.

Martinez MR (1984) Algae: biofertilizer for rice. *Philipp Counc Agric Res Resour Dev (PCARRD) Monirt* 12: 9-12.

Martin-Figueroa E, Navarro F, Florencio FJ (2000) The GS-GOGAT pathway is not operative in the heterocysts. Cloning and expression of *glsF* gene from the cyanobacterium *Anabaena* sp. PCC 7120. *FEBS Lett* 476: 282-286.

Mazur BJ, Chui CF (1982) Sequence of the gene coding for the f3-subunit of dinitrogenase from the blue-green alga *Anabaena. Proc Natl Acad Sci USA* 79: 6782-6786.

Meeks JC, Wolk CP, Thomas J, Lockau W, Shaffer PW, Austin SM, Chien WS, Galonsky A (1977) The pathways of assimilation of $^{13}NH_4^+$ by the cyanobacterium *Anabaena cylindrica*. *J Biol Chem* 252: 7894-7900.

Meeks JC (1998) Symbiosis between nitrogen-fixing cyanobacteria and plants. *Bioscience* 48: 266-276.

Meeks JC, Elhai J (2002) Regulation of cellular differentiation in filamentous cyanobacteria in free-living and plant-associated symbiotic growth states. *Microbiol Mol Biol Rev* 66: 94-121.

Merida A, Candau P, Florencio FJ (1991) Regulation of glutamine synthetase activity in the unicellular cyanobacterium *Synechocystis* sp. strain PCC 6803 by the nitrogen source: Effect of ammonium. *J Bacteriol* 173: 4095-4100.

Merida A, Flores E, Florencio FJ (1992) Regulation of *Anabaena* sp. strain PCC 7120 glutamine synthetase activity in *a Synechocystis* sp. strain PCC 6803 derivative strain bearing the *Anabaena glnA* gene and a mutated host *glnA* gene. *J Bacteriol* 174: 650-654.

Middleton EM, Teramura AH (1994) Understanding photosynthesis, pigment and growth responses induced by UV-B and UV-A irradiances. *Photochem Photobiol* 60: 38-45.

Montesinos ML, Muro-Pastor AM, Herrero A, Flores E (1998) Ammonium/ methylammonium permeases of a cyanobacterium. Identification and analysis of three nitrogen-regulated *amt* genes in *Synechocystis* sp. PCC 6803. *J Biol Chem* 273: 31463-31470.

Muro-Pastor AM, Valladares A, Flores E, Herrero A (2002) Mutual dependence of the expression of the cell differentiation regulatory protein *HetR* and the global nitrogen regulator *NtcA* during heterocyst development. *Mol Microbiol* 44: 1377-1385.

Muro-Pastor MI, Reyes JC, Florencio FJ (2001) Cyanobacteria perceive nitrogen status by sensing intracellular 2-oxoglutarate levels. *J Biol Chem* 276: 38320-38328.

Muro-Pastor MI, Reyes JC, Florencio FJ (2005) Ammonium assimilation in cyanobacteria. *Photosynth Res* 83: 135-150.

Murry MA, Horne AJ, Benemann JR (1984) Physiological studies of oxygen protection mechanisms in the heterocysts of *Anabaena cylindrica. Appl Environ Microbiol* 47, 449-454.

Murry MA, Wolk CP(1989)Evidence that the barrier to the penetration of oxygen into heterocysts depends upon two layers of the cell envelope. *Arch Microbiol* 151: 469-474.

Neilson AH, Doudoroff M (1973) Ammonia assimilation in blue-green algae. *Arch Microbiol* 89: 15-22.

Newton JW, Tyler DD, Slodki ME (1979) Effect of ultraviolet- B (280-320) radiation on blue-green algae (Cyanobacteria), possible biological indicators of stratospheric ozone depletion. *Appl Environ Microbiol* 37: 1137-1141.

Oelmueller R, Schuster C, Mohr H (1988) Physiological characterization of a plastidic signal required for nitrate induced appearance of nitrate and nitrate reductase. *Planta*, 174, 75-83.

Omata T, Andriesse X, Hirano A (1993) Identification and characterization of a gene cluster involved in nitrate transport in the cyanobacterium *Synechococcus* sp. PCC 7942. *Mol Gen Genet* 236: 193-202.

Orr J, Haselkorn R (1981) Kinetic and inhibition studies of glutamine synthetase from the cyanobacterium *Anabaena* 7120. *J Bioi Chern* 256: 13099-13104.

Orr J, Keefer LM, Keirn P, Nguyen TO, Wellems T, Heinrikson L, Haselkorn R (1981) Purification, physical characterization, and NH_2-terrninal sequence of glutamine synthetase from the cyanobacterium *Anabaena* 7120. *J BioI Chern* 256: 13091-13098.

Pace NR (1997) A molecular view of microbial diversity and the biosphere. *Science* 276: 734-740.

Parker BC, Heiskell LE, Thompson WJ, Zeller EJ (1978) Non-biogenic fixed nitrogen in Antarctica and some ecological implications. *Nature* 271: 651-652.

Pathak J, Richa, Rajneesh, Sonker AS, Kannaujiya VK, Sinha RP (2015a) Isolation and partial purification of scytonemin and mycosporine-like amino acids from biological crusts. *J Chem Pharm Res*7: 362-371.

Pathak J, Rajneesh, Richa, Kannaujiya VK, Sonker AS, Sinha RP (2015b) Diverse functions and applications of novel and unique cyanobacterial sheath pigment, scytonemin. In: Sinha RP, Richa, Rastogi RP, editors. *Biological Sciences: Innovations and Dynamics*: New India Publishing Agency, New Delhi, India. pp. 237-262.

Pathak J, Rajneesh, Ahmed H, Richa, Sinha RP (2017a) Metabolomic profiling of cyanobacterial UV-protective compounds. *Curr Metabolomics* 5(2): 138-163.

Pathak J, Rajneesh, Richa, Sonker AS, Kannaujiya VK, Sinha RP (2017b) Cyanobacterial extracellular polysaccharide sheath pigment, scytonemin: a novel multipurpose pharmacophore. In: Se-Kwon Kim, editor. *Marine Glycobiology: Principles and Applications*: Taylor and Francis Group, CRC Press, Boca Raton, Florida, USA, pp. 323-338.

Pathak J, Sonker AS, Richa, Rajneesh, Kannaujiya VK, Singh V, Ahmed H, Sinha RP (2017c) Screening and partial purification of photoprotective pigment, scytonemin from cyanobacterial crusts dwelling on the historical monuments in and around Varanasi, India. *Microbiol Res (Pavia)* 8: 6559.

Peterson RB, Friberg EE, Burris RH (1979) Diurnal variation in N_2 fixation and photosynthesis by aquatic blue-green algae. *Plant Physiol* 59: 74-80.

Popa R, Weber PK, Pett-Ridge J, Finzi JA, Fallon SJ, Hutcheon ID, Nealson KH, Capone DG (2007) Carbon and nitrogen fixation and metabolite exchange in and between individual cells of *Anabaena oscillarioides*. *ISME J* 1: 354-360.

Prasad V, Kumar A, Kumar HD (1998) Effect of UV-B on certain metabolic processes of the green alga, *Chlorella vulgaris*. *Internat J Env Sci* 55: 129-140.

Pratt BS, Thiel T (2014) Regulation of nitrogenase gene expression by transcript stability in the cyanobacterium *Anabaena variabilis*. *J Bacteriol* 196: 3609-3621.

Quesada A, Vincent WF, Lean DRS (1999) Community and pigment structure of arctic cyanobacterial assemblages: the occurrence and distribution of UV-absorbing compounds. *FEMS Microbiol Ecol* 28: 315-323.

Rai LC,Tyagi B, Rai PK, Mallick N (1998) Interactive effects of UV-B and heavy metals (Cu and Pb) on nitrogen and phosphorus metabolism of a N_2-fixing cyanobacterium, *Anabaena doliolum*. *Env Exp Bot* 39: 221-231.

Rajendiran K, Ramanujam, MP (2006) Interactive effects of UV-B irradiation and triadimefon on nodulation and nitrogen metabolism in *Vigna radiata* plants. *Biol plantarum* 50: 709-712.

Ramasubramanian TS, Wei TF, Golden JW (1994) Two *Anabaena* sp. strain PCC 7120 DNA-binding factors interact with vegetative cell-and heterocyst-specific genes. *J Bacteriol* 176: 1214-1223.

Ramirez ME, Hebbar PB, Zhou R,Wolk CP, Curtis SE (2005) *Anabaena* sp. strain PCC 7120 gene *devH* is required for synthesis of the heterocyst glycolipid layer. *J Bacteriol* 187: 2326-2331.

Ramos JL, Guerrero MG, Losada M (1982) Photoproduction of ammonia from nitrate by *Anacystis nidulans* cells. *Biochim Biophys Acta* 679: 323-330.

Ravishankara AR, Daniel JS, Portmann RW (2009) Nitrous oxide (N_2O): the dominant ozone-depleting substance emitted in the 21st century. *Science* 326: 123-125.

Resch C, Gibson J (1983) Isolation of the carotenoid-containing cell wall of three unicellular cyanobacteria. *J Bacteriol* 155: 345-350.

Richa, Kumari S, Kannaujiya VK, Mishra S, Sinha RP (2013a) Response of a Hot-spring Cyanobacterium *Scytonema* sp. strain HKAR-3 to Ultraviolet-B Radiation. *Int J Curr Biotechnol* 1: 32-36.

Richa, Kannaujiya VK, Kumari S, Mishra S, Sinha RP (2013b) Effects of ultraviolet-B radiation on a hot-spring cyanobacterium *Nostoc* sp. strain HKAR-2. *Acta Biologica Indica* 2: 265-276.

Richa, Sinha RP (2015) Biochemical characterization of sunscreening mycosporine-like amino acids from two *Nostoc* species inhabiting diverse habitats. *Protoplasma*. 252: 199-208.

Rowell P, Enticott S, Stewart WDP (1977) Glutamine synthetase and nitrogenase activity in the blue-green alga *Anabaena cylindrica*. *New Phytol* 79: 41-54.

Rubio LM, Herrero A, Flores E (1996) A cyanobacterial *narB* gene encodes a ferredoxin-dependent nitrate reductase. *Plant Mol Biol* 30: 845-850.

Sahoo A, Sarkar S, Singh RP, Kafatos M, Summers ME (2005) Declining trend of total ozone column over the northern parts of India. *Int J Remote Sens* 26: 33-40.

Sakamoto T, Inoue-Sakamoto K, Bryant DA (1999) A novel nitrate/ nitrite permease in the marine cyanobacterium *Synechococcus* sp. strain PCC 7002. *J Bacteriol* 181: 7363-7372.

Sandh G, El-Shehawy R, Diez B, Bergman B (2009) Temporal separation of cell division and diazotrophy in the marine diazotrophic cyanobacterium *Trichodesmium erythraeum* IMS101. *FEMS Microbiol Lett* 295: 281-288.

Saralabai YC, Thamizhchelvan P, Santhaguru K (1989) Influence of ultraviolet-B radiation on fixation and assimilation of nitrogen in *Crotalaria juncea* Linn. *Ind J Plant Physiol* 32, 65-67.

Sass L, Spetea C, Mate Z, Nagy F, Vass I (1997) Repair of UV-B induced damage of Photosystem II via *de novo* synthesis of the D1 and D2 reaction centre subunits in *Synechocystis* sp. PCC 6803. *Photosynth Res* 54: 55-62.

Schopf JW (1993) Microfossils of the early Archean chert: new evidence of the antiquity of life. *Science* 260: 640-646.

Schubert KR (1986) Products of biological nitrogen fixation in higher plants: synthesis, transport, and metabolism. *Annu Rev Plant Physiol* 37: 539-574.

Shrivastava AK, Chatterjee A, Yadav S, Singh PK, Singh S, Rai LC (2015) UV-B stress induced metabolic rearrangements explored with comparative proteomics in three *Anabaena* species. *J Proteomics* 127: 122-133.

Singh A (1997) Increased UV-B radiation reduces N_2-fixation in tropical leguminous crops. *Environ Pollut* 95: 289-291.

Singh G, Babele PK, Sinha RP, Tyagi MB, Kumar A (2013) Enzymatic and non-enzymatic defense mechanisms against ultraviolet-B radiation in two *Anabaena* species. *Process Biochem* 48: 796-802.

Singh SP, Häder DP, Sinha RP (2010) Cyanobacteria and ultraviolet radiation (UVR) stress: mitigation strategies. *Ageing Res Rev* 9: 79-90.

Sinha RP, Kumar HD, Kumar A, Häder DP (1995) Effects of UV-B irradiation on growth, survival, pigmentation and nitrogen metabolism enzymes in cyanobacteria. *Acta Protozool* 34: 187-192.

Sinha RP, Häder DP (1996) Photobiology and ecophysiology of rice field cyanobacteria. *Photochem Photobiol* 64: 887-896.

Sinha RP, Singh N, Kumar A, Kumar HD, Häder M, Häder DP (1996) Effects of UV irradiation on certain physiological and biochemical processes in cyanobacteria. *J Photochem Photobiol B* 32: 107-113.

Sinha RP, Singh N, Kumar A, Kumar HD, Häder, DP (1997) Impacts of ultraviolet-B irradiation on nitrogen-fixing cyanobacteria of rice paddy fields. *J Plant Physiol* 150: 188-193.

Sinha RP, Vaishampayan A, Häder DP (1998) Plant-cyanobacterial symbiotic somaclones as a potential bionitrogen-fertilizer for paddy agriculture: biotechnological approaches. *Microbiol Res* 153: 297-307.

Sinha RP, Häder DP (2000) Effects of UV-B radiation on cyanobacteria. *Recent Res Dev Photochem Photobiol* 4: 239-246.

Sinha RP, Dautz M, Häder DP (2001) A simple and efficient method for the quantitative analysis of thymine dimers in cyanobacteria, phytoplankton and macroalgae. *Acta Protozool* 40: 187-195.

Sinha RP, Häder DP (2002) UV-induced DNA damage and repair: a review. *Photochem Photobiol Sci* 1: 225-236.

Sinha RP, Kumari S, Rastogi RP (2008a) Impacts of ultraviolet-B radiation on cyanobacteria: photoprotection and repair. *J Sci Res* 52: 125-142.

Sinha RP, Rastogi RP, Ambasht NK, Häder DP (2008b) Live of wetland cyanobacteria under enhancing solar UV-B radiation. *Proc Natl Acad Sci India B* 78: 53-65.

Skopelitis DS, Paranychianakis NV, Paschalidis KA (2006) Abiotic stress generates ROS that signal expression of anionic glutamate dehydrogenases to form glutamate for proline synthesis in tobacco and grapevine. *Plant Cell* 18: 2767-2781.

Smith RC, Prézelin BB, Baker KS, Bidigare RR, Boucher NP, Coley T, Karentz D, MacIntyre S, Matlick HA, Menzies D, Ondrusek M, Wan Z, Waters KJ (1992) Ozone depletion: ultraviolet radiation and phytoplankton biology in Antarctic waters. *Science* 255: 952-959.

Srivastava PK, Singh VP, Prasad SM (2014) Low and high doses of UV-B differentially modulate chlorpyrifos-induced alterations in nitrogen metabolism of cyanobacteria. *Ecotoxicol Environ Saf* 107: 291-299.

Stacey G, Van Baalen C and Tabita FR (1979) Nitrogen and ammonia assimilation in the cyanobacteria: Regulation of glutamine synthetase. *Arch Biochem Biophys* 194: 457-467.

Stanier RY, Cohen-Bazire G (1977)Phototrophic prokaryotes: the cyanobacteria. *Ann Rev Microbiol* 31: 225-274.

Stewart WDP (1980) Some aspects of structure and function in N_2-fixing cyanobacteria. *Annu Rev Microbiol* 34: 497-536.

Sudaroli SJ, Rajendiran K (2013)Effect of elevated UV-B irradiation on the nodulation and nitrogen metabolism in *Sesbania grandiflora* (L.) *Pers. Int J Sci Nat* 4: 664-667.

Sudha JS, Rajendiran K (2014) Ultraviolet-B induced reduction in nodulation and nitrogen metabolism in *vigna mungo* L. Var. T-9 by elevated UV-B radiation. *Global J Biosci Biotechnol* 3: 370-373.

Tanigawa R, Shirokane M, Maeda Si S, Omata T, Tanaka K, Takahashi H (2002) Transcriptional activation of *NtcA* dependent promoters of *Synechococcus* sp. PCC 7942 by 2-oxoglutarate in vitro. *Proc Natl Acad Sci* 99: 4251-4255.

Tevini M, Teramura AH (1989) UV-B effects on terrestrial plants. *Photochem Photobiol* 50: 479-487.

Theil T (1993) Characterization of genes for an alternative nitrogenase in the cyanobacterium *Anabaena variabilis*. *J Bacteriol* 175: 6276-6286.

Thiel T, Lyons EM, Erker JC, Ernst AA (1995) A second nitrogenase in vegetative cells of a heterocyst-forming cyanobacterium. *Proc Natl Acad Sci USA* 92: 9358-9362.

Thiel T, Pratte B (2001) Effect on heterocyst differentiation of nitrogen fixation in vegetative cells of the cyanobacterium *Anabaena variabilis* ATCC 29413. *J Bacteriol* 183: 280-286.

Toepel J, Welsh E, Summerfield TC, Pakrasi HB, Sherman LA (2008) Differential transcriptional analysis of the cyanobacterium *Cyanothece* sp. strain ATCC 51142 during light-dark and continuous-light growth. *J Bacteriol* 190: 3904-3913.

Tyagi R, Srinivas G, Vyas D, Kumar A, Kumar HD (1992) Differential effect of ultraviolet-B radiation on certain metabolic processes in a chromatically adapting *Nostoc*. *Photochem Photobiol* 55: 401-407.

Vaishampayan A, Sinha RP, Häder DP (1998) Use of genetically improved nitrogen-fixing cyanobacteria in rice paddy fields: prospects as a source material for engineering herbicide sensitivity and resistance in plants. *Bot Acta* 111: 176-190.

Valladares A, Montesinos ML, Herrero A, Flores E (2002) An ABC-type, high-affinity urea permease identified in cyanobacteria. *Mol Microbiol* 43, 703-715.

Valladares A, Herrero A, Pils D, Schmetterer G, Flores E (2003) Cytochrome c oxidase genes required for nitrogenase activity and diazotrophic growth in *Anabaena* sp. PCC 7120. *Mol Microbiol* 47: 1239-1249.

Vazquez-Bermu´ dez MF, Herrero A, Flores E (2003) Carbon supply and 2-oxoglutarate effects on expression of nitrate reductase and nitrogen-regulated genes in *Synechococcus* sp. strain PCC 7942. *FEMS Microbiol Lett* 221: 155-159.

Vazquez-Bermu´dez MF, Herrero A, Flores E (2002) 2-Oxoglutarate increases the binding affinity of the *NtcA* (nitrogen control) transcription factor for the *Synechococcusgln A* promoter. *FEBS Lett* 512: 71-74.

Venkataraman GS (1981)Blue-green algae: a possible remedy to nitrogen scarcity. *Curr Sci* 50: 253-256.

Vincent WF, Neale PJ (2000) Mechanisms of UV damage to aquatic organisms. *In: The Effects of UV Radiation on Marine Ecosystems.* (Eds. de Mora SJ, Demers S, Vernet M), Cambridge Univ. Press, Cambridge, pp. 149-176.

Vincent WF, Roy S (1993) Solar ultraviolet-B radiation and aquatic primary production: damage, protection, and recovery. *Environ Rev* 1: 1-12.

Vosjan JH, Döhler G, Nieuwland G (1990) Effects of UV-B irradiance on the ATP content of microorganisms of the Weedell Sea (Antarctica). *Neth J Sea Res* 25: 391-393.

Walsby AE (1985)The permeability of heterocysts to the gases nitrogen and oxygen. *Proc R Soc Lond B* 226: 345-366.

Walsby AE (2007)Cyanobacterial heterocysts: Terminal pores proposed as sites of gas exchange. *Trends Microbiol* 15, 340-349.

Wang Q, Li H, Post AF (2000) Nitrate assimilation genes of the marine diazotrophic, filamentous cyanobacterium *Trichodesmium* sp. strain WH9601. *J Bacteriol* 182: 1764-1767.

Wei TF, Ramasubramanian TS, Pu F, Golden JW (1993) *Anabaena* sp. strain PCC 7120 *bifA* gene encoding a sequence-specific DNA-binding protein cloned by in vivo transcriptional interference selection. *J Bacteriol* 175: 4025-4035.

Williamson CE (1995) What role does UV-B radiation play in freshwater ecosystems? *Limnol Oceanogr* 40: 386-392.

Wolk CP (1973) Physiology and cytological chemistry of blue green algae. *Bacteriol Rev* 37: 32-101.

Wolk CP, Cai Y, Cardemil L, Flores E, Hohn B, Murry M, Schmetterer G, Schrautemeier B, Wilson R (1988) Isolation and complementation of mutants of *Anabaena* sp. strain PCC 7120 unable to grow aerobically on dinitrogen. *J Bacteriol* 170: 1239-1244.

Wolk CP (1992) Heterocyst formation. *Annu Rev Genet* 30: 59-78.

Wolk CP, Ernst A, Elhai J (1994) Heterocyst metabolism and development. *In: The molecular biology of cyanobacteria* (Eds. Bryant DA), Kluwer Academic Publishers, Dordrecht, pp. 769-863.

Wolk, CP, Zhu J, Kong R (1999) Genetic analysis of heterocyst formation. *In The Phototrophic Prokaryotes* (Eds. Peschek GA, Loeffelhardt W, Schmetterer G), Springer, New York, USA, pp. 509-515.

Wu H, Gao K, Villafañe VE, Watanabe T, Helbling EW (2005) Effects of solar UV radiation on morphology and photosynthesis of filamentous cyanobacterium *Arthrospira platensis*. *Appl Environ Microbiol* 71: 5004-5013.

Xue L, Li S, Zhang B, Shi X, Chang S (2011) Counter active action of nitric oxide on the decrease of nitrogenase activity induced by enhanced ultraviolet-B radiation in cyanobacterium. *Curr Microbiol* 62: 1253-1259.

Yoon HS, Golden JW (1998) Heterocyst pattern formation controlled by a diffusible peptide. *Science* 282: 935-938.

Yoon HS, Golden JW (2001) *Pat S* and products of nitrogen fixation control heterocyst pattern. *J Bacteriol* 183: 2605-2613.

Zhang CC, Laurent S, Sakr S, Peng L, Bedu S (2006)Heterocyst differentiation and pattern formation in cyanobacteria: A chorus of signals. *Mol Microbiol* 59: 367-375.

Zhang Y, Jiang HB, Qiu BS (2012) Effects of UV-B radiation on competition between the bloom-forming cyanobacterium *Microcystis aeruginosa* and the Chlorophyceae *Chlamydomonas microsphaera*. *J Phycol* 49: 318-328.

Zhou R, Wei X, Jiang N, Li H, Dong Y, Hsi KL, Zhao J (1998) Evidence that HetR protein is an unusual serine-type protease. *Proc Natl Acad Sci USA* 95: 4959-4963.

Zhu J, Kong R, Wolk CP (1998) Regulation of *hepA* of *Anabaena* sp. strain PCC 7120 by elements 5' from the gene and by *hepK*. *J Bacteriol* 180: 4233-4242.

In: Trends in Life Science Research
Editors: R. P. Sinha and U. P. Shrivastava

ISBN: 978-1-53613-241-0

Chapter 10

CROP RESPONSES OF BROCCOLI (*BRASSICA OLERACEA* L. *VAR. ITALICA*) TO SULFUR NUTRITION

Wafaa M. Shukry[1,*], Maha M. S. Abdallah[2], Omar A. Elshehaby[1] and Doaa A. Galilah[1]
[1]Department of Botany, Faculty of Science, Mansoura University, Mansoura, Egypt
[2]Department of Botany, National Research Center, Dokki, Giza, Egypt

ABSTRACT

A pot experiment was conducted in the green house during winter season to determine the quality of flowering head and to increase the nutritional values of Broccoli (*Brassica oleracea* L. var. *italica*). Plants were treated with industrial waste water (100%) (19.8 mg sulphate/L) which was collected as an effluent from El Delta company at a distance of about 1 km from the factory in Talkha region of El-Dakhlia governorate. K_2SO_4 (2.5mg /pot) was used as a sulfur source and the plants were grown in different soil types i.e., 100% clay soil, 1:1, 1:2, 1:3, 1:4 (clay:sand) (v/v) soil and 100% sand soil. Our results indicate that the above treatments increased the nutritional values of Broccoli plant in comparison to the control in terms of total carbohydrate and protein, enzymatic antioxidants (superoxide dismutase, SOD; catalase, CAT; ascorbate peroxidase, APX; glutathione reductase, GR) and non-enzymatic antioxidants (total phenols and flavonoids; ascorbic acid (vitamin C) and folic acid (vitamin B9) as well as glucosinolates), amino acids and mineral content. These findings suggest that application of waste water from a fertilizer factory improved the quality and nutritional values of Broccoli, (*Brassica oleracea* L. var. *italica*) plant grown in the soil type 1:4 (clay:sand) (v/v).

Keywords: broccoli, sulfur fertilization, nutritional values, yield

* Corresponding Author Email: Email: wafaa_12@yahoo.com.

INTRODUCTION

The entire world suffers from food shortage as a result of a population burst and huge loss of the agricultural soils due to desertification and erosion. Therefore, it is very essential to increase crop productivity by improving their ability to mitigate environmental stress conditions including low water availability, soil salinity and nutrients deprivation. The increase in agricultural food production worldwide over the past four decades has been associated with a remarkable increase in the use of fertilizers (Hirel et al. 2007).

Broccoli, (*Brassica oleracea* L. var. *italica*) is a cruciferous vegetable and a native to the Mediterranean region. The main edible parts of Broccoli are heads and florets named the inflorescence (head) (Finley 2005). Broccoli's inflorescence has been reported as good components of healthy diets based on their high contents of fiber, vitamin C, carotenoids and glucosinolates. In Egypt, Broccoli is still cultivated in a very limited and scattered region and the total cultivated area is not exactly known. Eating it in high amounts may have additional benefits, since it is a good source of many vitamins and minerals such as vitamin A and C, carotenoids, fiber, calcium and folic acid (Michaud et al. 2002). Broccoli has been known as one of main sources of natural antioxidants as phenolic compounds, vitamins and chemo-preventive compounds such as glucosinolates and their degradation product as isothiocyanates (Finley 2005). Broccoli is a rich source of sulforaphane, which has been shown to display potent anti-cacinogenic properties. Eating a few portions of Broccoli each week may help to decrease the risk of cancer. Nutrient and antioxidants may act together to reduce reactive oxygen species (ROS) level more effectively than single dietary antioxidant, because they can function as synergists (Trombino et al. 2004).

As for other cruciferous vegetable, its culture requires important amounts of sulphur (S) and nitrogen (N) fertilizers. Broccoli, as with most Brassicaceae, has greater S requirements than other large crop species such as wheat or maize (Abdallah et al. 2010 and 2014). Therefore, Broccoli is particularly sensitive to S deficiency or limitation, which reduces both head quality (De Pascale et al. 2008) and quantity. Sulfur (S) is an essential element for plant growth because it is present in major metabolic compounds such as amino acids (methionine and cysteine), glutathione, and proteins. S is a significant component of complete and balanced crop nutrition, and has justifiably gained more attention in recent years. Sulfur has long been recognized as a vital element for plant growth and development. Researchers observed crop responses to applied sulfur in a wide range of soils in many parts of the world (Brady and Weil 1996). Sulfur is available to plant mainly as sulfate (SO_4^{-2}) in the soil and as sulfur dioxide in the atmosphere. The roots of higher plants can only take up sulfur in the ionic form (SO_4^{-2}). Plants can take up SO_2 from the atmosphere through their leaves. Some sulfur compounds like free sulfuric acid, sulfites, sulfides and carbon bisulfide are toxic to plants (Yoshimoto et al. 2002). Sulfur has become more important as a limiting nutrient in crop production, clearly show a growing sulfur deficit in the soils of many regions of the world (Zhao et al. 2007) for several seasons. The main reasons for sulfur deficiency are (1) low organic matter content of the soil; (2) unfavorable soil environment cause decrease in mineralization rate of organic matter; (3) depletion of soil reverse due to intensive cultivation and the applied sulfur free fertilizers; (4) use of high rates of nitrogen and potassium fertilizers, which demand high rate of sulfur.

In the recent years, the limitation of water sources has attracted attention of researchers to proper use of unconventional water sources such as brine water and urban and industrial effluents. Due to urban development and increase in water consumption, a large volume of water is produced from raw wastewater and treated effluents (Naaz and Pandey 2010). Application of urban wastewaters in irrigation has been recommended as a rich source of fertilizing elements in many countries. Wastewater can compromise human health through incidence of bacteriological, viral, protozoan and parasitic diseases (Fattaet al. 2005; Qadiret al. 2014). In addition to providing large quantities of water, some effluents contain considerable amount of essential nutrients, which may prove beneficial for plants. Irrigation with wastewater is taken into consideration as an option to offset shortage of available water. In this case, it is necessary to assess impacts of using urban wastewater on physical and chemical properties to achieve sustainable development (Otokunefor and Obiukwu 2005).

The sandy soil in Egypt is considered as a part of the Sahara Desert of Northern Africa, is exposed to a combination of environmental stress conditions including low water availability, high irradiance, and nutrient deprivation. In these areas the water is the major factor limiting plant growth this water stress may be alleviated by irrigation whenever possible or using unconventional water sources such as industrial effluents, brine and urban water. Such stress may lead to reduce plant growth by affecting various physiological and biochemical processes, such as photosynthesis, respiration, translocation, ion uptake, carbohydrates, nutrient metabolism and growth promoters (Jaleel et al. 2008). So the aim of this work was to investigate the effects of sulfur fertilizers as K_2SO_4 (2.5 mg/pot) beside the use of industrial wastewater for irrigation on growth, some physiological aspects and nutritional values of the head of Broccoli (*Brassica oleracea* L. var. *italica*) plants grown in soil contained different ratios of clay and sand treatment.

In the current study preliminary experiments were carried out to test the ability of Broccoli seeds to germinate in different soil types collected from El-Mansoura district, Egypt according to different ratios from clay: sand v/v as follow: 100% clay soil, 1:1, 1:2, 1:3, 1:4 (clay: sand) (v/v) soil and 100% sand soil. At mature stage, after 98 days from transplanting, all Broccoli heads were harvested to record: head weight (g/plant), head diameter and head dry weight according to Gabalet al. (1985). Also, the analysis of the contents of total carbohydrates, total protein, total ascorbate, total glutathione, total phenols and flavonoids, ascorbic acid (vitamin C), folic acid, glucosinolates contents and amino acids as well as nutrients contents including K, Na, Ca, Mg, Fe and S was carried out.

MATERIALS AND METHODS

Plant Used and Growth Condition

This investigation was conducted at the green house of Faculty of Science, Mansoura University, Mansoura governorate, Egypt, during the winter season (2013-2014). Pure strain of Broccoli (*Brassica oleracea* L. var. *italica*) seeds was kindly supplied by the agriculture research center, ministry of agriculture, Giza, Egypt. Seed beds were well prepared, watered early in the morning with sprinkler till germination and seedling development. After 45 days, uniform plants were transplanted into plastic pots (35 cm in diameter) arranged in green house and filled with different types of soil. Each pot contains 10 kg soil. The area of the experimental plot was 25 m^2 consisted of many rows, each row consist of 5 pots. The plant distance was 50 cm a part on one side; pots were directly irrigated by normal irrigation system. Agricultural management, diseases and pest control program were followed according to the recommendation of the Egyptian ministry of agriculture. The following treatment was used as a source of S fertilization; industrial waste water as an effluents from El Delta Company for fertilizers of chemical industries (ASMEDA) at a distance of 1 Km from the factory in Talkha region from El-Dakhlia governorate as a source for irrigation, K_2SO_4 (2.5 mg/pot) and control (without treatment).

In the present study, preliminary experiments were carried out to test the ability of Broccoli seeds to germinate in different soil types collected from El-Mansoura district, Egypt according to different ratios from clay: sand v/v as follow:100% clay soil, 1:1, 1:2, 1:3, 1:4 (clay: sand) (v/v) soil and 100% sand soil (Table 1). At mature stage, after 98 days from transplanting, all Broccoli heads were harvested to record: head weight (g/plant), head diameter and head dry weight according to Gabalet al. (1985). Also, the analysis of the contents of total carbohydrates, total protein, total ascorbate, total glutathione, total phenols and flavonoids, ascorbic acid (vitamin C), folic acid, glucosinolates contents and amino acids as well as nutrients contents including K, Na, Ca, Mg, Fe and S was carried out.

Table 1. Some physical and chemical properties of the used soil with different ratios of clay to sand

Soil characters / Soil type	Soluble ions (mg/100g dry weight soil)												Available (mg\kg)		
	pH	EC	O.C (%)	Ca^{+2}	Mg^{++}	Na^{+}	K^{+}	TSS	$CaCO_3$	HCO_3^- (%)	Cl^- (%)	SO_4^{-2} (%)	N	P	K
100% clay	7.1	153.4	1.78	26.82	13.85	517.2	81.35	551.2	0.6	0.18	0.143	1.66	48.7	5.92	375
(1:1) (clay : sand v/v)	6.94	146.7	1.72	24.62	12.95	379.8	57.67	360.1	4.9	0.14	1.12	1.45	32.8	4.05	253
(1:2) (clay : sand v/v)	7.4	147.0	1.42	22.41	11.25	242.9	48.27	414.0	3.2	0.12	0.83	1.44	30.4	3.71	241
(1:3) (clay : sand v/v)	6.96	129.5	1.28	20.01	11.13	211.7	36.52	204.7	2.5	0.11	0.67	1.44	29.5	3.52	235
(1:4) (clay : sand v/v)	7.1	144.2	1.18	19.61	10.04	201.6	29.47	404.9	1.2	0.09	0.47	0.91	24.1	3.07	198
100% sand	7.3	133.6	0.88	18.81	9.52	191.2	22.91	311.5	0.8	0.06	0.28	0.87	22.9	2.93	194

Chemical Analysis

Total Soluble Carbohydrates

Total soluble carbohydrates were extracted by overnight submersion of dry tissue in 10 mL of 80% (v/v) ethanol at 25°C with periodic shaking according to the method described by (Homme et al. 1992). TSS were analyzed by reacting of 0.1 mL of ethanolic extract with 3.0 mL freshly prepared anthirone (150 mg anthrone+100 mL 72% H_2SO_4) in boiling water bath for ten minutes and reading the cooled samples at 625 nm using Spekol spectrocolorimeter VEB Carl Zeiss (Yemm and Willis1954).

Total Protein Extraction

According to the method as demonstrated by Bradford (1976), protein was extracted by macerating a known fresh weight of sample in a Mortar and Pestle with 2 mL of extraction buffer (0.2 M tris HCl (pH 6.8), 0.2% SDS, and 10% sucrose) then centrifuged at 4000 rpm for 15 min. Extraction buffer was prepared by adding 5g sucrose to 10 mL 1M tris HCl, 10 mL 10% SDS and completed to 50 mL with distilled water.

Assay of Non-Enzymatic Antioxidants

Ascorbate (AsA)

The water soluble AsA was determined by the methods of Kampfenkel et al. (1995). Total AsA was determined in a reaction mixture of 1.0 mL containing 50 μL apoplastic fluid, 100 μL phosphate buffer 0.2 M (pH 7.4), 50 μL dithiothreitol (DDT). After shaking and incubation for 20 min at 42ºC in water bath, 50 μL of 0.5% N-ethylmaleimide (NEM) was added, followed by incubation for 1 min at room temperature. AsA levels were determined in the same way with exception for the increasing of DDT and NEM which were substituted by water. Coloring was obtained by adding of 250 μL 10% trichloroacetic acid, 200 μL of 42% phosphoric acid, 200 μL of dipiridil dissolved in 70% ethanol and 100 μL of 3% phosphoric acid. After shaking, the mixture was incubated for 40 min at 42ºC and the absorbance values of the final colored solutions were recorded at 525 nm.

Glutathione (GSH)

About 0.2g of fresh Broccoli fruit was homogenized with 5 cm^3 of chilled buffer (50 mM potassium phosphate buffer, pH 7.5 and 1 mM EDTA). The homogenate was then centrifuged at 14,000 g for 20 min. The supernatant was used for GSH measurement. Determination of GSH was performed spectrophotometrically by the method of Beutler et al. (1963).

Total Phenolic and Flavonoid

Broccoli florets were ground into fine powder with liquid nitrogen, 0.5g was placed in a tube and 5 mL 80% methanol was added. Tubes were vortexed and left to stand for 20 min, centrifugation at 12000 rpm at 4°C for 5 min. The supernatant stored at 20°C and analyzed for determination of total phenolic and flavonoids within 2 weeks. The total phenolic content extract of Broccoli florets was determined according to the method modified by Atanassova et al. (2011). The total flavonoid content was measured with an aluminum chloride colorimetric assay described by Marinova et al. (2005).

Ascorbic Acid (Vitamin C) and Folic Acid

Ascorbic acid was quantified according to method of Harris and Ray (1935). Folic acid was determined according to simple spectrophotometric method as described by Rengstagoon and Hattanat (2012). One mL of the standard or sample solution of folic acid was mixed with 1 mL of 4M L-1hydrochloric acid, 1 mL of 1% (w/v) sodium nitrite, 1.0 mL of 1% (w/v) sulfamic acid and 0.1 mL of 1% (w/v) 3-aminophenol which result in an orange yellow complex. The absorption was measured at 460 nm using UV-visible spectrophotometer.

Glucosinolates

Glucosinolates have been estimated colorimetrically using the method of Gupta et al. (2012). Glucosinolates estimation based on the formation of a complex between hydrolytic products of glucosinolates and sodium tetrachloropalledate (II) and this complex is then analyzed colorimetrically. The amount of glucosinolate palladium complex formed or the rate of its formation is used to determine the quantity of glucosinolates. Heads were crushed in methanol and centrifuged at 5000 rpm. Sodium tetrachloropalledate (II) was added to the supernatant and glucosinolates content was read at 405 nm in ELIZA reader.

Assay of Enzyme Activities

Enzyme extracts were collected following the method described by Chen and Wang (2006). Leaf tissues were homogenized in ice-cold phosphate buffer (50 mM, pH 7.8), followed by centrifugation at 8,000 rpm and 4°C for 15 min. The supernatant was used immediately to determine the activities of enzymes.

Superoxide Dismutase

Superoxide dismutase, SOD (EC 1.12.1.1) activity was spectrophotometrically assayed at 560 nm by nitro-blue-tetrazolium (NBT) reduction method (Chen and Wang 2006). The reaction mixture (3 mL) contained 150 μL riboflavin (13μM), 2.5 mL methionine (13μM), 250 μL NBT (63 μM), 50 μL phosphate buffer (50 mM, pH 7.8) and 50 μL enzyme extract. One unit of SOD activity was defined as the amount of enzyme protein required for inhibition of the 50% reduction of NBT.

Catalase

Catalase, CAT(EC 1.11.1.6) activity was determined spectrophotometrically by following the decrease in absorbance at 240 nm (Chen and Wang 2006). The mixture (3 mL) contained 1.9 mL phosphate buffer (50 mM, pH7.0), 100 μL enzyme extract and 1 mL 0.3% H_2O_2. The reaction was initiated by adding enzyme extract. One unit of CAT activity was defined as the 0.01 deduction in absorbance at 240 nm per minute. The enzyme activities were calculated by Kong et al. (1999).

Ascorbate Peroxidase

Ascorbate peroxidase, APX (EC 1.11.1.11) activity was determined as described by Nakano and Asada (1981). The reaction mixture (1 mL) contained 50 mM potassium phosphate (pH 7.0), 1.2% sucrose, 1 mM hydrogen peroxide, 0.5 mM ascorbate and 10 mM 3-aminotriazole (an inhibitor of catalase). The hydrogen peroxide-dependent oxidation of ascorbate was followed by monitoring the decrease in absorbance at 290 nm, using the extinction coefficient of 2.8 mM cm^{-1}.

Glutathione Reductase

Glutathione reductase, GR (EC 1.6.4.2) activity was determined by the oxidation of NADPH at 340 nm with the extinction coefficient of 6.2 mM cm^{-1} as described by Lee and Lee (2000). The reaction mixture was composed of 100 mM potassium phosphate buffer (pH 7.8), 2 mM EDTA, 0.2 mM NADPH, 0.5 mM glutathione (oxidized form, GSSG) with an appropriate volume of enzyme extract in a 1 mL volume. The reaction was initiated by the addition of NADPH at 25°C.

Determination of Amino Acids

Free amino acids in the present investigation were determined using Eppendorf amino acid analyzer (LC 3000 Amino Acid Analyzer).

Detection and Quantification

Elutes from the column were detected spectrophotometerically. Eluent peaks with their areas and retention time were recorded by an attached integrator. The proper concentrations of free amino acids were obtained by comparing their respective peak areas on the chromatograms of the plant extracts with their corresponding areas obtained with the authentic samples (Rita Steed 2010).

Determination of Ion Contents

Samples were dried in an oven at 80°C till constant dry weight and dry weights of samples were recorded. The dried matter was digested in concentrated HNO_3 and made up to volume with deionized distilled water as described by Motsara and Roy (2008). K^+ and Na^+ concentrations in the plants were measured by Flame-Emission Spectrophotometer. Ca^{+2}, Mg^{+2} and Fe^{+2} concentrations were measured by atomic absorption spectrophotometry (PHF 80B Biology Spectrophotometer). Meanwhile, sulfur content was determined as described by Soon et al. (1996). Approximately 0.5g of plant sample was pre-digested in 7 mL of HNO_3 for 1h. Then 3.5 mL of 70% $HClO_4$ were added and the contents of the closed Teflon vessel digested at 100% power for 3 min and at 40% for a further 20 min. The samples were made up to 30 mL volume, filtered through a Whatman No.1 filter paper and analyzed by an ARL 3580 ICP spectrometer. Digestion was *accomplished* in a CEM microwave oven (model MDS 2000). Data were calculated as μmol g^{-1} dry weight.

Statistical Analysis

The data were statistically analyzed on complete randomized design system according to Snedecor and Cochran (1980). Combined analysis of the two growing seasons was carried out. Means were compared by using least significant difference (LSD) at 5% levels of probability.

RESULTS

Growth Parameters

Many parameters of head of Broccoli plant grown in different soil types and treated with either wastewater or K_2SO_4 (2.5 mg/pot), were determined including head diameters and fresh and dry weight as illustrated in Figure 1. The used treatments had a significant increase in all the measured parameters of Broccoli head in all soil types than that of control plants. In 100% clay soil, the diameter of head was significantly increased with both treatments comparing with the control. The treatment with waste water recorded the highest value relative to K_2SO_4 treatment. Fresh and dry weights of head responded in a similar manner to that of head diameter. Treatment with waste water follow this sequence for both fresh and dry weight of head: (1:2) > 100% clay soil > (1:1) > (1:4) > (1:3) (clay: sand v/v) soil for all tested parameters. Meanwhile, for K_2SO_4 (2.5 mg/pot) treatment and according to soil type it follow this sequence: 100% clay soil >1:1 > 1:2 ≥ 1:3 ≥ 1: 4 (clay: sand v/v) soil. In 100% sand soil (Figure 1) all head traits (head diameter, head fresh and dry weight), subjected to significant increase in treatment with waste water comparing with the control. Meanwhile in case of K_2SO_4 application at concentration (2.5mg/pot); there was a complete inhibition of plant productivity of inflorescences.

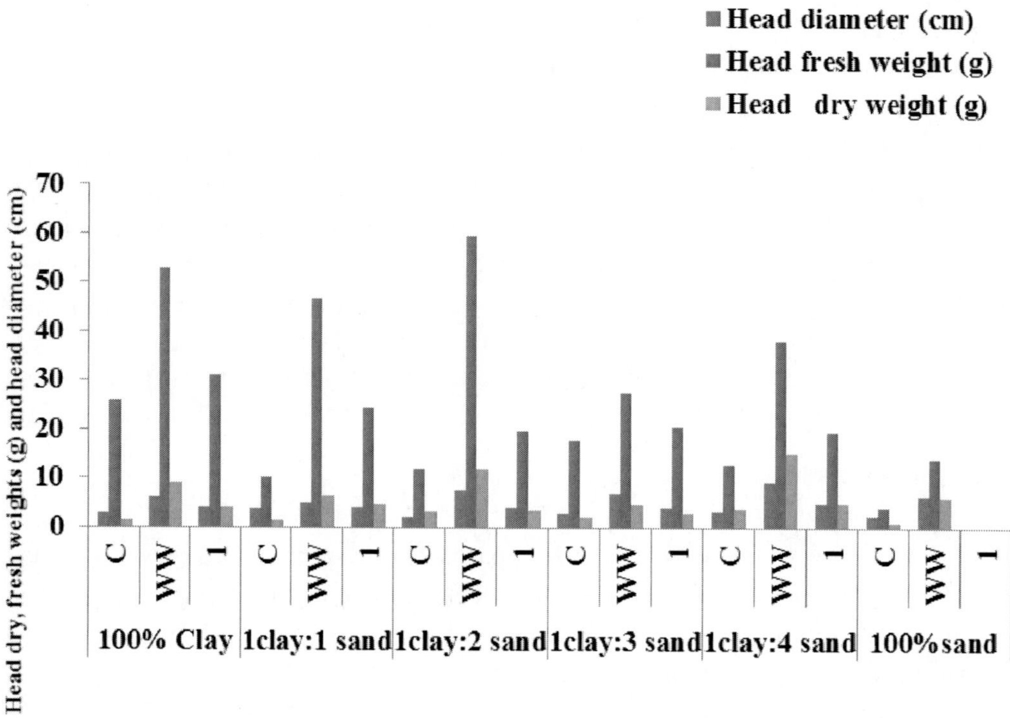

Figure 1. Effect of inorganic fertilizer, K_2SO_4 (2.5 mg/pot), and waste water on head diameter, fresh and dry weight of *Brassica oleracea* in different soil composition (C: control; WW: waste water; 1: K_2SO_4 (2.5 mg/pot) *Non- significant change at 5% level.

Total Carbohydrates

Total carbohydrates content of Broccoli head is demonstrated in Figure 2, which indicates significant differences in the content of total carbohydrates according to either the use of waste water or the application of K_2SO_4 (2.5 mg/pot). Also, the mixing of clay soil with sand soil in different ratios achieved a significant variance in total carbohydrate content of head of Broccoli. Both treatments achieved a significant increase in all types of soil meanwhile, the highest content is recorded in soil type (1:1) (clay: sand v/v) amounted to 142.63 and 122.28 mg/g dry weight for treatment with waste water and K_2SO_4 (2.5 mg/pot) respectively. The pattern was similar upon growth in other soils under investigation with a magnitude of response in the following order: (1:3) then (1:4) and (1:2) (clay: sand v/v) respectively, in comparison to control values.

Total Protein

Total protein was determined in different soil types under effect of either waste water or K_2SO_4 (2.5 mg/pot). Data tabulated for total protein in Broccoli head (Figure 2) showed that, there was significant increase in the total protein content in all soil types in case of treatment with waste water following this sequence: 1:1 > 1:3 > 1:2 > 1:4 (clay: sand v/v) soil > 100% clay soil > 100% sand soil. Treatment with waste water achieved a higher content of total protein than in treatment with K_2SO_4 (2.5 mg/pot). The application of K_2SO_4 (2.5 mg/pot) induced a significant increase and decrease in soil type (100% clay, 1:1 and1:3 (clay: sand v/v)) and (1:2 and 1:4 (clay: sand v/v)) respectively. Concerning the control values, the highest overall content of total protein was obvious in (1:1) (clay: sand v/v) soil which enhanced the highest overall content of total protein between all soil types used.

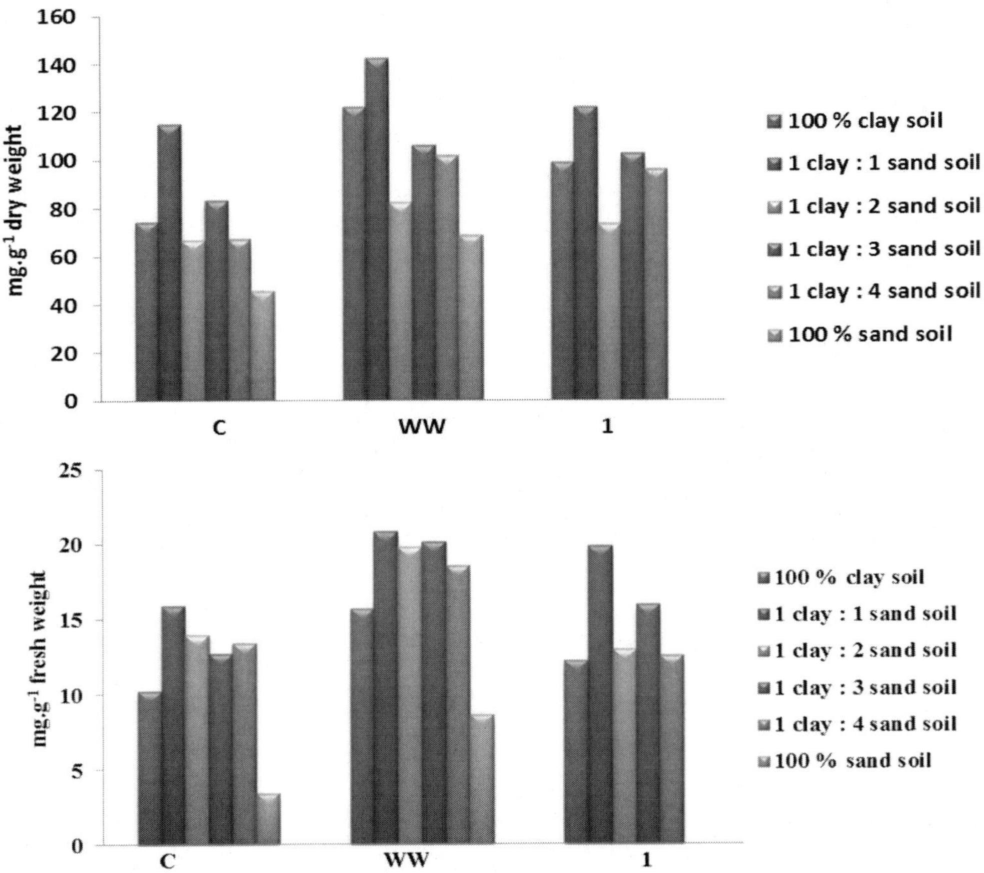

Figure 2. Effect of inorganic fertilizer, K_2SO_4 (2.5 mg/ pot) and waste water on the total carbohydrate and protein contents of *Brassica oleracea* grown in different types of soil. Data are represented as mg/g dry weight for total carbohydrates and $\mu g/g^{-1}$ fresh weight for total proteins (C: control; WW: waste water; 1: K_2SO_4 (2.5 mg/pot). *Non- significant change at 5% level.

Antioxidant System

Non Enzymatic Antioxidants

Total Ascorbate (AsA) and Glutathione (GSH)

The changes in total AsA and reduced GSH content in response to the various treatments and different soil types are shown in Figure 3. It is apparent that, ASA content differed according to different treatments including waste water or K_2SO_4 (2.5 mg/pot). Treatment with waste water followed this sequence: 100% clay soil > (1:4) > (1:2) > (1:3) > (1:1) (clay: sand v/v) soil > 100% sand soil. Generally, treatment with K_2SO_4 at the selected concentration achieved significant different values for the variously treated samples and obeyed the following sequence: 100% clay soil > (1:2) > (1:1) (clay: sand v/v) soil, meanwhile at soil type 1:4 and 1:3 (clay: sand v/v) a non-significant change was observed. As for the reduced GSH content, a slight increase was maintained higher than that of AsA in response to both experimental treatments, however, the magnitude of increase appeared to follow more or less similar trend in treatment with waste water according to different type of soil. Meanwhile the content of GSH in different type of soil according to treatment with K_2SO_4 follow this sequence: 1:4 > 1:3 ≥ 1:1 ≥ 1: 2 > 100% clay soil > (clay: sand v/v) soil.

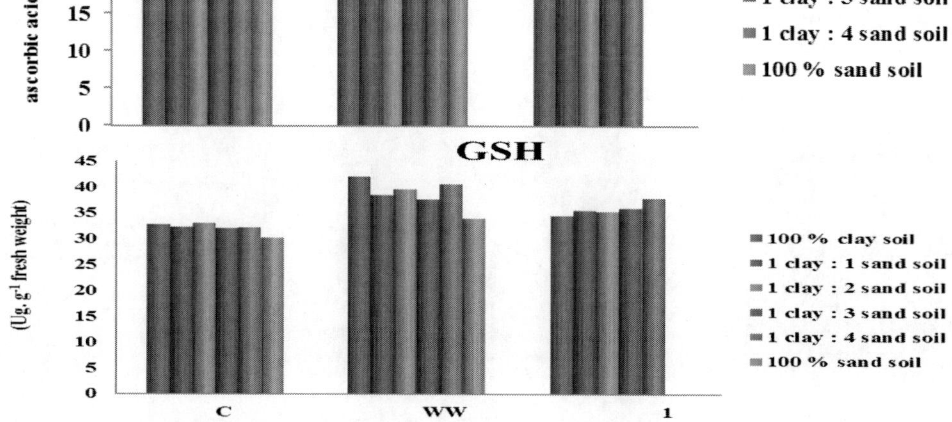

Figure 3. Effect of inorganic fertilizer K_2SO_4 (2.5 mg/pot) and waste water on activity of antioxidant compound of *Brassica oleracea* inflorescence grown in different soil type (C: control; WW: waste water; 1: K_2SO_4 (2.5 mg/pot) *Non- significant change at 5% level.

Total Phenols and Flavonoids

Many recent studies showed a close relationship between different treatments (fertilization) and bioactive components in the yield of plants, from these compounds total phenols and flavonoids. The changes in content of total phenols and flavonoids are given in Figure 4 which showed, in general, a significant increase in such contents by effect of waste water or K_2SO_4 (2.5 mg/pot). This sequence was displayed with respect to treatment with waste water: 100% clay > (1:4) > (1:2) > (1:1) > (1:3) (clay: sand v/v) > 100% sand soil. Waste water achieved a higher significant increase in different types of soil than in treatment with K_2SO_4 (2.5 mg/pot). Application of K_2SO_4 (2.5 mg/pot) induced a significant progressive increase in total phenols in (1:1) and (1:4) (clay: sand v/v) and a non-significant increase in 100% clay soil, (1:2) and (1:3) (clay: sand v/v) in head of Broccoli plants.

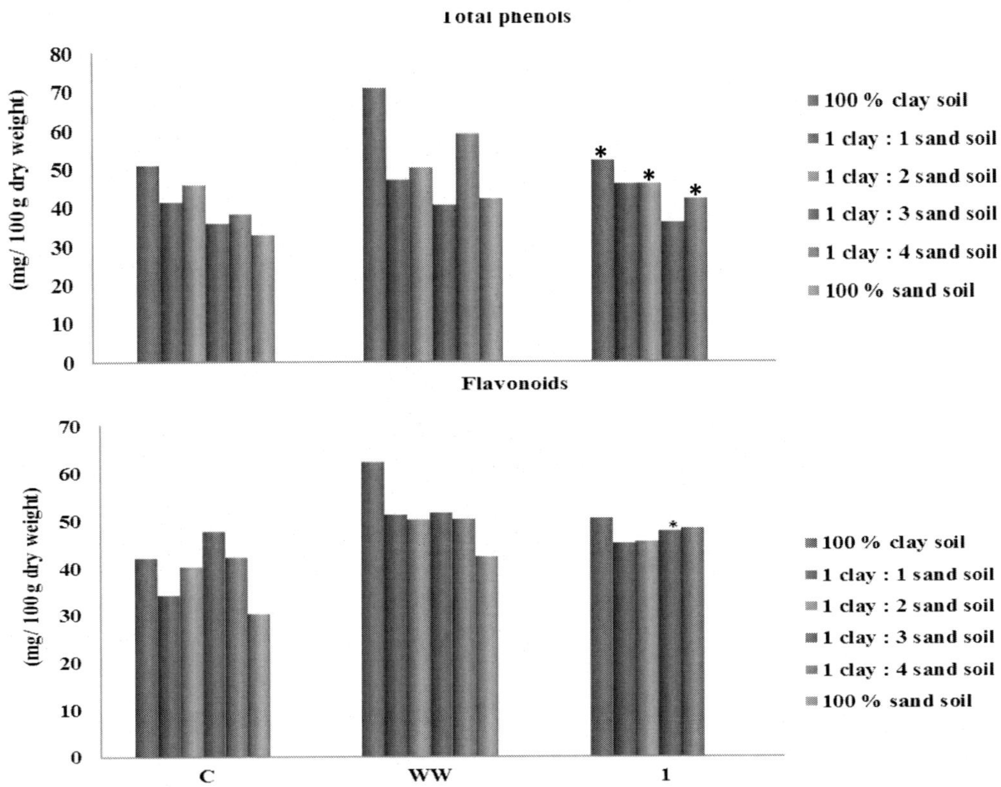

Figure 4. Effect of inorganic K_2SO_4 (2.5 mg/ pot) and waste water fertilizer on total phenols and flavonoids content in inflorescence of *Brassica oleracea* plant grown in different types of soil, data are represented as mg/100 g dry weight (C: control; WW: waste water; 1: K_2SO_4 (2.5 mg/pot)) *Non-significant change at 5% level.

Concerning flavonoids content, the data revealed a various changes in the content of flavonoids according to different types of soil. According to this, flavonoids content followed this sequence: (1:3) > (1:4) ≥ 100% clay soil > (1:2) > (1:1) (clay: sand v/v) >

100% sand soil, with respect to control values. Using waste water achieved a high significant increase in flavonoids contents in inflorescence than that in treatment with K_2SO_4 (2.5 mg/pot) in comparing with the control values. For using waste water, the flavonoids content recorded the highest value in 100% clay soil than (1:3) ≥ (1:1) > (1:4) > (1:2) (clay: sand v/v) > 100% sand soil. On the other side, on application of K_2SO_4 (2.5 mg/pot) to different type of soils, it induced a significant increase in all soil types except in (1:3) (clay: sand v/v), which induced a non- significant increase in flavonoids content (Figure 4).

Ascorbic Acid and Folic Acid

The data in Figure 5 shows the the bioactive compounds of Broccoli head such as vitamins. Selected samples taken for analysis included: samples treated with wastewater or inorganic fertilizer K_2SO_4 (2.5 mg/pot) in addition to use tap water as a control sample. Concerning to vitamins, vitamin C (ascorbic acid) and vitamin B9 (folic acid) in Broccoli head, were subjected to analysis. From the data obtained, it was obvious that the content of ascorbic acid in plants treated with waste water is higher than that of K_2SO_4 (2.5 mg/pot) comparing with control samples and obey the following sequence: 100% clay> (1:4) > (1:2) > (1:1) > (1:3) (clay: sand v/v) >100% sands soil in treatment with waste water.

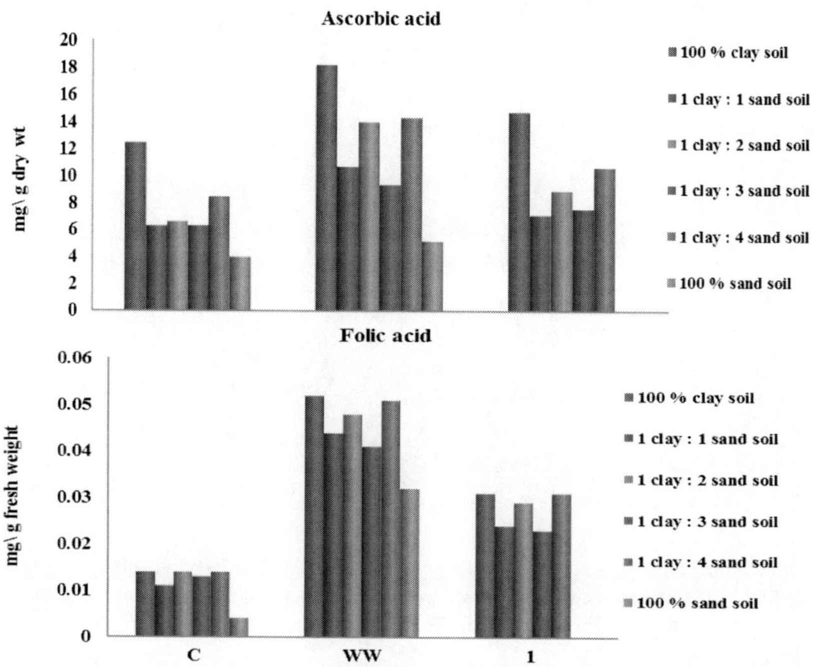

Figure 5. Effect of inorganic fertilizer K_2SO_4 (2.5 mg/ pot) and waste water on the content of ascorbic acid (vitamin C) and folic acid (vitamin B9) in inflorescence of *Brassica oleracea* plant grown in different types of soil (C: control; WW: waste water;1: K_2SO_4 (2.5 mg/pot)) *Non- significant change at 5% level.

The present investigation was conducted to assess the possible effects of waste water and inorganic fertilizers K_2SO_4 (2.5 mg/pot) on contents of folic acid of the selected samples. Treatment with waste water induced a significant increase in the content of folic acid than those of K_2SO_4 treatment comparing with control values. The content of folic acid varies from soil type to another and followed this arrangement: 100% clay, (1:4) > (1:2) > (1:1) > (1:3) (clay: sand v/v).

Glucosinolates (GLS)

Total GLS content were detected and shown in Figure 6, which indicates different soil types with different treatments either inorganic K_2SO_4 (2.5 mg/pot) or using waste water. All soil types achieved a significant increase in glucosinolates content except in (1:3) (clay: sand v/v) in which both treatments achieved a non-significant increase in GLS content. The highest overall content was achieved in soil type (1:1) (clay: sand v/v). According to waste water, such treatment enhanced higher values than the other treatment in all soil types, the following sequence was displayed with respect to the control value: (1:1) > (1:4) >100 clay > (1:2) > (1:3) (clay: sand v/v) > 100% sand soil. Again, exogenous application of K_2SO_4 stimulated an increase in total GLS and followed the pattern as follows: 1:1 > 100% clay > 1:4 > 1:3 > 1:2 (clay: sand v/v) in comparing to control values.

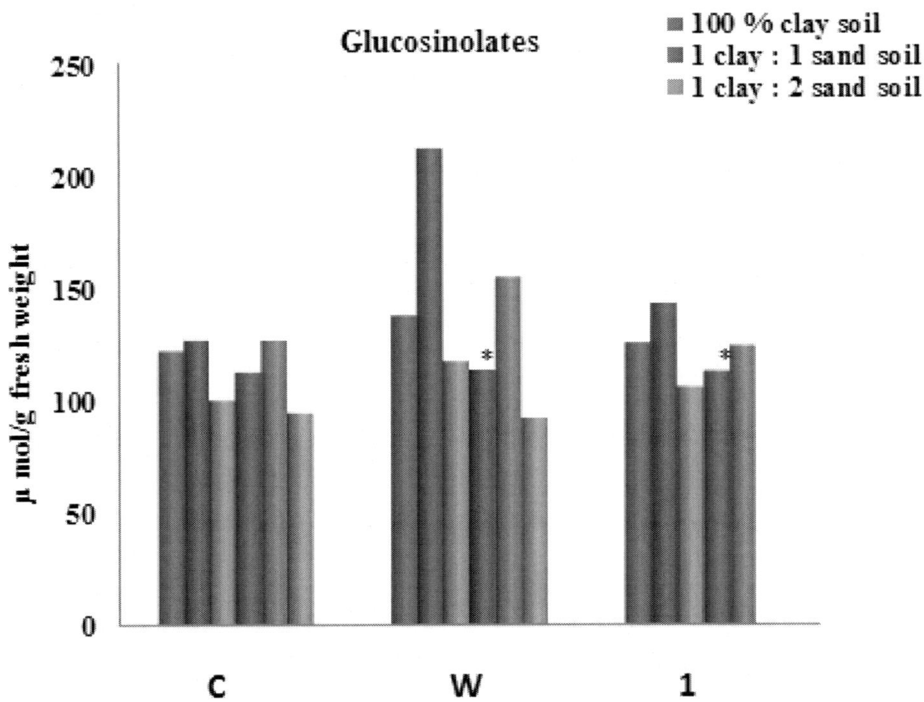

Figure 6. Effect of inorganic fertilizer K_2SO_4 (2.5 mg/pot) and wastewater on content of glucosinolates in inflorescence of *Brassica oleracea* plant grown in different soil types (C: control; WW: waste water; 1: K_2SO_4 (2.5 mg/pot)) *Non-significant change at 5% level.

Changes in Activities of Antioxidant Enzymes

SOD is regarded as the primary scavenger for ROS where it converts superoxide free radical anion (O_2) to H_2O_2 and O_2 (Rahnama and Ebrahim Zadeh 2005), and CAT, APX and GR detoxify hydrogen peroxide in peroxisomes, cytosol and chloroplast. Their activities were measured as representing enzymes involved in antioxidative metabolism and altered upon the interactive effect of different sources of sulfur in Broccoli head. The present study was aimed to investigate the changes in the antioxidant system. Thus, selected samples of the variously treated Broccoli heads were taken for analysis of antioxidant compounds and antioxidant enzymes activities. Figure 7 shows the activity of SOD, CAT, APX and GR enzymes in heads of the variously treated Broccoli plants, either with waste water or inorganic fertilizer K_2SO_4 (2.5 mg/pot) and grown in different types of soil.

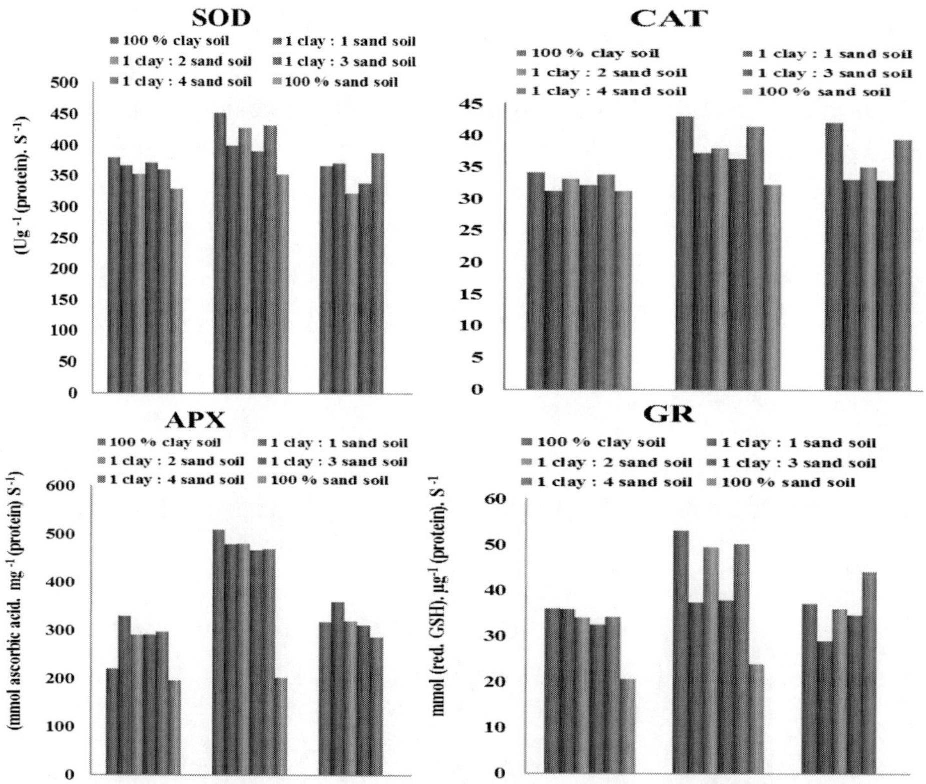

Figure 7. Effect of inorganic fertilizer K_2SO_4 (2.5 mg/ pot) and waste water on the activity of antioxidant enzyme of *Brassica oleracea* inflorescences grown in different soil type (C: control; WW: waste water; 1: K_2SO_4 (2.5 mg/pot)). *Non-significant change at 5% level.

Superoxide Dismutase (SOD)

SOD specific activities (unit/mg proteins) are represented in Figure 7. The data revealed that the activity of SOD in Broccoli heads increased significantly in waste water treatment than the treatment with K_2SO_4 as compared with the absolute control value in all types of soils. The highest value of SOD activity was observed in 100% clay soil in both control and waste water treatment, meanwhile the highest value in treatment with K_2SO_4 (2.5 mg/pot) was at soil type (1:4) (clay: sand v/v).

Catalase (CAT)

Perusal of the data in Figure 7 revealed a significant increase in CAT activity in Broccoli head owing to treatment with or with K_2SO_4 (2.5 mg/pot) and comparing to control value. The magnitude of increase being more obvious when treatment with waste water in 100% clay and (1:4) (clay: sand v/v). In Broccoli head, the following sequence of treatments: 100% clay soil > (1:4) > (1:2) > (1:1) > (1:3) (clay: sand v/v) was displayed generally with respect to the activity of catalase enzyme.

Ascorbate Peroxidase (APX)

In relation to control values, treatment with waste water orK_2SO_4 (2.5 mg/pot) induced a significant increase in the activity of APX in the head of Broccoli (Figure 7). Furthermore, application of waste water induced the highest increase in APX activity. An activity value comparable to that of control and other types of soil being increased in case of using 100% clay soil when treated with waste water and in (1:1) (clay: sand v/v) when treated with K_2SO_4 (2.5 mg/pot). Meanwhile, the observed decline in the APX activity, in response to treatment with waste water appeared when application of 100% sand soil.

Glutathione Reductase (GR)

GR activity of Broccoli head were found to increase significantly and variably in the variously treated Broccoli plants, the magnitude of increase being highest upon the use of waste water Figure 7. The highest values were achieved in application of 100% clay soil and (1:4) (clay: sand v/v) with respect to the activity level of GR followed with the other ratios of soil. On the other hand, a sharp decrease in the actively level of GR was observed in case of 100% sand soil when treated with waste water. However, in treatment with K_2SO_4 (2.5 mg/pot), there was a significant increase in GR activity comparing with the control except in soil type (1:1) (clay: sand v/v), where it recorded a significant decrease in GR activity. On the other side, the highest value of GR activity was recorded at (1:4) (clay: sand v/v) in treatment with K_2SO_4 (2.5 mg/pot).

Amino Acids Content

The data in Table 2, revealed the changes in the amino acids composition of the variously yielded inflorescences of Broccoli plants. With respect to amino acids composition of Broccoli inflorescences of plants grown in different types of soil and treated with K_2SO_4 (2.5 mg/pot) or irrigated with waste water as a source of sulfur is presented in Table 2. The level of amino acids generally showed a significant difference between different treatments and different types of soil.

In soil type 100% clay, essential amino acid histidine (His) and L-valine (Val) recorded two times than the control value in using K_2SO_4(2.5 mg/pot) as inorganic sulfur fertilizer, meanwhile L-arginine (Arg) recorded three times than the control. For treatment with waste water, there was increase in all essential amino acids comparing with the control. For non- essential amino acids, there was an increase in its contents in either treatments with K_2SO_4 (2.5 mg/pot) or with waste water comparing to the control. Glutamine (Glu) recorded a bi-fold increase in treatment with waste water than the control. So, the total amino acids in treatment with waste water was higher than the treatments with K_2SO_4, meanwhile the essential/ non-essential (Ess/Non-Ess) ratio was higher in treatment with K_2SO_4 than with waste water. In soil type (1:1) (clay: sand v/v), L- arginine (Arg) and L-lysine (Lys) were recorded a higher value than the treatment with K_2SO_4. In general, non-essential amino acids recorded a less values in either treatments with K_2SO_4 or waste water than the control. Also, all non-essential amino acids except tyrosine (Tyr) recorded a high value in sulfur treatment with K_2SO_4 than with waste water. So, the total amino acid in Broccoli inflorescences was high in treatment with K_2SO_4 than with waste water, meanwhile the (Ess/ Non-Ess) was high in waste water than with K_2SO_4. As shown in Table 2, the treatment of soil type (1:2) (clay: sand v/v) with either waste water or K_2SO_4 induced a variable changes in essential amino acids of Broccoli inflorescences comparing with the control where L-methionine (Meth) disappeared in either treatments with waste water or K_2SO_4. The use of K_2SO_4 fertilizer induced a drastic decrease in essential amino acids; Histidine (His) and L-phenylalanine (Phe). On the other side, amino acids L- lysine (Lys), L-arginine (Arg) and L-valine (Val) were recorded in bi-fold amount than the control.

Sulfur fertilization with waste water induced an increase in essential amino acids His, Ile, Leu, Arg and Val, meanwhile Lys recorded a three times higher value than the control. For non- essential amino acids in treatment with K_2SO_4, there were more or less similar to the control values. So, the waste water recorded a high value of total amino acid and low value for Ess/ Non-Ess ratio. On the other side, this ratio Ess/Non-Ess was more or less similar in treatment with K_2SO_4 and control. The results represented in Table 2, revealed the following main changes in essential amino acids for (1:3) (clay: sand v/v) soil. The changes in essential amino acids in treatment with K_2SO_4 revealed that His recorded a threefold increase comparing with the control value and bi-fold increase was observed for Leu and Phe. Moreover, there was *de novo* synthesis of essential amino acid Meth which disappeared in inflorescence of Broccoli plants in treatment with waste water and in control. Also, in waste water treatments, there was disappearance of essential amino acid His and Ile. For non-essential amino acids, in general, there were an increase in Gly, Pro, Tyr, Ala, and decrease in Glu in treatment with K_2SO_4 compared with the control and the waste water. So, the total amino acids recorded a high value in treatment with K_2SO_4 than the treatment with waste water and the same trend was observed for Ess/Non-Ess ratio.

Table 2. Effect of K_2SO_4 (2.5 mg/ pot) and waste water on contents of amino acids of *Brassica oleracea* plants grown in different types of soil

Soil type	Treat-ments	Essential amino acids									Non-Essential amino acids							Ess. a.as.	Non-Ess. a.as	Total amino acids	Ess\Non-Ess
		HIS	ILE	LEU	LYS	PHE	THR	METH	ARG	VAL	GLY	PRO	SER	TYR	ALA	ASP	GLU				
100% clay	Control	10.11	6.09	30.33	15.10	17.62	467.5	-	23.30	16.18	10.09	7.27	18.77	7.83	32.56	35.33	227.75	127.9	339.6	467.5	0.38
	Waste water	15.09	8.83	42.66	24.82	23.05	836.6	-	61.59	27.77	12.75	17.49	28.27	15.04	48.24	47.04	450.06	217.7	618.9	836.6	0.35
	K_2SO_4 (2.5 mg/pot)	21.98	9.63	45.74	21.19	29.85	734.0	0.76	69.92	31.89	14.30	18.49	25.19	17.61	51.41	45.75	318.17	243.1	490.9	734.0	0.50
1 clay: 1 sand	Control	16.46	9.32	43.74	18.40	25.86	770.4	-	51.49	23.62	12.56	17.20	24.90	19.33	47.23	43.02	404.11	202.0	568.4	770.4	0.36
	Waste water	14.68	9.50	35.22	19.27	15.93	592.7	2.21	68.78	25.64	10.15	15.96	14.86	13.44	39.26	33.11	267.92	198.0	394.7	592.7	0.50
	K_2SO_4 (2.5 mg/pot)	13.54	8.21	34.83	16.51	16.61	669.9	-	42.27	25.41	11.37	19.58	22.76	8.77	42.35	41.09	355.68	168.3	501.6	669.9	0.34
1 clay: 2 sand	Control	13.20	8.96	37.43	11.81	16.21	559.2	1.15	31.91	28.32	11.82	13.29	23.12	10.65	43.30	39.67	256.94	160.4	398.8	559.2	0.40
	Waste water	18.60	10.99	45.54	33.35	9.17	1399.5	-	52.64	42.89	11.66	-	19.69	25.58	48.92	40.22	1030.35	223.1	1176.4	1399.5	0.19
	K_2SO_4 (2.5 mg/pot)	6.86	11.80	52.89	23.92	5.01	747.6	-	60.40	55.26	12.85	13.95	23.13	19.79	57.25	41.19	351.50	227.9	519.7	747.6	0.44
1 clay: 3 sand	Control	7.45	11.81	46.79	24.67	29.95	868.2	-	51.56	62.38	12.56	20.21	22.83	31.83	45.87	44.17	444.17	246.6	621.6	868.2	0.40
	Waste water	-	-	38.34	11.40	14.91	532.1	-	26.87	30.14	6.22	5.10	10.95	26.58	22.53	20.62	314.19	125.9	406.2	532.1	0.31
	K_2SO_4 (2.5 mg/pot)	21.57	18.10	71.08	14.20	67.57	847.5	16.32	54.41	38.04	22.55	31.74	24.10	55.68	92.09	42.57	266.71	312.1	535.4	847.5	0.58

Table 2. (Continued)

Soil type	Treat-ments	Essential amino acids									Non-Essential amino acids							Ess. a.as.	Non-Ess. a.as	Total amino acids	Ess\ Non-Ess
		HIS	ILE	LEU	LYS	PHE	THR	METH	ARG	VAL	GLY	PRO	SER	TYR	ALA	ASP	GLU				
1 clay: 4 sand	Control	-	4.47	23.31	12.91	2.57	463.82	-	20.50	26.49	5.81	12.66	9.94	11.97	23.47	22.53	283.32	94.12	369.7	463.82	0.25
	Waste water	13.83	7.60	31.11	17.81	15.64	623.2	1.11	36.89	24.21	9.37	15.61	19.62	7.85	37.28	35.96	339.78	157.7	465.5	623.2	0.34
	K_2SO_4 (2.5 mg/pot)	-	5.44	23.98	10.31	5.36	453.2	-	29.47	13.34	8.02	13.39	15.01	-	30.20	28.62	263.17	95.3	358.4	453.2	0.27
100% sand	Control	8.50	8.87	43.14	47.02	24.19	668.1	1.97	49.51	26.15	11.97	10.79	23.72	17.23	43.21	40.79	321.95	222.2	445.9	668.1	0.50
	Waste water	23.01	10.49	51.93	22.35	34.14	1051.4	2.05	70.37	33.15	15.87	15.20	29.73	23.92	58.27	49.87	624.17	264.1	787.3	1051.4	0.34
	K_2SO_4 (2.5 mg/pot)	-	-	-	-	-	-	-	-	-	-	-	-	-	-	-	-	-	-	-	-

The supplementation of soil type (1:4) (clay: sand v/v) with waste water induced a *de novo* appearance of essential amino acids His and Met which disappeared in control and in treatment with K_2SO_4, so, the total essential amino acids with this treatment was higher than the control and K_2SO_4treatments. All non-essential amino acids recorded a high value in treatment with waste water, except Tyr which disappeared in treatment with K_2SO_4. So, the sulfur fertilization with waste water induced an increase in total amino acids and in the ratio of Ess/Non-Ess than the treatment with K_2SO_4. For the content of essential amino acids, in general, the supplementation of soil with waste water induced appreciable increase in total essential amino acids and a bi-fold increase in His, Lys, Meth and Arg. Also, there was an increase in all non-essential amino acids in comparison to the control. So, the content of the total amino acids in treatment with waste water was higher than in the control, but the ratio of Ess/Non-Ess was lower than the control.

Ionic Contents

It is of great importance to refer to that from the data concerned with major growth attributes of inflorescence (Figure 1-6) indicated that plants cultivated in sandy soil treated with K_2SO_4 failed to complete its life cycle and/ or to give yield of significant value. Data tabulated in Table 3 shows the mineral content of head of Broccoli plants grown under treatment with waste water or inorganic fertilizer K_2SO_4 (2.5 mg/pot). It is apparent from the data that, many elements are quantified in Broccoli separated head such as K^+, Na^+, Ca^{+2}, Fe^{+3}, Mg^{+2} and S. Concerning K^+ ion content in Broccoli head, it is apparent that, treatment with inorganic fertilizer in the form of K_2SO_4 (2.5 mg/pot) increased the content of K^+ more than that of waste water and followed this sequence:(1:1) > (1:4) > 100% clay soil > (1:3) > (1:2) (clay: sand v/v) soil. From presented data in, it is obvious that, content of Na^+ ion in head varied with different treatments applied to different types of soil. In general, content of Na^+ ion is higher in plants treated with waste water compared to either with control or K_2SO_4 (2.5 mg/pot) in all types of soils and follow this trend: (1:1) > (1:3) > (1:2) > (1:4) (clay: sand v/v) soil > 100% clay soil, finally 100% sandy soil for waste water treatment only. On the other hand, K/Na ratio recorded a non-significant value in all used treatment and all types of soils. The changes of Ca^{+2} ion content showed that, in all types of soil, treatment of waste water significantly increased content of Ca^{+2} ion compared with inorganic fertilizer K_2SO_4 (2.5 mg/pot). In addition, both treatments followed the same sequence: (1:1) > (1:2) > (1:3) > (1:4) (clay: sand v/v) soil > 100% clay soil. It is apparent that, content of Mg^{+2} ion changes in different soil types, treated with either waste water or K_2SO_4 (2.5 mg/pot). Treatment with waste water significantly increased Mg^{+2} ion content of head of Broccoli plants grown in different soils according to this sequence: (1:1) > (1:2) > (1:4) > (1:3) (clay: sand v/v). Iron content of head was significantly and non-significantly increased in response to waste water treatment or K_2SO_4, respectively and depends on investigated soil type. The data obtained showed that, content of sulfur differs according to either treatments or the soil types. In general, its content is higher in case of treatment with waste water which induced a significant increase than in case of inorganic fertilizer K_2SO_4. For the sulfur content, according to soil type, it followed following order in treatment with waste water: (1:1) > (1:3) > 100% clay > (1:4) > (1:2) (clay: sand v/v) and 100% sand.

Table 3. Effect of inorganic fertilizer K_2SO_4 (2.5 mg/pot) and waste water on mineral ion contents in the heads of *Brassica oleracea* plant grown in different types of soils. Each value is the mean of three samples. Data represented as µM or mM/g dry weight

Soil type	Treatment	K^+	Na^+	K/Na	Ca^{++}	Mg^{++}	Fe^{++}	S^{++}
100% clay	Control	784.2	277.8	2.82	294.9	206.5	7.54	335.9
	Waste water	888.3	340.0	2.61*	327.7	244.2	8.20*	431.6
	K_2SO_4 (2.5 mg/pot)	923.0	306.1	3.02*	314.0	232.1	7.90*	372.8
	LSD at 5%	4.6	3.1	2.4	1.6	2.2	3.0	1.4
(1:1) (clay : sand v/v)	Control	920.6	393.0	2.34	444.3	316.0	8.60	321.4
	Waste water	967.9	466.9	2.07*	540.0	475.9	10.71	452.6
	K_2SO_4 (2.5 mg/pot)	1240.3	411.2	3.02*	499.4	394.7	9.44*	429.6
	LSD at 5%	6.2	1.6	1.6	3.0	1.9	1.7	2.4
(1:2) (clay : sand v/v)	Control	799.2	300.0	2.66	357.0	292.0	9.32	158.7
	Waste water	809.5	379.1	2.14*	461.1	312.7	12.02	365.8
	K_2SO_4 (2.5 mg/pot)	865.7	343.4	2.52*	384.4	295.9	9.54*	318.0
	LSD at 5%	5.9	3.5	1.6	2.7	1.9	1.8	2.5
(1:3) (clay : 1 sand v/v)	Control	766.4	343.9	2.23	330.9	231.0	7.12	304.3
	Waste water	821.0	374.6	2.19*	352.1	246.0	9.84	437.6
	K_2SO_4 (2.5 mg/pot)	874.1	351.7	2.49*	344.7	237.2	7.92*	400.9
	LSD at 5%	3.8	1.9	3.3	2.2	2.9	2.1	3.8
(1:4) (clay : 1 sand v/v)	Control	814.2	272.9	2.98	310.3	215.3	8.71	274.3
	Waste water	858.7	350.7	2.45*	337.6	297.0	10.90	424.0
	K_2SO_4 (2.5 mg/pot)	961.4	319.4	3.01*	329.5	282.6	9.87*	314.9
	LSD at 5%	2.4	2.3	1.9	1.7	3.0	1.4	1.8
100% sand	Control	614.9	98.7	6.23	167.9	120.9	6.00	162.0
	Waste water	695.0	113.2	6.14	184.2	131.2	6.71	211.6
	K_2SO_4 (2.5 mg/pot)	-	-	-	-	-	-	-
	LSD at 5%	2.2	1.4	1.0	1.3	2.0	1.7	2.6

*Non-significant change at 5% level.

DISCUSSION

Growth Parameter

In the sample of Broccoli plants as affected by S fertilizers both waste water and K_2SO_4 (2.5 mg/pot) gave a significant increase effect in the yield parameters i.e., head diameter, fresh and dry weight as compared to the control. Data showed that inorganic fertilizer (K_2SO_4) at concentration of (2.5 mg/pot) gave the lowest head yield compared with wastewater. The results were in agreement with that mentioned by Chatterjee et al. (2005), where, addition of wastewater increased the total head yield and gave the superior yield of Broccoli. The observations are consistent with the previous results obtained by Gad and Kandil (2012). It was found that biomass yields recorded under recycled effluent usage are equal to and in many cases higher than those obtained for crops grown elsewhere under more favorable conditions (Parameswaran 1999).

Our results are consistent with the findings of Charfiet al. (1997) who recorded that leaf area as well as weights of fruit and pulp of olive trees irrigated with effluent was higher than those of the control. The greater growth of the olive trees obtained by effluent application may be due to the sufficient availability of essential nutrients in the effluent which might have enhanced growth parameters and did not reach a point causing a decrease in the diffusion pressure gradient between the plant and the medium or ion toxicity and ion imbalance or a combination of any of these adverse effects (Greenway and Munns 1980).

Total Carbohydrates and Total Protein

In the present study, increasing level of sand ratio achieved a significant variance in total carbohydrate and total protein content in head of Broccoli in response to K_2SO_4 (2.5 mg/pot) treatment and irrigation with waste water. The highest content was recorded in (1:1) (clay: sand v/v) soil for both treatments. Present results are in good harmony with Konwar and Jha (2010) who established that, waste effluents contain organic and inorganic nutrients like Mg^{++}, Ca^{++}, and S^{++}, which promoted formation of chlorophylls, with its role in carbohydrate metabolism.

The type and concentration of free sugars influence the flavor of *Brassica* products where fructose, glucose, and sucrose are the major soluble sugars found in *Brassica*. Abdallah et al. (2015) mentioned that, the substantial increase in carbohydrate contents may be due to the activation of photosynthetic machinery, as a result of the stimulatory effects of the S-containing fertilizers on photosynthetic process and carbohydrate metabolism. Von Uexkull (1986) found that sulfur availability may influence photosynthetic rate since ferredoxin and acetyl-CoA contain S and play a significant role in the reduction of CO_2 and production of organic compounds. Also, sulfur is necessary for enzymatic reactions, chlorophyll formation, synthesis of certain amino acids and vitamins, hence, it helps to obtain a good vegetative growth leading to a high yield (Marschner 1995).

Dietary proteins are the major source of amino acids. Most proteins are broken down by enzymes into amino acids. (Malhiand Gill 2007) stated that consumption of Broccoli has been highly recommended based on the presence of secondary plant metabolites with health protective effects. A part from these compounds, Broccoli might report high levels of proteins, which however, are likely to be affected by cultivar, environment, treatments and type of inflorescence (Kopriva 2006).

In the present study, different sulfur treatments affected greatly the protein content of Broccoli head in which there was a significant increase in total protein in all used soils in case of treatment with waste water which achieved a higher content of total protein than in treatment with K_2SO_4 (2.5 mg/pot). The highest overall content of total protein was obvious in (1:1) (clay: sand v/v) soil comparing with the control values. In this respect, protein concentration in seed varied significantly among the four *Brassica* species cultivers. Protein concentration in seed increased significantly with increasing S rate (Malhi and Gill 2007). Earlier studies with rape seed have shown large increase in protein content in seed with application of sulfate-fertilizer on S-deficient soil**.** The rape seed (*B. napus*) meal contains napin and cruciferin as storage proteins and oleosin as a structural protein associated with oil bodies (Kumar et al. 2012).

Total Ascorbate and Glutathione

Glutathione (GSH) is an important antioxidant in plants and is one of the main S-storage compounds in plants. It is a major non-enzymatic scavenger of ROS in plants exposed to stresses acting as precursor of phytochelatins (PCs) sulfur (S)-rich peptide. Preventing damage to important cellular components caused by ROS radicals and peroxides (Pompella et al. 2003). Ascorbate (AsA) eliminates ROS through coordination with the production of GSH in ascorbate-glutathione cycle (Noctor et al. 2002).

Results obtained, indicated that, reduced GSH content, slightly increased than that of total AsA in response to K_2SO_4in both experimental tested treatments, however, the magnitude of increase in response to K_2SO_4 appeared to follow more or less similar to that induced by waste water the same trend in treatment with waste water according to different used soils. Meanwhile the content of GSH in different used soils under K_2SO_4 treatment follow this sequence: $1:4 > 1:3 \geq 1:1 \geq 1:2$ (clay: sand v/v) soil> 100% clay soil.

The previous results indicated that, S application increased the GSH and AsA content of head. In this respect, Thangavel et al. (2007) stated that, sulfur is incorporated into organic molecules in plants and is located in thiol (-SH) groups in proteins (cys-residues) or non-protein thiols (GSH). These results are in good agreement with that obtained by Rausch and Wachter (2005) who found that, supplementary S-fertilization to high S-loving crops such as brassica and leguminous crops could enhance defense operations, *via* its incorporation into the antioxidants, GSH and AsA.

It is interesting to mention that, the marked increase of AsA and GSH contents particularly in response to plant growth in the soil (1:4, clay: sand) is related to the effect of soil water availability, the water holding capacity the availability of mineral nutrition to plants which are prevailed in sandy soils and hence subject plants to water and nutrient deficiency stresses. In this connection, AsA is the major, antioxidant buffer in the apoplast, it is a key primary antioxidant that reacts with OH radicals, O_2^-, and H_2O_2 (Chen and Gallie 2004). AsA indirectly eliminates H_2O_2 through the activity of APX. In the ascorbate-glutathione pathway AsA together with GSH affects plant tolerance to ROS by participating in the detoxification of ROS in plant cells (Anjumet al. 2008a).

The positive correlation was observed between the contents of ASA and GSH and the rate of photosynthesis and plant dry mass in broccoli suggests that there exists a concomitant relation between ASA and GSH contents, net photosynthesis and plant dry mass. This conclusion may be in agreement with Anjum et al. (2008b).

Total Phenols and Flavonoids

Phenolic molecule is characteristic of a plant species or even of a particular organ or tissue of the plant. They are closely related to the secondary and nutritional quality of foods derived from plant resources. Phenolic compounds may act as antioxidant and protect food from oxidative deterioration (Karakaya 2004). They have a great importance in terms of the nutritional and commercial properties of agricultural products through their contribution to sensory properties such as color and flavour (Pérez-Lópezet al. 2007). Morton et al. (1997) stated that, flavonoids have wide range of biological activities, so they can be used in numerous medical treatments connected to cancer prevention.

The results showed that the use of waste water resulted in higher values of total phenols and flavonoids compared to inorganic fertilizers. The amount of total phenols was between 33 and 59 mg/100 g DW and flavonoids were found between 30 and 51.5 mg/100g DW. These results are in agreement with those obtained by Liu et al. (2008) who found that the total phenolics of fennel 70% methanolic extract was 7.01 mg/g DW and the total flavonoids content of the same extract was 3.29 mg/g DW. Waste water treatment containing organic residues has a stimulating effect on the biosynthesis of phenolics when compared with control and bioorganic fertilizers. The antioxidant activity of phenolics is mainly due to their redox properties, which allow them to act as reducing agents, hydrogen donors, and singlet oxygen quenchers (Rice-Evans et al. 1997). This is in agreement with the study conducted in China by Juan et al. (2008) on effects of S on total phenolics and antioxidant activity of leaf Mustard. Vallejo et al. (2003b) reported that the main phenolic constituents of Broccoli are flavonols (quercetin and kaempferol derivates) and phenolic acids (hydroxyl cinnamic acids derivates). Phenolic compounds also contribute to the health properties of these vegetables. The importance of the flavonoids was known to possess significant antimicrobial activities and was utilized as natural plant protectants. It could be suggested that flavonoids content may be an alternative to conventional fungicides in the control of storage grains against some fungi (Vallejo et al. 2003a).

Ghasemzadeh et al. (2012) reported the strong positive relationship between total phenols and antioxidant activity that appears to be the trend in many plant species. The importance of phenolic compounds as antioxidant in the maintenance of health and protection against coronary heart disease and cancer is also of raising interest among scientists, food manufactures and consumers. This trend leads the future toward functional food with specific health effects. Carbonaroet al. (2002) also suggested that flavonoids and phenolics have a preventive rolein the development of cancer and heart disease. Consumption of controlled diets high in the fruits and vegetables increased significantly the antioxidant capacity of plasma (Carbonaro et al. 2002). Moreover, epidemiological studies showed that significant negative correlation was found between the intake of fruits and vegetables and heart disease mortality. Floridi (2003) showed that the presence of phenolic in food is particularly important for their oxidative stabilityand antimicrobial protection. From clinical point of view, there are numerous evidences indicating the efficiency of cruciferous plants (as sweet fennel) in disease alleviation.

Folic and Ascorbic Acid

The content of folic and ascorbic acids were significantly increased over the control value with all the used treatments. Irrigation with waste water was the more effective treatment where the content of folic acid and ascorbic acid was higher as compared with treatment with K_2SO_4. Moreover growth of Broccoli in the soil 1:4 (clay:sand v/v) achieved the highest content of folic and ascorbic acids content of Broccoli head, compared with other types of soils (Figure 5).

Regarding the values of vitamin C, it is interesting to mention that, the vitamin C values of Broccoli plant treated with waste water or inorganic fertilizers K_2SO_4 (2.5 mg/pot), Broccoli plant contained drastic values (0.044:0.052) mg/gFW, in response to the mentioned treatments. Badawi et al. (2005) reported that vitamin C content ranged from 21.83 to 48.04 mg/100 gFW mineral fertilization was applied. Koudela and Petříková (2008) determined the content of vitamin C of three cultivars of sweet fennel and found that the values of vitamin C content ranged from 87 to 347 mg/kg fresh matter. High levels of vitamin C have been reported in Chinese cabbage, Broccoli, cauliflower, and cabbage. The content of this vitamin in different cultivars of cabbage (*Brassica oleraceae* L.) ranges from 12 to112.5 mg/100 g (Bahorun et al. 2004).

In this respect, Brassica vegetables were reported to contain high levels of vitamins (Heimler et al. 2006), including carotenes, vitamin C, and folic acid. Broccoli (*Brassica oleracea*) extracts are protective role against ROS presumably due to the presence of vitamin C, α-tocopherol, γ-tocopherol, and ß-carotene (Eberhardt et al. 2005).

On the other hand, folic acid contents of Broccoli plant treated with waste water and inorganic fertilizers K_2SO_4 (2.5 mg/pot) are illustrated in Figure 5, and indicated that Broccoli plant contained drastic values in response to the mentioned treatments. Folic acid is known to reduce the risk of neural tube defects and may be associated with the reduced risk of vascular disease and cancer (Cornel et al. 2005), while low-folate intake has been identified as a main cause of anemia (Bollheimer et al. 2005).

Glucosinolates Content

A large proportion of the known sulfated metabolites play various roles in plant defense against biotic and abiotic stress. A well-studied example of such compounds is glucosinolates (GSLs), which in addition to the above role; they are also responsible for taste and flavor of cruciferous vegetables, and some of their degradation products, isothiocyanates, possess an anti-cancer activity (Fahey et al. 2001).

According to the present study, the treatment with different sulfur fertilizers, or use of waste water in irrigation (Figure 6) confirmed that sulfur fertilization generally resulted in an increase in glucosinolates content of head Broccoli plants as compared with control. The total glucosinolates concentrations ranged from 91.89 to 211.92 μmole/gFW. The higher content was found in the head of plants grown in (1:1) (clay: sand v/v soil) and treated with waste water, meanwhile the lowest one was recorded in the heads of plants grown in 100% sand soil for the same treatment. It is possible to explain that the decrease in sulfur supply encountered in sandy soil may lead to this reduction of the total glucosinolates concentrations.

With their high-sulfur content and role in defense, glucosinolates might be thought to accumulate less under conditions of sulfur deficiency if plants place a higher priority on the use of sulfur for protein synthesis and essential functions other than defense. Glucosinolates may act as an expendable reservoir of sulfur in times of sulfur deprivation (Fahey et al. 2001). Also, Shahbaz et al. (2014) reported that, sulfur treatment of *Brassica* generally resulted in enhanced levels of water-soluble, non-protein thiol compounds in the shoot, which could be ascribed to an accumulation of cysteine and glutathione, which also function as reduced sulfur donors in the synthesis of glucosinolates.

Supporting explanation, it was reported by many investigators that, glucosinolates content is related to sulfur availability in growth media. On this contrary, *Brassica napus* seed's glucosinolates content varied from 9 to 24 μmol/g seed in ten different sites across England varying in used soils, despite the plots receiving the same amount of sulfur fertilizer (Withers and O'Donnell 1994).

Sulfur fertilization usually causes an increase in glucosinolates content ranging from 25% to more than 50-folds, depending on the plant species, amount of sulfur applied, and type of treatment (Falk et al. 2007). In this respect Schonhof et al. (2007) documented that, an insufficient sulfur supply led to reduction of total glucosinolates that had received an optimal N supply. Other studies on Brassicaceae like rape seed, the alkyl glucosinolates concentration decreased significantly with an insufficient sulfur supply in combination with optimal N supply. The so called aliphatic GSLs are derived mainly from methionine, but also from alanine, leucine, isoleucine and valine, while aromatic GSLs are derived from tryptophan (Fismeset al. 2000). In seeds of oil seed rape, Zhao et al. (1999) reported that when S is limiting, most of it is incorporated into proteins, and there for, less sulfur is available for glucosinolates synthesis. Staley et al. (2010) reported that GSLs in cabbage (*Brassica oleracea* L. var. *capitata*) were three times more when grown on organic manure than when it was grown on chemical fertilizers. Schonhof et al. (2007) and Pradhan et al. (2007) reported that Broccoli crop yield product appearance as well as glucosinolates concentration was influenced by S supply which is of a crucial importance in terms of crop management. Furthermore, sulfur application on Broccoli was essential to produce higher yield with good quality curds (Elwan and El-Hamed 2011) because sulfur-containing compounds can provide a highly effective antioxidant system (Al-Malaika 1984).

Antioxidant Enzymes

For scavenging ROS, plant possesses effective systems that protect them from destructive oxidative reaction. Part of this system is identified as "antioxidative enzymes." They are key elements in the defense mechanisms. Antioxidants are defined as compounds that can delay, inhibit, or prevent the oxidation of oxidizable materials by scavenging free radicals and diminishing oxidative stress (Foyer and Halliwell 1976).

According to the present study, the activity of antioxidant enzyme in terms of SOD, CAT, APX, and GR in heads of the Broccoli plants treated, either with waste water irrigation or inorganic fertilizers K_2SO_4 (2.5 mg/pot) recorded a significant increase in response to treatments applied in all the investigated soils. The use of waste water achieved the highest value than that of K_2SO_4treated plants.

In this regard, Foyer and Halliwell (1976) suggested that the increase in the activities of antioxidant enzymes detected in *L. latifolium*is not a direct consequence of H_2O_2 accumulation, but is mediated through GR in intracellular environment of *L. latifolium*. Yoshimoto et al. (2003) illustrated that, sulfur application was the best in improving antioxidant enzymes especially GSH in head of Broccoli plants. This is may be related to the role of S in plants which is found in amino acids (Cys, Met), vitamins and cofactors (biotin) and in a variety of secondary products (glucosinolates in Crucifereae).

In this investigation, the results obtained demonstrated that Broccoli crop yield appearance as well as antioxidant system especially for glucosinolates concentration was influenced by S supply which is of a crucial importance in terms of crop management. These results are compatible also with the findings of Schonhof et al. (2007). Improvement in antioxidant with the use of organics in Broccoli was also reported by Sanwal et al. (2006).

Regarding the effect of the used soils, the highest values were recorded in 1:4 (clay: sand v/v) soil except for APX activity where the highest value was recorded in (1:1) (clay: sand v/v) soil, soil water availability affects the mineral nutrition of plants. Soil water availability which is closely related to its composition exhibited an increase in SOD, APX and CAT activities in Broccoli as compared to soil 1:1 (clay: sand v/v) soil. These results are in agreement with AbdElgawad et al. (2014). Higher levels of enzyme activities in Broccoli plant grown under water deficiency because of rapid leaching of water by the sandy soil may play here crucial role against drought stress. Adverse environmental stresses increase CAT and APX activities in several cycles of physiological processes. At stress conditions higher content of hydrogen peroxide is detoxified by CAT and glutathione peroxidase (Datet al. 2000).

Amino Acids Content

About 70% of sulfur in plants is present as amino acids. Among the two sulfur containing amino acids (SAAs), methionine is considered essential, while cysteine is considered a non-essential for humans (Fürst and Stehle 2004). According to the results of this study, sulfur application either in the form of K_2SO_4 or by waste water irrigation in different used soils caused a considerable variation in amino acid contents of essential or non-essential as well as total amino acids and in the ratio of essential/non-essential amino acids. In this connection, it was reported that, Broccoli is a very good source of amino acids (Kmiecik et al. 2010). This may be due to conversion of methionine to other compounds like glucosinolates which is in agreement with Mikkelsen et al. (2002) that amino acids such as methionine, tryptophan or phenylalanine are the precursors for glucosinolates production. The prevalent non-essential amino acids in the head of Broccoli plants are L-glutamic acid (Glu) which might be converted directly into protein and chlorophyll (Gomes and Rosa 2001). Tisdal et al. (1986) found that, both cysteine and methionine contents of alfalfa could be increased up to certain point on sulfur application to the plant. Regarding the changes in the ratio of essential/non-essential amino acids, the highest value was recorded in response to treatment with K_2SO_4 in soil (1: 3) (clay : sand v/v) owing to the highest value of essential amino acids with special regard to the methionine comparing with the other treatments. This is in accordance with the results obtained by Gomes and Rosa (2001) who detected a total of 17amino acids in the inflorescences of eleven Broccoli cultivars. The major amino acid was L-glutamine, which represented an average between 39.5%and 55.5% of the total amino acids content among cultivars, followed by L-glutamic acid with a variation between 12.1% and 17.4%. For most of the amino acids there were significant differences between cultivars.

According to the present results, the amino acid cysteine was not be detected in all treatments. This may be, due to its presence in very small amounts, or it may be incorporated into proteins or peptides and glutathione (GSH) (Lewandowska and Sirko 2008) which is the most abundant low molecular weight tripeptide which plays role inplant stress defense, redox regulation, and sulfur storage and transport. Alternatively, cysteine is further metabolized to serve as a donor of reduced sulfur for synthesis of methionine (Nikiforova et al. 2003). An intermediate of the sulfate assimilation pathway, sulfite, is metabolized into sulfo-lipids, essential components of chloroplast membrane. Sulfate reduction is localized to plastids only. Cysteine is synthesized in all three compartments capable of protein synthesis: plastids, mitochondria and cytosol (Leustek et al. 2000).

Ionic Content

The change in ionic contents (K, Na, Ca, Mg, Fe and S) of heads of Broccoli showed different variations in response to inorganic fertilizer (K_2SO_4) and industrial waste water. The predominant element in all samples was potassium. The potassium ion content recorded a high amount in response to K_2SO_4 than that detected in response to waste water. Meanwhile a reverse situation was obtained in Na ion content, where its content was high in treatment with waste water than in treatment with K_2SO_4. Consequently, there was non-significant change in K/Na ratio in response to both $K_2SO_4$and waste water treatments in comparison to the control. These data are in accordance with those obtained by Dos Santos et al. (2013) which recorded the high concentrations of this element in food plants. For further evaluation of the obtained results, the readily soluble minerals in use of chemical synthetic fertilizers are allowed (Wang et al. 2004; Oplanićet al. 2009).

Likewise, high Ca, Mg, Fe and S content was found in treatment with waste water as compared with K_2SO_4 treatment. With regarding that, a non-significant increase in Fe content was observed with K_2SO_4compared with the control, so the use of S-fertilizers either in form of K_2SO_4 or waste water will maintain a high ionic content of Broccoli head. High S, Ca, Na, Mg and Fe ionic contents in the industrial waste water samples was more than conventional farming. In this respect, wastewater may supply organic matter and mineral nutrients to soil that are beneficial to crop production (Naeem et al. 2006). Inorganic fertilizer is considered as a major source of plant nutrients. According to many authors, due to important functions in plant metabolism played by sulfur and due to its positive effect on the yield size and quality (Cazzato et al. 2012) the element cannot be disregarded in crop fertilization.

CONCLUSION

From the results obtained it can be concluded that Broccoli plant (*Brassica oleracea* L. var. *italica*) can be cultivated in different types of soils irrigated with industrial waste water, where the vigorous growth and yield of heads was obtained as compared to the K_2SO_4 treatment and control. For Broccoli plant (*Brassica olereaca* L. var. *italica*) in general, the best quality of metabolic changes and nutritive value of head (inflorescence) were obtained when *Brassica oleracea* was cultivated in 1:4 (clay: sand v/v) soil and supplemented with industrial waste water than the other treatments.

ACKNOWLEDGMENT

This work was supported by Mansoura University, Faculty of Science, Mansoura governorate, Egypt.

REFERENCES

Abdallah M, Magdeldin M, Mahmoud S (2015) Effect of nitrogen and sulfur on yield, yield components, some chemical composition and nutritional quality of canola plant grown in saline soil condition. *Res J Pharm Biol Chem Sci* 6: 1055-1064.

Abdallah M, Dubousset L, Meuriot F, Etienne P, Avice JC, Ourry A (2010) Effect of mineral sulfur availability on nitrogen and sulfur uptake and remobilization during the vegetative growth of *Brassica napus*. *J Exp Bot* 61: 2635-2646.

Abd Elgawad H, Peshev D, Zinta G, Van den Ende W, Janssens IA, Asard H (2014) Climate extreme effects on the chemical composition of temperate grassland species under ambient and elevated CO_2: a comparison of fructan and non-fructan accumulators. *PLoS One* 9: 1-13.

Al-Malaika S (1984) A novel application of sulfur containing antioxidants for the stabilization of polyolefins. *J Brit Polym* 16: 301-310.

Anjum NA, Umar S, Ahmad A, Iqbal M, Khan NA (2008a) Sulfur protects mustard (*Brassica campestris* L.) from cadmium toxicity by improving leaf ascorbate and glutathione. *J Plant Growth Regul* 54: 271-279.

Anjum N, Umar S, Singh S, Nazar R, Khan NA (2008b) *Sulfur assimilation and cadmium tolerance in plants Sulfur assimilation and abiotic stress in plants*, Springer, pp. 271-302.

Atanassova M, Georgieva S, Ivancheva K (2011) Total phenolic and total flavonoid contents, antioxidant capacity and biological contaminants in medicinal herbs. *J U Chem Technol Metall* 46: 81-88.

Badawi M, El-Magd MA, Hassan H, El-Sharky M (2005) Effect of bio-gerilization, nitrogen sources, nitrogen levels and their interactions on the vegetative growth, chemical content and oil yield of sweet fennel. *Egypt J Appl Sci* 20: 567-591.

Bahorun T, Luximon-Ramma A, Crozier A, Aruoma OI (2004) Total phenol, flavonoid, proanthocyanidin and vitamin C levels and antioxidant activities of Mauritian vegetables. *J Sci Food Agric* 84: 1553-1561.

Beutler E, Duron O, Kelly BM (1963) Improved method for the determination of blood glutathione. *J lab Clin Med* 61: 882-888.

Bollheimer LC, Buettner R, Kullmann A, Kullmann F (2005) Folate and its preventive potential in colorectal carcinogenesis: How strong is the biological and epidemiological evidence. *J Crit Rev Oncol Hematol* 55: 13-36.

Bradford MM (1976) A rapid and sensitive method for the quantitation of microgram quantities of protein utilizing the principle of protein-dye binding. *J Anal Biochem* 72: 248-254.

Brady NC, Weil RR (1996) *The nature and properties of soils*: Prentice-Hall Inc. Nyle C. and Weil, Published by MacMillan.

Carbonaro M, Mattera M, Nicoli S, Bergamo P, Cappelloni M (2002) Modulation of antioxidant compounds in organic vs conventional fruit (peach, *Prunus persica* L., and pear, *Pyrus communis* L.). *J Agric Food Chem* 50: 5458-5462.

Cazzato E, Laudadio V, Stellacci A, Ceci E, Tufarelli V (2012) Influence of sulphur application on protein quality, fatty acid composition and nitrogen fixation of white lupin (*Lupinus albus* L.). *J EurFood Res Technol* 235: 963-969.

Charfi D, Trigui A, Medhioub K (1997) *Effects of irrigation with treated wastewater on olive trees cvchemlali of sfax at the station of elhajeb (sfax-tunisia).* Paper presented at the III International Symposium on Olive Growing 474.

Chaterjee B, Ghanti P, Thapa U, Tripathy T (2005) Effect of organic nutrition in sprouting broccoli (*Brassica oleracea* L. var. *italica* Plenck). *J Veg Sci* 32: 51-54.

Chen JX, Wang XF (2006) *Plant physiology experimental guide*. Higher Education Press, Beijing, pp. 55-56.

Chen Z, Gallie DR (2004) The ascorbic acid redox state controls guard cell signaling and stomatal movement. *J Plant Cell* 16: 1143-1162.

Cornel M, De Smit D, De Jong-van den Berg L (2005) Folic acid-the scientific debate as a base for public health policy. *J Reprod Toxicol* 20: 411-415.

Dat J, Vandenabeele S, Vranová E, Van Montagu M, Inzé D, Van Breusegem F (2000) Dual action of the active oxygen species during plant stress responses. *J Cell Mol Life Sci* CMLS 57: 779-795.

De Pascale S, Maggio A, Orsini F, Bottino A, Barbieri G (2008) Sulfur fertilization affects yield and quality of friarielli (*Brassica rapa* l. subs sylvestris L. Janch. Var. esculenta Hort.) grownon a floating system. *J Hortic Sci Biotechnol* 83: 743-748.

Dos Santos WNL, Ferreira SLC, Silva LOB, de Santana FA, Junior MMS, Matos GD (2013) A review of reflux systems using cold finger for sample preparation in the determination of volatile elements. *Microchem J* 106: 307-310.

Eberhardt MV, Kobira K, Keck A-S, Juvik JA, Jeffery EH (2005) Correlation analyses of phytochemical composition, chemical, and cellular measures of antioxidant activity of broccoli (*Brassica oleracea* L. var. *italica*). *J Agric Food Chem* 53: 7421-7431.

Elwan M, El-Hamed KA (2011) Influence of nitrogen form, growing season and sulfur fertilization on yield and the content of nitrate and vitamin C of broccoli. *J Sci Hortic* 127: 181-187.

Fahey JW, Zalcmann AT, Talalay P (2001) The chemical diversity and distribution of glucosinolates and isothiocyanates among plants. *J Phytochem* 56: 5-51.

Falk K, Tokuhisa J, Gershenzon J (2007) The effect of sulfur nutrition on plant glucosinolate content: physiology and molecular mechanisms. *J Plant Biol* 9: 573-581.

Fatta D, Anayiotou S, Papadopoulos I (2005) An overview of the water and wastewater management practices in Cyprus. *J Crop* 35: 133-136.

Finley JW (2005) Proposed criteria for assessing the efficacy of cancer reduction by plant foods enriched in carotenoids, glucosinolates, polyphenols and seleno compounds. *J Ann Bot* 95: 1075-1096.

Fismes J, Vong PC, Guckert A, Frossard E (2000) Influence of sulfur on apparent N-use efficiency yield and quality of oilseed rape (*Brassica napus* L.) grown on a calcareous soil. *Eur J Agron* 12: 127-141.

Floridi L (2003) Two approaches to the philosophy of information. *J Minds Mach* 13: 459-469.

Foyer CH, Halliwell B (1976) The presence of glutathione and glutathione reductase in chloroplasts: a proposed role in ascorbic acid metabolism. *J Planta* 133: 21-25.

Fürst P, Stehle P (2004) What are the essential elements needed for the determination of amino acid requirements in humans. *J nutr* 134: 1558S-1565S.

Gabal M, Abdallah I, Abed I, Hassan F, Hassanen S (1985) Evaluation of some American tomato cultivars grown for early summer production in Egypt. *Anna Agric Sci* 1: 5-17.

Gad N, Kandil H (2012) Effects of inorganic and organic fertilizers on growth and production of brocoli (*Brassica Oleracea* L.). *Soil Forming Factors and Processes from the Temperate Zone* 8: 61-69.

Ghasemzadeh A, Azarifar M, Soroodi O, Jaafar HZ (2012) Flavonoid compounds and their antioxidant activity in extract of some tropical plants. *J Med Plants Res* 6: 2639-2643.

Gomes MH, Rosa E (2001) Free amino acid composition in primary and secondary inflorescences of 11 broccoli (*Brassica oleracea* var., *italica*) cultivars and its variation between growing seasons. *J Sci Food Agric* 81: 295-299.

Greenway H, Munns R (1980) Mechanisms of salt tolerance in non-halophytes. *J Ann Rev Plant Physiol* 31: 149-190.

Gupta S, Sangha MK, Kaur G, Atwal AK, Kaur P, Kumar H, Banga SS (2012) Variation in glucosinolate and tocopherol concentrations in a germplasm collection of *Brassica juncea* (L.) Czern. Coss. *Biochem Anal Biochem* 1: 1-121.

Harris L, Ray S (1935) Diagnosis of vitamin-C sub nutrition by urine analysis: with a note on the antiscorbutic value of human milk. *J Lancet* 225: 71-77.

Heimler D, Vignolini P, Dini MG, Vincieri FF, Romani A (2006) Antiradical activity and polyphenol composition of local Brassicaceae edible varieties. *J Food Chem* 99: 464-469.

Hirel B, Le Gouis J, Ney B, Gallais A (2007) The challenge of improving nitrogen use efficiency in crop plants: towards a more central role for genetic variability and quantitative genetics within integrated approaches. *J Exp Bot* 58: 2369-2387.

Homme PM, Gonzalez B, Billard J (1992) Carbohydrate content, frutane and sucrose enzyme activities in roots, stubble and leaves of rye grass (*Loliumperenne* L.) as affected by sources/link modification after cutting. *J Plant Physiol* 140: 282-291.

Jaleel CA, Sankar B, Murali PV, Gomathinayagam M, Lakshmanan GMA Panneerselvam R (2008) Water deficit stress effects on reactive oxygen metabolism in *Catharanthus roseus*; impacts on ajmalicine accumulation. *Colloids Surf B Biointerfaces* 62: 105-111.

Juan L, Zhujun Z, Jóska G (2008) Effects of nitrogen and sulfur on total phenolics and antioxidant activity in two genotypes of leaf mustard. *J Plant Nutr* 31: 1642-1655.

Kampfenkel K, Van Montagu M, and Inze′D (1995) Extraction and determination of ascorbate and dehydroascorbate from plant tissue. *Anal Biochem* 225: 165-167.

Karakaya S (2004) Bioavailability of phenolic compounds. *J Crit Rev Food Sci Nutr* 44: 453-464.

Kmiecik W, SŁUPSKI J, Lisiewska Z (2010) Comparison of amino acid content and protein quality in raw broccoli and in broccoli after technological and culinary processing. *J Food Process Preserv* 34, 639-652.

Kong FX, Hu W, Chao SY, Sang WL, Wang LS (1999) Physiological responses of mexicana to oxidative stress of SO_2. *Environ Exp Bot* 42: 201-209.

Konwar D, Jha D (2010) Response of rice (*Oryza sativa* L.) to contamination of soil with refinery effluents under natural conditions. *Assam U J Sci Technol* 5: 14-22.

Kopriva S (2006) Regulation of sulfate assimilation in *Arabidopsis* and beyond. *J Anna Bot* 97: 479-495.

Koudela M, Petříková K (2008) Nutritional compositions and yield of sweet fennel cultivars-*Foeniculum vulgare* Mill. ssp. vulgare var. azoricum (Mill.). *J Hortic Sci* 35: 1-6.

Kumar V, Budki P, Rani S (2012) Persistence of circannual rhythms under constant periodic and aperiodic light conditions: sex differences and relationship with the external environment. *J Exp Biol* 215: 3774-3785.

Lee DH, Lee CB (2000) Chilling stress-induced changes of antioxidant enzymes in the leaves of cucumber in gel enzyme activity assays. *Plant Sci* 159: 75-85.

Leustek T, Martin MN, Bick JA, Davies JP (2000) Pathways and regulation of sulfur metabolism revealed through molecular and genetic studies. *J Ann Rev Plant Biol* 51: 141-165.

Lewandowska M, Sirko A (2008) Recent advances in understanding plant response to sulfur-deficiency stress. *J Acta Biochem* 55: 457-471.

Liu J, Zhu Z, Gerendas J (2008) Effects of nitrogen and sulfur on total phenolics and antioxidant activity in two genotypes of leaf mustard. *J Plant Nutr* 31: 1642-1655.

Malhi S, Gill K (2007) Interactive effects of N and S fertilizers on canola yield and seed quality on S-deficient Gray Luvisol soils in northeastern Saskatchewan. *Can J Plant Sci* 87: 211-222.

Marinova D, Ribarova F, Atanassova M (2005) Total phenolics and total flavonoids in Bulgarian fruits and vegetables. *J U chem technol Metall* 40: 255-260.

Marschner H (1995) Functions of mineral nutrients: macronutrients. In: *Mineral nutrition of higher plants 2nd Edition* (Ed. Marschner H). Academic Press, NY, pp. 299-312.

Michaud J, Wu F, Osato M, Cottles GM, Yanagida M, Asou N, Raskind WH (2002) *In vitro* analyses of known and novel mutations in dominant familial platelet disorder with predisposition to acute myelogenous leukemia: implications for mechanisms of pathogenesis. *J Blood* 99: 1364-1372.

Mikkelsen MD, Petersen BL, Olsen C, Halkier B (2002) Biosynthesis and metabolic engineering of glucosinolates. *J Amino Acids* 22: 279-295.

Morton MS, Matos-Ferreira A, Abranches-Monteiro L, Correia R, Blacklock N, Chan PS, Cheng C, Lloyd S, Chieh-ping W, Griffiths K (1997) Measurement and metabolism of isoflavonoids and lignans in the human male. *J Cancer Lett* 114: 145-151.

Motsara M, Roy RN (2008) *Guide to laboratory establishment for plant nutrient analysis*: Food and Agriculture Organization of the United Nations Rome. Vol. 19.

Naaz S, Pandey S (2010) Effects of industrial waste water on heavy metal accumulation, growth and biochemical responses of lettuce (*Lactuca sativa* L.). *J Environ Biol* 31: 273-276.

Naeem M, Iqbal J, Bakhsh M (2006) Comparative study of inorganic fertilizers and organic manures on yield and yield components of mung bean (*Vignaradiata*L.). *J Agric Soc Sci* 2: 227-229.

Nakano Y, Asada K (1981) Hydrogen peroxide is scavenged by ascorbate-specific peroxidase in spinach chloroplasts. *J Plant Cell Physiol* 22: 867-880.

Nikiforova V, Freitag J, Kempa S, Adamik M, Hesse H, Hoefgen R (2003) Transcriptome analysis of sulfur depletion in *Arabidopsis thaliana*: interlacing of biosynthetic pathways provides response specificity. *J Plant* 33: 633-650.

Noctor G, Gomez L, Vanacker H, Foyer CH (2002) Interactions between biosynthesis, compartmentation and transport in the control of glutathione homeostasis and signalling. *J Exp Bot* 53: 1283-1304.

Oplanić M, Ban D, Ilak-Peršurić A, Žnidarčič D (2009) Profitability of leek (*Allium porrum* L.) in three production systems. *Int J food Agric Environ* 7: 376-381.

Otokunefor T, Obiukwu C (2005) Impact of refinery effluent on the physicochemical properties of a water body in the Niger delta. *J Appl Ecol Environ Res* 3: 61-72.

Parameswaran M (1999). Urban waste water use in plant biomass production. *J Resour Conserv Recy* 27: 39-56.

Pérez-López AJ, del Amor FM, Serrano-Martínez A, Fortea MI, Núñez-Delicado E (2007) Influence of agricultural practices on the quality of sweet pepper fruits as affected by the maturity stage. *J Sci Food Agric* 87: 2075-2080.

Pompella A, Visvikis A, Paolicchi A, De Tata V, Casini AF (2003) The changing faces of glutathione, a cellular protagonist. *J Biochem Pharmacol* 66: 1499-1503.

Pradhan SK, Nerg A, Sjoblom A, Holopainen JK, Heinonen-Tanski H (2007) Use of human urine fertilizer in cultivation of cabbage (*Brassica oleracea*var capitata)-impact on chemical, microbial and flavor quality. *J Agric Food Chem* 55: 8657-8663.

Qadir M, Quillérou E, Nangia V, Murtaza G, Singh M, Thomas RJ, Noble AD (2014) *Economics of salt-induced land degradation and restoration*. Paper presented at the Natural Resources Forum. 38: 282-295.

Rahnama H, Ebrahimzadeh H (2005) The effect of NaCl on antioxidant enzyme activities in potato seedlings. *J Biol Plant* 49: 93-97.

Rausch T, Wachter A (2005) Sulfur metabolism: a versatile platform for launching defense operations. *J Trend Plant Sci* 10: 503-509.

Rengstagoon W, Hattanat N (2012) *Simple spectrophotometric method for determination of folic acid.* The 4th Annual Northeast Pharmacy Profession in harmony. The 4th Annual Northeast Pharmacy Research Conference of 2012, Pharmacy Profession in Harmony.

Rice-Evans CA, Miller NJ, Paganga G (1997) Structure-antioxidant activity relationships of flavonoids and phenolic acids. *J Free Radical Biol Med* 20: 933-956.

Sanwal SK, Laximinarayana K, Yadav DS, Rai N, Yadav RK (2006) Growth yield and dietary antioxidants of broccoli as affected by fertilizer type. *J Veg Science* 12: 13-26.

Schonhof Li S, Krumbein A, Li L, Stützel H, Schreiner M (2007) Glucosinolate concentration in turnip (*Brassica rapa* sp. *rapifera* L.) roots as affected by nitrogen and sulfur supply. *J Agric Food Chem* 55: 8452-8457.

Shahbaz M, Stuiver C, Posthumus F, Parmar S, Hawkesford M, Kok L (2014) Copper toxicity in Chinese cabbage is not influenced by plant sulfur status, but affects sulfur metabolism-related gene expression and the suggested regulatory metabolites. *J Plant Biol* 16: 68-78.

Soon Y, Kalra Y, Abboud S (1996) Comparison of some methods for the determination of total sulfur in plant tissues. *J Commun Soil Sci Plant Anal* 27: 809-818.

Staley JT, Stewart-Jones A, Pope TW, Wright DT, Leather SR, Hadley P, Rossiler JT, van Emden HF, Poppy GM (2010) Varying responses of herbivores to altered plant chemistry under organic and conventional treatments. *Proc Royal Soc B* 227: 779-786.

Steed, Rita (2010) *Analysis of Amino Acids by HPLC*, Agilent Technologies, Inc. 800: 227-9770.

Thangavel P, Long S, Minocha R (2007) Changes in phytochelatins and their biosynthetic intermediates in red spruce (*Picea rubens* Sarg.) cell suspension cultures under cadmium and zinc stress. *J Plant cell Tissue Organ Cult* 88: 201-216.

Tisdale S, Reneau R, Platou J (1986) Atlas of sulfur deficiencies. *J Sulfur Agric* 27: 295-322.

Trombino S, Serini S, Di Nicuolo F, Celleno L, Andò S, Picci N, Palozza P (2004) Antioxidant effect of ferulic acid in isolated membranes and intact cells: synergestic interactions with α-tocopherol, β-carotene, and ascorbic acid. *J Agric Food Chem* 52: 2411-2420.

Vallejo F, Tomas-Barberan FA, García-Viguera C (2003a) Effect of climatic and sulphur fertilisation conditions, on phenolic compounds and vitamin C, in the inflorescences of eight broccoli cultivars. *Eur Food Res Technol J* 216: 395-401.

Vallejo F, Tomás-Barberán FA, Benavente-García AG, García-Viguera C (2003b) Total and individual glucosinolate contents in inflorescences of eight broccoli cultivars grown under various climatic and fertilization conditions. *J Sci Food Agric* 83: 307-313.

Von Uexkull H (1986) Sulfur interactions with other plant nutrients. In: *International Symposium on Sulfur in Agricultural Soil*s, Dhaka (Bangladesh), pp. 20-22.

Wang LI, Giovannucci EL, Hunter D, Neuberg D, Su L, Christiani DC (2004) Dietary intake of cruciferous vegetables, glutathione S-transferase (GST) polymorphisms and lung cancer risk in a Caucasian population. *J Cancer Cause Control* 15: 977-85.

Withers PJA, O'Donnell FM (1994) The response of double-low winter oilseed rape to fertilizer sulfur. *J Sci Food Agric* 66: 93-101.

Yemm E, Willis A (1954) The estimation of carbohydrates in plant extracts by anthrone. *Biochem J* 57: 508-514.

Yoshimoto N, Inoue E, Saito K, Yamaya T, Takahashi H (2003) Phloem-localizing sulfate transporter, Sultr1; 3, mediates re-distribution of sulfur from source to sink organs in *Arabidopsis*. *J Plant physiol* 131: 1511-1517.

Yoshimoto N, Takahashi H, Smith FW, Yamaya T, Saito K (2002) Two distinct high-affinity sulfate transporters with different inducibilities mediate uptake of sulfate in *Arabidopsis* roots. *Plant J* 29: 465-473.

Zhao F, Hawkesford M, McGrath S (1999) Sulphur assimilation and effects on yield and quality of wheat. *J Cereal Sci* 30: 1-17.

Zhao X, Chambers E, Matta Z, Loughin TM, Carey EE (2007) Consumer sensory analysis of organically and conventionally grown vegetables. *J Food Sci* 72: 87-91.

In: Trends in Life Science Research
Editors: R. P. Sinha and U. P. Shrivastava
ISBN: 978-1-53613-241-0

Chapter 11

ARBUSCULAR MYCORRHIZAE IMPROVES THE IONIC STATUS OF *PHASEOLUS VULGARIS* L.

***Wafaa M. Shukry*[1,*], *Gamal M. Abdel-Fattah*[1], *Mahmoud M. B. Shokr*[2] *and Mai A. Ahmed*[2]**
[1]Botany Department, Faculty of Science, Mansoura University, Mansoura, Egypt
[2]Horticultural Research Institute, Agricultural Research Center, Cairo, Egypt

ABSTRACT

The effects of arbuscular mycorrhizal (AM) fungi with different levels of NPK fertilizers was studied on the growth, photosynthetic pigments and ionic contents of common bean plants (*Phaseolus vulgaris* L.) throughout the three stages of growth. Common bean plants were subjected to five levels of NPK fertilizers (0, 25, 50, 75, 100%). Application of AMF significantly increased the growth and yield parameters (fresh and dry weights of roots and shoots, shoot length, number of leaves, leaf area, number of flowers as well as yield) and ionic contents (N, P, K, Mg, Ca and Na) of common beans with minimized levels of NPK as compared to equivalent non-mycorrhizal ones. The results obtained reveal that inoculation with AMF and with the 50 and 75% concentrations of NPK along with AMF were greater than the other concentrations and non-mycorrhizal plants. Mycorrhizal common bean plants had significantly higher shoot (fresh and dry weight), root (fresh and dry weight), shoot length, leaf area and photosynthetic pigments (Chl *a*, C *b*, carotenoids, and total pigments) in leaves of mycorrhizal and non-mycorrhizal common bean plants at the three stages. The same trend was recorded with the number of flowers. Ionic contents (N, P, K, Ca, Mg, and Na) in leaves and seeds, shoot, and root were significantly increased in mycorrhizal plants at different NPK levels as compared to those of non-mycorrhizal plants particularly at 50% and 75% concentrations of NPK, but lower Na^+ concentration in mycorrhizal common bean seeds than those of non-mycorrhizal were recorded.

* Corresponding Author Email: wafaa_12@yahoo.com.

Keywords: arbuscular mycorrhiza, (*Phaseolus vulgaris* L.), growth, ionic contents, NPK fertilizers

INTRODUCTION

Since the middle of the 20th century, food demand has increased at an unprecedented pace. With human numbers rising to 6.70 billion today (Southgate 2009). In recent years, the consumption of legumes, particularly dry beans, has increased in some West European countries and the United States. This is due to an increased realization of consumers about the nutritional characteristics in foods. Egypt suffers from food shortage problem as a result for a huge increment of population and the huge loss of agricultural soils due to desertification and erosion problems (Abd Allah et al. 2015). According to this view, increasing crop production for fulfilment of food requirements of the day by day with increasing population, so the systems are relying exclusively on the use of chemical fertilizers. Whereas, it is proved that use of chemical fertilizers and other pesticides are causing tremendous harm to the environment by pollution and contamination in water and soil and common men are suffering by many means, especially their health has been affected very much because of the situation (Weisenburger 1993). Therefore, the need to increase production of crops, to preserve soil fertility and to protect the environment from detrimental, agronomic techniques has brought about a revision of productive systems in agriculture. Recently, the employment of beneficial microorganisms has gained popularity (Parial et al. 2014).

Common bean was domesticated more than 7,000 years ago in ancient Mesoamerica (Mexico and Central America) and the Andean region, which is now grown worldwide. It is said to be the "poor man's meat" and is particularly important as the diet of the underprivileged. Now, for more than 300 million of the world's people, an inexpensive bowl of common beans is the centerpiece of the daily diet. It is the most important legume worldwide for direct human consumption, far outdistancing chickpeas, faba beans, lentils, and cowpeas (CIAT 2007).Common bean is a warm season crop, where average temperatures range between 15°C and 23°C. Legume growing better under subtropical and temperate conditions. It can found in tropical areas but does not well under very wet conditions that cause fungal attacks and flower drop (Romero-Arenas et al. 2013). It can grow under higher temperatures (35°C) but this may hamper seed production. It has slight frost tolerance but growth stops below 10°C and frost hinders yield at various stages of growth (Wortmann 2006).

The common bean grows well on large variable soils with pH ranging from 4 to 9. It grows better on well-drained, sandy loam, silt loam or clay loam soils, rich in organic matter content. Dry beans production (theoretically only *Phaseolus* species) was about 23 million ton in 2012, cultivated on 29 million ha (FAO 2013).

In Africa, common bean is noted for its versatility and diversity. It is adapted to varied climatic and agronomic conditions, and exhibits considerable variation in growth habit and seed type. It is an important food to people of all income categories, especially to the poor as a source of dietary protein. It is grown primarily by small scale farmers who have limited resources and usually produce the crop under adverse conditions such as low input use, marginal lands, and intercropping with competitive crops (CIAT 2004). In Recent days, consumption of legumes particularly dry beans, has increased in some West European countries and the United States, due to an increased realization of consumers about the nutritional characteristics in foods (Peksen and Artik 2005). The common bean characterized as a nearly perfect food because of its high protein content and generous amounts of fiber, complex carbohydrates, and other dietary necessities (López et al. 2005). When consumed as seed, beans constitute an important source of dietary protein (22% of seed weight) and are a very good source of cholesterol-lowering fiber, as are most other beans. In addition it has high fiber content prevents blood sugar levels from rising too rapidly after a meal, making these beans an especially good choice for individuals with diabetes, insulin resistance or hypoglycemia (Celleno 2007). Common beans are an excellent source of the trace mineral, molybdenum, an integral component of the enzyme sulfite oxidase, which is responsible for detoxifying sulfites, and iron, an integral component of hemoglobin, which transports oxygen from the lungs to all body cells, and is also part of key enzyme systems for energy production and metabolism (Queiroz et al. 2002). In general, dry beans contain 12% moisture, 22.9% protein, 60.6% carbohydrates, 1.3% fat, 3.2% minerals, 260 mg Ca, 410 mg P, 5.8 mg iron, and 346 calories/100 g. The vitamin contents of the dried beans are: thiamine, 0.6; riboflavin, 0.2; nicotinic acid, 2.5; and ascorbic acid, 2 mg/100g (de Almeida Costaet al.2006).

Food reduction problem of Egypt may occurs due to a noticed shortage in fertile cultivated soils in the old Nile Valley and Delta, which represent about 3-4% of the total area of Egypt (Zaki and Radwan 2006). Due to alkalinity of Egyptian soils, the obtainable phosphorus in the added fertilizer reduces sharply after a short period since application rapidly and transformed to tricalcium phosphate which is unavailable to the plants (khalil 2013).

Today fertilizer has become important to modern agriculture to feed the growing population. Chemical fertilizers are used extensively in modern agriculture in order to improve crop yield (Rai et al. 2014). However, the production and productivity of crops has increased by chemical fertilizers, the increased use of chemical fertilizers has posed many environmental and health problems. Biofertilizer can be used to substitute organics and chemical fertilizer, which does not source of pollution like chemical fertilizer and give faster effect compare to organic fertilizer, it should be put at the first place (Norhazimah 2009).

So, an attention was direct towards using micro-organisms to enhance it through their activities and providing most of the essential nutrients required to crop productivity. Most of the plants form symbiotic relationship with a group of fungi called mycorrhiza (Cardon and Whitbeck 2007). AM fungi may increase the effectiveness of absorbing capability of surface host root as much as ten times. Ions such as P, Zn and Cu do not diffuse readily through soil Because of this poor diffusion, roots deplete these immobile soil nutrients from the zone immediately surrounding the root. The increase in plant growth resulting from AM symbiosis is usually associated with increased nutrient uptake by the hyphae from the soil. Mycorrhizal fungal hyphae extend into the soil, penetrating into nutrient depletion zone and increased the effectiveness of immobile elements (Sadhana 2014).Several researches have shown that vesicular arbuscular rnycorrhizal (VAM) fungi can improve immobile nutrients uptake in plants, particularly Phosphorus (Safapour et al. 2011). Therefore, it is likely that plants infected with VAM fungi can significantly benefit from rock phosphate application in these soils.

In addition, Abdel-Fattah et al. (2014) also reported that mycorrhizal inoculation significantly increased plant growth responses, P and N concentrations in shoot and root tissues, acid and alkaline phosphatase activities, and total soluble proteins in root tissues compared with the non-mycorrhizal plants. These stimulations were related to the level of the mycorrhizal colonization in the root tissues which reflected on the yield and growth of plant. According to mycorrhizae function lower the rhizosphere pH due to selective uptake of NH_4^+ (ammonium-ions) and release of H^+ ions, that led to decrease soil pH and increases the solubility of phosphorus precipitates. The hyphal uptake of NH_4^+ also increases the flow of nitrogen to the plant as NH_4^+ is absorbed to the soil's inner surfaces and must be taken up by diffusion (Hamel 2004). Enhanced nutrient uptake by AM fungi is often associated with dramatic increase in dry matter yield, typically amounting to several-fold increases for plant species having high dependency on mycorrhiza (Pharudi 2010). Researchers have proved that AMF inoculation is effective in increasing crop growth and yield under sterilized soil (Grantetal. 2005).

MATERIALS AND METHODS

Experimental Design

The experiment was conducted using a randomized complete block design with two mycorrhizal treatments (+AMF, -AMF) combined with 5 concentrations of NPK fertilizers (0, 25, 50, 75, 100%) and were replicated 8 times (3 plants/pot) to give 80 pots with (35cm in diameter) containing 10kg sterilized soil (clay : sand 2:1 v/v). All plants were grown for 3 months and maintained under natural conditions at the Experimental Research Station, Faculty of Agriculture, Mansoura University.

Inoculum Preparation of Arbuscular Mycorrhizal Fungi

The inoculum of arbuscular mycorrhizal species including *Glomusclarum, Glomusmosseae, Glomusintraradices, Gigaspora margarita and Gigaspora gigantean* were originally isolated from rhizosphere soil of common bean plants using the wet sieving and decanting technique (Gerdemann and Nicolson 1963), and then identified by the author.

The identified arbuscular mycorrhizal spores were left to multiply for 6 months on sudangrass plants (*Sorghum halepense* L.) under natural conditions. Plants were irrigated with tap water as needed and the nutrient solutions (Long-Ashton without phosphorus) of plants were provided. A mixture of soil and infected sudangrass root, which contained spores were used as mycorrhizal inoculum. The AM fungal inoculums containing 5 g of rhizosphere soil (approx. 725 spores) and 0.5 g of infected sudangrass root fragments with an infection level of 88.6% was inoculated to each mycorrhizal pot.

Plant and Growth Conditions

Seeds of Common bean (*Phaseolus vulgaris* L.) (Nebraska), obtaining from the Horticultural Research Center, Egypt and used throughout this investigation. Seeds were surface sterilized in 0.0l% mercuric chloride for 3 min, subsequently washed with sterilized water. The results of soil analysis were recorded in (Table 1).

Table l. Physical and chemical analysis of soil used in this study.

Clay (%)	13.85
Silt (%)	37
Sand (%)	49.15
Texture	Loam
Organic matter (%)	1.31
P (ppm)	18.6
K (ppm)	10.2
N (ppm)	59.5
Ca^{2+} (meq/l00 g)	0.5
K^{+} (meq/l00 g)	0.22
Mg^{2+} (meq/l00 g)	0.4
Na^{+} (meq/l00 g)	0.35
pH	8.1
EC (dsm^{-1})	0.294
TSS (%)	0.094

For mycorrhizal treatments (AMF), each pot was inoculated with 5g of rhizosphere soil (approx. 275 spores/g soil) and 0.5g of chopped AM fresh sudan grass roots (M = 88.6%). The inoculum was placed 3-5cm below the soil surface of common bean plants to produce mycorrhizal plants. In the non-mycorrhizal treatments (-AMF), each pot received 10 ml filter leaching (Whatman No. 1) from infected roots and sterilized equal amount of soil inoculum to provide the same microorganisms other than mycorrhizal propagules. Each pot received Super phosphate (15% P_2O_5) was used as a source of phosphorus fertilizer, while nitrogen fertilizer used ammonium sulphate (20.5%) and potassium sulfate (48% K) as a source of potassium.

Measurements

Shoot fresh and dry weight, root fresh and dry weight, shoot length, and Leaves area throughout three stages of growth(vegetative at 35 days; flowering at 45 days and fruiting stage at 55 days) as well as number of flowers during flowering stage. Mycorrhizal dependency (MD) was defined as a percentage of a plant growth subjected to AMF application (Mengeet al. 1978), and calculated from the following formula as $\{(M-NM)/NM\}\times100$, where M is parameter value of mycorrhizal plants and NM is parameter value of non-mycorrhizal plants. Photosynthetic pigments (Chl *a*, Chl *b*, carotenoids, and total pigments) in leaves of mycorrhizal and non mycorrhizal common bean plants for three stages by Harborne (1984). Weight of 100 seed (g), weight offers hand dry weight of pods (g/plant), Length of pod (cm), number of pods/plant, and harvest index calculated according to (Beadle 1993) as follow seed weight (g) plant^{-1}/straw weight (g) plant^{-1}.After digestion by acid, analysed of N, P, K, Ca, Na and Mg contents. Total nitrogen (N) was measured by using Kjeldahl method (Nelson and Sommers 1973). Phosphorus (P) was determined following ammonium molybdate blue method (Chapman and Pratt 1961). Potassium (K) and sodium (Na) were assayed using a flame spectrophotometer (Corning 400, UK), and calcium (Ca) and magnesium (Mg) were determined using atomic absorption (PerkinElmer, Model 2308, USA) according to (Allen 1989).

Statistical Analysis

Data were subjected to statistical analysis using two-factor analysis of variance (ANOVA) and the least significant LSD at the significance level of $P \leq 0.05$ method using the Costat software (Cohort, Berkeley, CA, USA).

RESULTS

Growth Parameters

Results of effect of mycorrhizal colonization on growth parameters are presented in (Tables 2). In general, all growth parameters were significantly increased in common bean plants colonized with mycorrhizal fungi particularly at lower concentration of NPK than non-mycorrhizal plants in all harvests of plant growth. In this connection, AMF plants, in most cases, had significantly higher shoot fresh and dry weight, root fresh and dry weight, shoot length, at 50% and 75% NPK chemical fertilizers than non-inoculated plants, on the other hand, it observed that 100% NPK concentration decreased significantly and the lowest values were found in control treatments in the three stages respectively.

Plate 1. Effect of different levels of NPK fertilizers on growth of mycorrhizal and non-mycorrhizal common bean plants.

Table 2. Growth parameters (fresh and dry weights of roots and shoots, leaf area, number of leaves, and shoot length) of mycorrhizal (+AMF) and non-mycorrhizal (-AMF) common bean plants grown under different concentrations of NPK

Growth stage	Treatments		Fresh weight (g/plant)		Dry weight (g/plant)		Leaves area (cm^2/plant)	No. of leaves/plant	Shoot length (cm/plant)
	NPK conc. (%)	AMF status	Shoot	Root	Shoot	Root			
Vegetative	0	−AMF	38.96i	3.15i	5.47j	0.69f	193.83j	7.33f	15.33f
		+AMF	39.07i	3.32h	5.56i	0.71f	205.56i	8.00ef	17.76e
	25	−AMF	40.52h	3.41h	5.66h	0.74f	216.86h	8.66de	18.83de
		+AMF	41.36g	3.65g	5.78g	0.82e	230.10g	9.00cde	19.60cde
	50	−AMF	42.28f	3.91f	5.91f	0.85de	243.4f	9.33cd	20.06cd
		+AMF	45.49b	4.66b	6.35b	1.03ab	295.83b	11.33a	23.13ab
	75	−AMF	43.07e	4.13e	6.02e	0.91cd	256.50e	9.66bcd	21.50bc
		+AMF	46.33a	4.78a	6.47a	1.05a	308.93a	11.66a	24.10a
	100	−AMF	43.93d	4.23c	6.14d	0.94c	264.00d	10.00bc	22.56ab
		+AMF	44.73c	4.47d	6.28c	0.97bc	282.23c	10.66ab	23.00ab
Flowering	0	−AMF	41.78j	3.39j	5.73j	0.77g	219.50j	7.66g	23.66g
		+AMF	42.5i	3.56i	5.84i	0.82fg	231.70i	8.33fg	24.66g
	25	−AMF	43.25h	3.75h	5.93h	0.83f	241.46h	9.00efg	25.00fg
		+AMF	44.48g	3.99g	6.08g	0.90e	254.70g	9.66def	27.66ef
	50	−AMF	45.15f	4.17f	6.18f	0.93de	275.43f	10.33cde	29.66de
		+AMF	48.10b	5.00b	6.69b	1.14ab	327.53b	12.00ab	37.66b
	75	−AMF	6.08e	4.41e	6.33e	0.97d	286.43e	10.66bcd	31.33cd
		+AMF	48.64a	5.19a	6.82a	1.17a	340.56a	12.66a	38.33a
	100	−AMF	46.87d	4.61d	6.44d	1.05c	301.80d	11bcd	33.66bc
		+AMF	47.65c	4.82c	6.54c	1.09bc	312.46c	11.33abc	35.66ab
Fruiting	0	−AMF	37.86j	3.92j	5.05g	0.86f	207.76j	8.66f	30.66g
		+AMF	38.21i	4.05i	5.12g	0.92ef	221.80i	9.00ef	31.33fg
	25	−AMF	38.78h	4.20h	5.26f	0.96e	293.13d	12.0bc	36.16bcd
		+AMF	39.14g	4.40g	5.31ef	1.03d	301.46c	12.66ab	37.66abc
	50	−AMF	39.64f	4.56f	5.37e	1.10c	278.53e	11.66bcd	35.66bcde
		+AMF	41.81b	5.27b	5.75b	1.23ab	309.43b	13.00ab	38.66ab
	75	−AMF	40.13e	4.72e	5.48d	1.12c	264.46f	11.33bcd	34.66cdef
		+AMF	42.37a	5.44a	5.83a	1.29a	315.53a	14.00a	41.33a
	100	−AMF	40.73d	4.87d	5.57c	1.18b	236.26h	10.00def	32.00efg
		+AMF	41.27c	5.12c	5.67b	1.22b	250.80g	10.66cde	33.66defg

Values in each column followed by the same letters are not significantly different at $P \leq 0.05$.

On the other hand, no significant differences were observed in the number of leaves between the mycorrhizal and non-mycorrhizal plants grown at different concentration of NPK fertilizers. However, the reduction in most growth criteria were more distinct in non-mycorrhizal than in mycorrhizal plants (Plate 1). Number of flowers of mycorrhizal common bean plants was significantly higher than non-mycorrhizal ones (Figure 1).

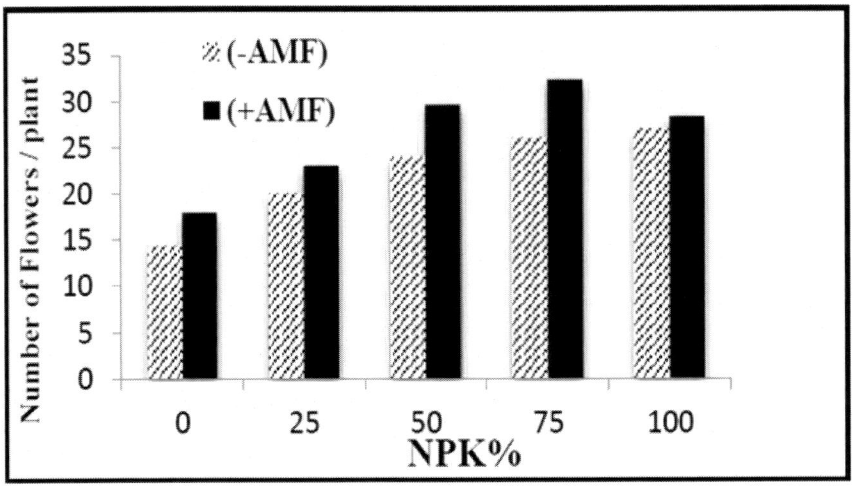

Figure 1. Effect of mycorrhizal colonization with different concentrations of NPK on number of flowers (flowers/plant) of common bean plants at flowering stage of plant growth.

Photosynthetic Pigments

The obtained results from the content of photosynthetic pigments (Chl *a*, Chl *b*, carotenoids, and total pigments) in leaves of mycorrhizal and non-mycorrhizal common bean plants are recorded in (Table 3). In general, the content of photosynthetic pigments was increased with plant growth but the contents of photosynthetic pigments in inoculated mycorrhizal plants were significantly greater than those of non-mycorrhizal plants at all stages of plant growth. In this interaction, the content of photosynthetic pigments in leaves of mycorrhizal plants was significantly higher than non-mycorrhizal plants, the greatest values were recorded at 50% and 75% NPK fertilizers with mycorrhizal infection as compared to 100% NPK concentration alone, and significantly reduced in 100% NPK fertilizers with mycorrhizal fungi. In fruiting stages, the photosynthetic pigments were lower than the vegetative and flowering stages but the mycorrhizal plants is still greater than non-mycorrhizal plants. The effects of mycorrhizal fungi on increasing of chlorophyll content in plants may be due to increased uptake of phosphorous and photosynthetic activity of the plants, the mycorrhizal fungi can be attributed to increased Mg and Fe uptake which are essential for chlorophyll biosynthesis. The increased levels of Chlorophyll in mycorrhizal common bean plants presumably would have contributed to achieve higher photosynthetic rates in the whole plant, thus benefiting both growth of host plants (Plate 2) and the development and functionality of the symbiosis.

Table 3. Photosynthetic pigments (mg.g^{-1} fresh weight) of mycorrhizal (+AMF) and non-mycorrhizal (-AMF) common bean plant leaves grown under different concentrations of NPK

Growth stage	Treatments		Chl. *a* (mg g-^{1} plant)	Chl. *b* (mg g^{-1} plant)	Carotenoids (mg g-^{1} plant)	Total pigments (mg g-^{1} plant)
	NPK conc. (%)	AMF status				
Vegetative	0	–AMF	0.761h	0.532h	0.711h	2.004h
		+AMF	0.772g	0.542g	0.723g	2.037g
	25	–AMF	0.782f	0.549g	0.733fg	2.061fg
		+AMF	0.790ef	0.557f	0.741f	2.088f
	50	–AMF	0.798e	0.568e	0.752e	2.118e
		+AMF	0.836b	0.603b	0.792b	2.234b
	75	–AMF	0.808d	0.575e	0.762de	2.145de
		+AMF	0.849a	0.612a	0.805a	2.267a
	100	–AMF	0.817cd	0.584d	0.770cd	2.172cd
		+AMF	0.825c	0.592c	0.780c	2.198fc
Flowering	0	–AMF	0.797f	0.562g	0.726f	2.055h
		+AMF	0.806e	0.563g	0.735ef	2.104g
	25	–AMF	0.813e	0.575f	0.744def	2.133ef
		+AMF	0.827de	0.582ef	0.756f	2.156e
	50	–AMF	0.833cde	0.588e	0.765cde	2.186d
		+AMF	0.867ab	0.618b	0.813ab	2.299ab
	75	–AMF	0.844bcd	0.596d	0.776bcd	2.217cd
		+AMF	0.878a	0.630a	0.823a	2.331a
	100	–AM F	0.854abcd	0.597d	0.788abc	2.230cd
		+AMF	0.859abc	0.610c	0.794abc	2.264bc
Fruiting	0	–AMF	0.692g	0.487h	0.612j	1.791j
		+AMF	0.704f	0.493h	0.622i	1.82i
	25	–AMF	0.713e	0.503g	0.633h	1.85h
		+AMF	0.723d	0.512f	0.644g	1.88g
	50	–AMF	0.73d	0.522e	0.653f	1.905f
		+AMF	b0.766	0.555b	0.688b	2.01b
	75	–AMF	0.741c	0.532d	0.662e	1.935e
		+AMF	a0.776	0.565a	0.698a	2.04a
	100	–AMF	0.748c	0.536d	0.672d	1.956d
		+AMF	0.76b	0.548c	0.679c	1.987c

Values in each column followed by the same letters are not significantly different at $P \leq 0.05$.

Ionic Contents

Nitrogen Content

Nitrogen concentration in common bean was improved by mycorrhizal inoculation (Table 4). Mycorrhizal inoculation significantly increased the nitrogen content in leaves, shoots and roots as compared to the non-inoculated common bean plants at all harvests. The higher values were markedly observed at 50% and 75% concentration of NPK chemical fertilizers when compared to other treatments.

Plate 2. Effects of AMF on the growth of common bean plants. Note the leaves of mycorrhizal plants were more green than that of non-mycorrhizal common bean plants.

Phosphorus Content

Mycorrhizal inoculation significantly affected the phosphorus concentration in plant tissue (Table 4). Mycorrhizal inoculation significantly increased the phosphorus content of the leaves, shoots and roots as compared to the non-inoculated common bean plants. The phosphorus concentration in leaves showed highest value than shoot and root. The higher values were observed at 50% and 75% concentration of NPK chemical fertilizers with mycorrhizal colonization than 100% NPK concentration of the increase in absorption of phosphorus by VA mycorrhizal plants.

Potassium Content

Generally, potassium concentration in leaves, shoots, and roots of AMF common bean plants were significantly higher than non-mycorrhizal common bean plants grown at different concentration of NPK fertilizers (Table 4). These effects were more pronounced at 50% and 75% of NPK fertilizers. Such increases in potassium contents in response to the mycorrhizal effects were highly associated with the levels of mycorrhizal colonization.

Table 4. Effects of mycorrhizal colonization under different concentrations of NPK on N, P, and K concentration (%) in leaf, shoot and root of common bean plants

Growth stage	Treatments		N(%)			P%			K%		
	NPK conc. (%)	AMF status	Leaf	Stem	Root	Leaf	Stem	Root	Leaf	Stem	Root
Vegetative	0	−AMF	3.33g	0.52g	1.35h	0.39 g	0.078h	0.135h	2.26g	0.43g	0.66i
		+AMF	3.42f	0.58fg	1.40gh	0.40f	0.081h	0.146g	2.36f	0.47fg	0.78h
	25	−AMF	3.48ef	0.62f	1.45g	0.40f	0.091g	0.148g	2.43ef	0.54f	0.85g
		+AMF	3.55e	0.71e	1.58f	0.42e	0.103f	0.156f	2.49e	0.63e	0.93f
	50	−AMF	3.66d	0.78d	1.66ef	0.42d	0.111ef	0.165e	2.61d	0.72d	1.06e
		+AMF	3.98b	1.03b	1.98b	0.46b	0.135ab	0.196b	2.96b	0.98ab	1.33b
	75	−AMF	3.73d	0.83d	1.72e	0.43d	0.116de	0.176d	2.66d	0.78cd	1.10e
		+AMF	4.06a	1.10a	2.06a	0.47a	0.143a	0.204a	3.09a	1.04a	1.47a
	100	−AMF	3.81c	0.89c	1.80d	0.44c	0.124cd	0.182cd	2.79c	0.84c	1.17d
		+AMF	3.92b	0.99b	1.89c	0.45bc	0.129bc	0.186c	2.89b	0.92b	1.23c
Flowering	0	−AMF	3.602i	0.39g	1.07h	0.414g	0.069h	0.119i	2.48j	0.37g	0.61g
		+AMF	3.70h	0.40g	1.11h	0.419g	0.075h	0.125i	2.60i	0.46f	0.70f
	25	−AMF	3.77g	0.46f	1.17g	0.431f	0.085g	0.131h	2.69h	0.53ef	0.77ef
		+AMF	3.82g	0.54e	1.24f	0.441e	0.086fg	0.137g	2.80g	0.54e	0.83e
	50	−AMF	3.95f	0.58de	1.32e	0.451d	0.094ef	0.147f	2.87f	0.66d	0.91d
		+AMF	4.28b	0.77b	1.56b	0.487b	0.122ab	0.175b	3.34b	0.94a	1.22b
	75	−AMF	4.03e	0.62cd	1.35de	0.457d	0.102de	0.154e	3.04e	0.72cd	0.99c
		+AMF	4.43a	0.83a	1.65a	0.502a	0.126a	0.181a	3.42a	0.98a	1.33a
	100	−AMF	4.11d	0.66c	1.39d	0.466c	0.109cd	0.163d	3.12d	0.76c	1.04c
		+AMF	4.21c	0.75b	1.50c	0.482b	0.114bc	0.17c	3.22c	0.85b	1.17b
Fruiting	0	−AMF	2.94i	0.67f	1.49h	0.369h	0.088f	0.133g	2.03j	0.47i	1.02d
		+AMF	3.02h	0.74f	1.55g	0.377g	0.092f	0.136g	2.13i	0.54h	1.11cd

Growth stage	Treatments		N(%)			P%			K%		
	NPK conc. (%)	AMF status	Leaf	Stem	Root	Leaf	Stem	Root	Leaf	Stem	Root
	25	–AMF	3.10g	0.81e	1.62f	0.384g	0.099e	0.146f	2.22h	0.59g	1.18cd
		+AMF	3.20f	0.91d	1.66f	0.393 f	0.107d	0.153e	2.29g	0.67f	1.22c
	50	–AMF	3.27e	0.97d	1.74e	0.403e	0.111d	0.160de	2.36f	0.72e	1.27bc
		+AMF	3.60b	1.25a	1.93ab	0.437b	0.137a	0.185ab	2.71b	1.02b	1.58a
	75	–AMF	3.33e	1.04c	1.81d	0.409e	0.118c	0.162d	2.47e	0.82d	1.42ab 1.42ab 1.42ab
		+AMF	3.71a	1.32a	1.98a	0.446a	0.143a	0.192a	2.81a	1.09a	1.60a
	100	–AMF	3.43d	1.14b	1.84cd	0.418d	0.124c	0.173c	2.54d	0.9c	1.52a
		+AMF	3.52c	1.18b	1.89bc	0.427c	0.130b	0.181b	2.62c	0.91c	1.54a

Values in each column followed by the same letters are not significantly different at $P \leq 0.05$.

Table 5. Effects of mycorrhizal colonization under different concentrations of NPK on Ca, Mg, and Na concentration (%) in leaf, shoot and root of common bean plants

Growth stage	Treatments		Ca %			Mg %			Na %		
	NPK conc. (%)	AMF status	Leaf	Stem	Root	Leaf	Stem	Root	Leaf	Stem	Root
Vegetative	0	–AMF	2.33h	0.58h	1.22h	1.45i	0.31h	0.59h	1.92a	0.25i	0.53j
		+AMF	2.42g	0.65g	1.32g	1.50h	0.42g	0.67g	1.82b	0.37h	0.61i
	25	–AMF	2.49f	0.74f	1.33g	1.61g	0.50f	0.76f	1.76b	0.43g	0.67h
		+AMF	2.57e	0.84e	1.44f	1.69f	0.60e	0.84e	1.68c	0.52f	0.72g
	50	–AMF	2.69d	0.92d	1.49e	1.77e	0.65e	0.93d	1.57d	0.6e	0.83f
		+AMF	3.01b	1.21b	1.80b	2.12b	0.98b	1.21b	1.23h	0.89b	1.13b
	75	–AMF	2.74d	0.98d	1.58d	1.89d	0.77d	0.97d	1.49e	0.67d	0.88e
		+AMF	3.11a	1.32a	1.93a	2.18a	1.06a	1.33a	1.16h	0.97a	1.22a
	100	–AMF	2.80c	1.07c	1.61d	1.94d	0.85c	1.07c	1.42f	0.73c	0.96d
		+AMF	2.97b	1.17b	1.72c	2.03c	0.93b	1.13c	1.34g	0.86b	1.08c

Table 5. (Continued)

Growth stage	Treatments		Ca %			Mg %			Na %		
Flowering	0	−AMF	2.61g	0.72i	1.34i	1.64j	0.51i	0.78j	2.09a	0.38j	0.63j
		+AMF	2.68g	0.83h	1.44h	1.72i	0.64h	0.91i	1.98b	0.52i	0.73i
	25	−AMF	2.77f	0.90h	1.54g	1.81h	0.77g	1.02h	1.92b	0.61h	0.81h
		+AMF	2.86e	1.03g	1.62f	1.98g	0.83g	1.12g	1.77c	0.71g	0.93g
	50	−AMF	2.96d	1.15f	1.65f	2.09f	0.98f	1.19f	1.7d	0.82f	0.99f
		+AMF	3.33b	1.55b	2.02b	2.38b	1.41b	1.65b	1.39f	1.22b	1.44b
	75	−AMF	3.03d	1.25e	1.75e	2.26e	1.10e	1.33e	1.65d	0.91e	1.12e
		+AMF	3.46a	1.73a	2.09a	2.48a	1.53a	1.76a	1.28g	1.32a	1.55a
	100	−AMF	3.12c	1.34d	1.85d	1.91d	1.21d	1.43d	1.52e	0.98d	1.22d
		+AMF	3.26b	1.48c	1.94c	2.19c	1.32c	1.54c	1.44f	1.13c	1.35c
Fruiting	0	−AMF	2.43j	0.89h	1.47g	1.37g	0.65h	0.90g	1.65g	0.51h	0.73h
		+AMF	2.50i	0.98g	1.54f	1.45f	0.72g	0.98f	1.57f	0.54h	0.78g
	25	−AMF	2.59h	0.98g	1.62e	1.48f	0.77g	1.07e	1.51f	0.61g	0.86f
		+AMF	2.70g	1.06f	1.66de	1.57e	0.86f	1.12de 1.12de	1.47e	0.70f	0.93e
	50	−AMF	2.77f	1.12ef	1.71d	1.62d	0.93e	1.15d	1.40d	0.76ef	0.95e
		+AMF	3.10b	1.42b	1.95b	1.89b	1.12ab	1.43a	1.17b	0.93ab	1.19b
	75	−AMF	2.84e	1.17e	1.79c	1.67d	0.97de	1.21c	1.37d	0.82de	1.02d
		+AMF	3.17a	1.51a	2.02a	1.97a	1.19a	1.46a	1.09a	0.98a	1.28a
	100	−AMF	2.93d	1.25d	1.83c	1.77c	1.02cd	1.25c	1.30c	0.86cd	1.09c
		+AMF	3.02c	1.33c	1.92b	1.80c	1.08bc	1.35b	1.21bc	0.90bc	1.17b

Values in each column followed by the same letters are not significantly different at $P \leq 0.05$.

Magnesium Calcium and Sodium Contents

The results of Mg and Ca concentrations are presented in (Table 5). These results show that, colonization of common bean plants with mycorrhizal fungi significantly increased the nutrient contents in leaves, shoot and root compared to those of equivalent non-mycorrhizal plants at all harvests particularly at 50% and 75% concentration of NPK fertilizers with mycorrhizal colonization in three stages as a consequence of better growth. Such increases in nutrient contents were highly related with the intensity of mycorrhizal colonization for each treatment. Table 5 showed that Na contents in leaves, shoots and roots, in mycorrhizal colonization tended to reduce Na in leaves concentration than non-mycorrhizal common bean plants. AM-inoculated plants exhibited a reduction in Na concentration in shoot tissues compared to non-inoculated control plants at all harvests. The rate of reduction in Na content was related to the infection of mycorrhizal fungi.

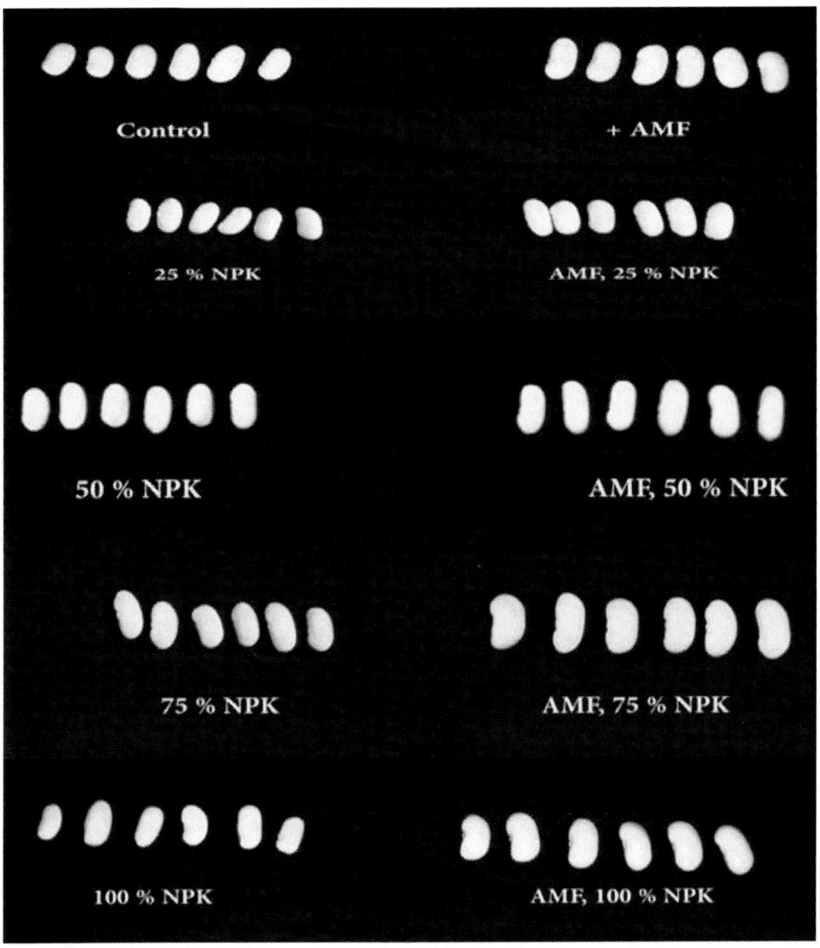

Plate 3. Effects of mycorrhizal colonization under different concentration of NPK fertilizers on the seeds yield of common bean plants.

Table 6. Yield parameters of mycorrhizal (+AMF) and non-mycorrhizal (-AMF) common bean plants under different concentrations of NPK fertilizers

Treatments		Number of pods/plant	Pod length (cm/pod)	Pods weight (g/plant)	Number of seeds/ pod	100 seeds weight (g)	Weight of seed (g/plant)	Harvest index %
NPK conc. (%)	AMF status							
0	–AMF	6.00j	8.23j	17.31i	3.00g	43.6j	12.18j	398.99g
	+AMF	7.33i	9.00i	25.20h	4.00f	45.03i	17.53i	544.17f
25	–AMF	8.66h	9.43h	29.70gh	4.66f	46.6h	21.69h	652.89e
	+AMF	9.66g	10.00g	33.08g	5.66e	48.76g	24.47g	685.54de
50	–AMF	11.00f	10.40f	39.28f	6.00de	49.96f	27.49f	687.83de
	+AMF	19.33b	12.40b	90.33b	7.66b	57.43b	65.50b	1183.34b
75	–AMF	12.33e	10.73e	45.45e	6.66cde	52.16e	31.13e	712.89d
	+AMF	21.66a	13.56a	111.15a	9.00a	59.36a	80.06a	1374.83a
100	–AMF	14.66d	11.06d	58.75d	6.33cd	53.26d	38.26d	835.54c
	+AMF	16.66c	11.40c	72.40c	7.00bc	55.7c	56.27c	1186.51b

Values in each column followed by the same letters are not significantly different at $P \leq 0.05$.

Yield Parameters

Yield parameters (number, length, and weight of pods, number of seed/pod, weight of 100 seeds, weight of seed/plant and harvest index) of common bean plants were greatly affected by each mycorrhizal colonization and NPK fertilizers alone or in combination (Table 6 and Plate 3). In this connection, these parameters were significantly higher in mycorrhizal plants than those in the non-mycorrhizal plants, particularly in 50% and 75% of NPK fertilizers.

Concentration of Nutrients in Seeds of Common Bean Plants

The concentrations of P, N, K, Mg, and Ca contents in seeds of common bean plants decreased with increasing NPK concentration (Table 7). Reduction in nutrient contents as a result of NPK concentration was more pronounced in non-mycorrhizal than in mycorrhizal plants. Colonization of common bean plants with mycorrhizal fungi significantly increased the nutrient contents in seeds compared to those of non-mycorrhizal plants particularly at 50% and 75% NPK concentration. Such increases in nutrient contents were highly associated with the intensity of mycorrhizal colonization for each treatment. On the other hand, the effects of mycorrhizal colonization tended to reduce Na in seeds of common bean plants. Inoculated common bean plants with mycorrhizal fungi reduce the concentration of Na than non-mycorrhizal common bean plants. The lowest value was recorded at 75% concentration of fertilizers with mycorrhizal colonization, and the higher value was observed at control treatments.

Table 7. Concentrations (%) of N, P, K, Ca, Mg and Na in mycorrhizal (+AMF) and non-mycorrhizal (-AMF) in seeds of common bean plants

Treatments		N	P	K	Ca	Mg	Na
NPK conc. (%)	AMF status						
0	−AMF	2.43j	0.298j	2.08j	1.99j	1.32j	1.67a
	+AMF	2.47i	0.311i	2.18i	2.09i	1.39i	1.58b
25	−AMF	2.53h	0.318h	2.26h	2.19h	1.53h	1.51b
	+AMF	2.59g	0.332g	2.34g	2.27g	1.59g	1.40c
50	−AMF	2.65f	0.340f	2.46f	2.37f	1.69f	1.32cd
	+AMF	2.91b	0.383b	2.83b	2.71b	2.07b	0.97f
75	−AMF	2.72e	0.354e	2.57e	2.43e	1.79e	1.23d
	+AMF	2.97a	0.397a	2.92a	2.83a	2.13a	0.89f
100	−AMF	2.78d	0.364d	2.65d	2.52d	1.88d	1.13e
	+AMF	2.84c	0.371c	2.76c	2.62c	1.96c	1.08e

Values in each column followed by the same letters are not significantly different at $P \leq 0.05$.

Mycorrhizal Dependency

Results of mycorrhizal dependency (MD) for improving each plant growth (shoot fresh weight, leaf area, total pigments, and some of nutrients) of common bean plants grown under different concentrations of NPK fertilizers are shown in (Tables 8). Mycorrhizal dependency values of common bean plants in responding to AMF inoculation were significantly higher at 50% and 75% concentration of NPK fertilizers than control conditions among all calculated parameters ($P \leq 0.05$). Improved growth of common bean plants depends highly on arbuscular mycorrhizal fungi, indicating that AM colonization can enhance the uptake of nutrients which reflected on the plant growth and yield components.

Table 8. Mycorrhizal dependency (MD)* for several parameters of common bean plants in flowering stage grown under different NPK concentration

Parameters	0	25	50	75	100	LSD
Shoot fresh weight	1.72d	2.84c	6.52a	5.55b	1.67d	0.41
Leaves area	5.55b	5.47b	18.91a	18.89a	3.53c	1.01
Total pigments	2.36b	1.51b	5.16a	5.14a	1.33b	1.37
N	2.39b	3.75b	12.84a	15.13a	4.93b	3.23
P	2.64b	2.51b	13.48a	13.29a	3.74b	2.85
K	12.69b	4.41c	24.07a	20.76a	6.43c	5.9

Values in each row followed by the same letters aren't significantly different at $P \leq 0.05$.

* MD = [(M-NM)/NM] × 100, where M is parameter value of mycorrhizal plants and NM is parameter value of non-mycorrhizal plants.

Discussion

Growth Parameters

In this study, the results demonstrated that colonization of bean plants with AM fungi significantly increased growth parameters (shoot and root fresh and dry weights, number of leaves, shoot length, and leave area), as compared to non-mycorrhizal plants that have NPK recommended fertilizers only particularly at 50% and 75% NPK concentration with mycorrhizal colonization than non-mycorrhizal plants. Arbuscular mycorrhizal fungi are the most widespread root fungal symbiont and are associated with the vast majority of higher plants. AMF have been shown to improve soil structure (Miller and Jastrow 2000) and have a great importance due to their great capability to increase the plant growth and yield through efficient nutrient uptake (Turkmen et al. 2008). In relation to plant growth and grain yield, our results clearly showed the beneficial effect of AM colonization on plant growth compared with non- mycorrhizal plants. It is widely known that the AM fungi are the most important microbial symbioses for the majority of plants improving their nutrient content, growth, and productivity (Grover et al. 2011; Berendsen et al. 2012). Enhanced growth effects on mycorrhizal plants are often related to improvement of P and N by leading their hyphal networks to plants. This pattern of response to mycorrhizal infection in low P soils is entirely consistent with previous studies (Abdel-Fattah and Asrar 2011; Smith et al. 2011; Sheng et al. 2013). This result also supports the previous findings by (Abdel-Fattah et al. 2014), who indicated that increased soil phosphorus generally reduced AMF development and consequently the mycorrhizal benefits. The fertilizer effect on plant growth was much more pronounced after inoculation of AMF and its combination with beneficial bacteria (Wu et al. 2005).

In the present study, our results show that inoculating common bean plants with vesicular-arbuscular mycorrhiza significantly increased shoot biomass and root biomass production could be attributed to enhance inorganic nutrition absorption and greater rates of photosynthesis. These data are in agreement with those of Abdel-Fattah et al. (2014), who reported that the arbuscular mycorrhizal inoculation significantly increased growth parameters (dry matter, leaf area and shoot height) of soybean plants. Also, the effects of mycorrhiza symbiosis on some crops such as Sugarcane, Mungbean, Wheat and Tomato confirmed that mycorrhiza affect uptake capacity and vegetative growth (Bayani et al. 2015). Also, mycorrhizal root have been known to absorb phosphorus faster than non-mycorrhizal plants, this could be resulted in a higher root biomass. The higher biomass and fruit yield observed in inoculated plants is consistent with the findings of some earlier studies (El-Shaikh and Mohammed 2009) and Okon (2014), who explained that due to the enhanced nutrient absorption caused by the *Glomus mosseae* whose extra-radical hyphae must have increased root absorbing surface area leading to more nutrient absorption for enhanced growth.

The results are also in good agreement with these of Abdullahi and Sheriff (2013) who also reported that increased plant growth parameter in plant shoot and root % colonization when applied with 60-30-50 kg ha^{-1}NPK to inoculated plants and this was comparable to un-inoculated plants treated with high doses of fertilizer (120-60-50kg ha^{-1}NPK) and result also revealed low level of colonization under un-inoculated plants and using AMF could be reduced the amount of fertilizer needed to produce on ion that leads to increase plant growth parameters, nutrients concentration in plant shoot and root % colonization were obtained when 60-30-50 kg ha^{-1}NPK was applied to inoculated plants and this was comparable to un-inoculated plants treated with high chemical fertilizers.

In consequence, many researchers have proved the importance of phosphorus in plant vegetative growth. According to Patton et al. (2002); Omotoso and El-Waraky (2014) lead to increasing in the plant heights of inoculated plants comparing by non-inoculated plants, this may be due to the mycorrhiza fungus producing a number of growth regulators that contribute for improvement of the components of different growth (Barker and Tagu 2000).

Photosynthetic Pigments

The results reported here, showed that plant photosynthetic pigments (Chl *a*, *b*, carotenoids, Chl *a*/*b* and total pigments) in leaves of common bean plants were significantly increased in the mycorrhizal plants with mycorrhizal fungi as compared to non-inoculated ones at all levels of NPK fertilizers. Such increases in photosynthetic pigments of mycorrhizal common bean plants were directly proportional to the respective level of mycorrhizal colonization. No significant differences in Chl a/b ratio were observed between mycorrhizal and non-mycorrhizal common bean plants grown in NPK-fertilized soil. In contrast, Chl *a*/*b* ratio of mycorrhizal plants was significantly lower than in non-mycorrhizal plants, presumably due to the greater pigments content of the mycorrhizal plants.

The effects of mycorrhizal fungi on increasing of chlorophyll content in plants may be due to increase the uptake of phosphorous and photosynthetic activity of the plants (Karthikeyan et al. 2009), or may be attributed to increased biosynthesis of pigments and/or decreased of their degradation (Shukry 2001). Some species of mycorrhizae caused an increase of the Chl *a*, *b* and total chlorophyll in *Pogostemon patchouli* (Selvaraj et al. 2009).

Common bean plants inoculated with mycorrhizal fungi showed a significant increase in photosynthetic pigments content, this could be attributed to greater rate of photosynthesis in inoculated plants due to the increased content of nitrogen and magnesium (major components of chlorophyll molecules) in mycorrhizal plants.

According to Krishna et al. (2005) have been stated that the mycorrhizal fungi can be attributed to increase Mg and Fe uptake which are essential for chlorophyll biosynthesis. These results are in agreements with those obtained by Shukry (2001) on wheat and bean plants and Bavaresco and Foghar (1996) who found that, the inoculation with VAM increased Fe and chlorophyll concentrations in leaves of the grapes.

Ionic Contents

Mycorrhizal colonization has been reported to increase the nutrients contents and results in more vigorous plants. The enhanced plant of most nutrient contents were pronounced at 50% and 75% levels of NPK fertilizers due to AM colonization. Nitrogen is absorbed and transported to plant shoots by AM fungi when present in the soil. Whereas, nitrogen is obtained by the extra radical hyphae of AM fungi in different forms ranging from amino acids, peptides, ions (NO_3^-or NH_4^+) to recalcitrant organic nitrogen forms. Of particular interest in this study, the concentrations of N, P, K, Mg and Ca in leaves of common bean plants were significantly decreased with increasing NPK concentrations in soil (Richardson et al. 2009). These results are in a good agreement with Bhattacharjee and Dutta Sharma (2012) who supported the results of our present findings that increasing of the nitrogen and phosphorus contents of pigeon pea. According to Krishna et al.(2005) have been stated that the mycorrhizal fungi can be attributed to increased Mg and Fe uptake which are essential for chlorophyll biosynthesis. A large proportion of nutrients (potassium, nitrogen, phosphorus, calcium, magnesium and sodium) absorbed by mature bean plants are found in different ratios in different organs of bean plants, where these amounts are in the order of leaves > roots> stems in all plants of mycorrhizal and non-mycorhizal under all concentrations of NPK fertilizers, where these amounts increased with increasing concentration of fertilizers. In this context, Xian-bin et al. (2000) reported that, the potassium circling amount between root and shoot increased as potassium supply levels increased. The potassium distribution proportion in organs of flue-cured tobacco was highest in leaf, lower in stem and root under different potassium supply levels. The highest amount of elements in roots is higher than that of the stem is in accordance with those obtained by Armstrong and Kirkby (1979) who stated that, the elements are carried away from the roots by the transpiration stream to the aerial portions of the plant, where they are utilized, water evaporates, and some of these elements enter the phloem and return to the root.

In this connection, Jia (2004) reported that, inoculation with AM fungi promoted biomass production and photosynthetic rates in *Vicia faba* because of the enhanced P supply due to AM fungi inoculation (Bucher 2006). The positive effect of AM fungi on phosphate uptake has been attributed to:

An exploration of a larger soil volume by the extra radical mycelium.

i. The small hyphal diameter leading to an increase of phosphate absorbing area and, compared to non-mycorrhizal roots, higher phosphate in flux rates per surface unit.
ii. The formation of polyphosphates by mycorrhizal fungi and so lower internal inorganic phosphate concentrations.
iii. The production of organic acids and phosphatases that catalyse the release of phosphate from organic complexes. In addition to phosphate, AM fungi enhance uptake of nitrogen, potassium, calcium, copper, manganese, magnesium, iron, and zinc (Clark and Zeto 2000).

On other hand, the concentration of Na^+ in mycorrhizal plants was significantly reduced than those non-inoculated plants. Alternatively, it appears that the role of mycorrhizal inoculation is partly to prevent Na absorption by the root and translocation to shoot tissues and the symbiosis can control uptake of Na when it becomes toxic to plants these results were confirmed by (Giri et al. 2007). Also (Abdel-Fattah and Asrar 2011) reported that AM plants exhibited low Na^+ and increased K^+ and Ca^+ contents in shoot tissues compared to non-mycorrhizal plants of wheat.

Yield Parameters

Yield parameters of the present results (number of pods, length of one pod, pods weight, number of seed/pod, 100 seeds weight, weight of seed/plant and harvest index) revealed that colonization of common bean plants with AM significantly increased yielded parameters as compared to control plants at all levels of NPK fertilizers. The effect of mycorrhizal inoculation, NPK fertilization and their interaction on common bean seeds yield, indicated that inoculation with mycorrhiza have a significant effect on seed yield compared to non-inoculation treatment. These results supported by the previous results of Ishac et al.1994 and Sadek 2005 who found that, the yield and yield attributed of faba bean tended to be higher in mycorrhizal colonized plants than non-mycorrhizal plants. And also Safapour et al. (2011) reported that inoculated red bean plants with mycorrhizal fungi showed a high seed yield.

Arbuscular mycorrhizal fungi colonized host roots uses the extra radical mycelium to explore a larger volume of soil, and transfer nutrients from soil to the plants more efficiently and increasing photosynthesis, which reflected on the improvement of plant growth and enrichment of fruit nutrition and increasing of seed yield (Tanwar et al. 2013; Abdel-Fattah et al. 2014). Also, these observations are also in good agreement with those of Utkhede (2006) who indicated that inoculation of tomato plants by AMF produced significantly higher fruit number and fruit yield compared to the non-inoculated plants.

In addition, the present results are also coincided those of with (Parial et al. 2014) who found that 75% NPK fertilizer and AM fungi treated maize plant was found significantly higher than the treated with 75% NPK fertilizer alone and almost similar to the maize plant treated with 100% fertilizer alone. Seedlings treated with 50% NPK fertilizer and AM fungi also showed good results as compared to 50% NPK fertilizer alone. From this data it can be stated that the use of AM fungi may reduce about 25% of chemical fertilizer usage. According to this observation, inoculation of seedlings with VAM at nursery stage and in the mainfield recorded leads to an additional yield and saved 25% of recommended dose of NPK with improvement in soil health, vegetable yield and quality (Singh and Kalloo 2000). This may be because mycorrhizal fungi help to increase nutrient uptake by increasing the surface area of the plant absorptive system (roots) that was evident from the present study and exploring soil by extra radical hyphae beyond the root hair and P-depletion zone and thus facilitates the absorption of the main elements in the soil (N, P, and K), and increasing its production of dry matter (Johri et al.2015).

Concentration of Nutrients Content in Seeds

The present results indicated that, increasing in nutrient contents in seeds yield in response to the mycorrhizal effects were highly associated with the level of mycorrhizal infection. Mycorrhizal common bean plants had higher contents of N, P, K, Ca, and Mg contents than those of non-mycorrhizal plants particularly at 50% and 75% NPK concentration of fertilizers. Our results are in good agreement with those of Abdullahi and Sheriff (2013) who found that using AMF could be reduced the amount of fertilizer needed to produce onion since increased plant growth parameters and nutrients concentration. Azcón et al. (2003) described that mycorrhizal lettuce plants with high availability of N and P in the soil reduced the content of macro and micronutrients in their tissues. However, when plants were fertilized with low level of both, P and N, nearly all the macro and micronutrients (P, K, Ca, Mg, Cu, Fe, Mn, Zn and S) were enhanced by AMF colonization (Baslam et al. 2013).

These results are confirmed by Khalil (2013) who reported that nutrients were taken up by the hyphae to the plant, which lead to a very efficient mobilization and uptake of phosphate, nitrogen, potassium, magnesium, copper, zinc, boron, sulphur and other elements that were transported to the plant. On the other hand, the mycorrhizal colonization lowering the Na concentrations, the non-mycorrhizal common bean plants were higher in Na as compared to mycorrhizal plants. This prevents Na ions that are toxic from interfering in metabolic pathways of growth and yield (Abdel-Fattah and Asrar 2011; Evelin et al. 2011).

Mycorrhizal Dependency

Bhattacharjee and Sharma (2012) reported that mycorrhizal infection alters the physiology of plants by increasing photosynthetic rates due to the presence of large number of chloroplasts in the bundle sheath of the leaves, changing the position of photosynthates in shoots and roots and changing them in eral uptake from soil, which consequently changes the nutritional composition of host tissues. This change in tissues in turn brings structural and biochemical changes in root cells, thus, the quality and quantity of root exudates. In the present study, vesicles and arbuscules were observed from root so finoculated treatment decreased with increasing adding NPK fertilizer, indicating that the highest value for root infection achieved. It can be concluded that the high rate of chemical phosphorus fertilizers application, led to antagonistic interaction with mycorrhiza. In low amount of added chemical phosphorus fertilizer, mycorrhiza fungi were able to significantly increase the root infection (Abdel-Fattah 1991).

In addition, the present results are in agreement with those of (Koide 1991; Soleimanzadeh, 2010) who showed that very high phosphorus levels may reduce mycorrhizal colonization. And also these can be explained by Valentine et al. 2001 and Gryndler et al. 2005 who stated that, the excessive chemical fertilizers have a negative effect on AMF colonization. In the present study AM fungi reduced the use of chemical fertilizer by 25%. It was also observed that when chemical fertilizer was used in combination with mycorrhiza, the plant growth occurs highest. Similar observations were recorded by Toussaint et al. (2007). Many of the compounds produced by plants in response to AM fungal colonization function as antioxidants in the plant and also in our diet. In consequence, mycorrhizal symbiosis and mycorrhiza dependency as results showed the symbiosis of plant roots with AM fungi was induced in 50% and 75% NPK fertilizers and 50% NPK fertilizers, dependency effects of AM symbiosis on mineral absorption increase with reducing root area (Porcel 2004).

Plant dependency on AMF was varied extremely among crops (Al- Karaki et al. 2004). Therefore, plant response to AMF has affected on management in cropping systems (Bouamri et al. 2006). Mycorrhizal efficiency is important for improving plants growth, but under high levels of chemical fertilization, mycorrhiza dependency has been slowed down (Ezawaetal. 2000).

CONCLUSION

Use of NPK fertilizer is a common practice among farming community throughout the world but application of NPK to legumes is limited especially in the developing countries. The results of this study have shown that inoculation with mycorrhizal fungi can enhance growth quality of common bean plants as compared to other treatments.

Application of NPK fertilizer at 50% and 75% along with VAM gave a significant increase in growth and components like shoot (fresh and dry weight), root (fresh and dry weight), shoot length, number of leaves, leaf area and number of flowers, and photosynthetic pigments (Chl *a*, Chl *b*, carotenoids, and total pigments) in leaves of mycorrhizal and non mycorrhizal common bean plants at three stages. Concentrations of ionic nutrients such as N, P, K, Ca, Mg, and Na, which are the components of grain, increased as compared to non-mycorrhizal ones. These have led to the increase in the economic value of common bean for sustainable agricultural purposes and healthy food production.

Acknowledgment

The authors gratefully acknowledge the Botany Department, Faculty of Science, Mansoura University, Egypt where this research was undertaken.

References

Abd Allah MMS, El-Bassiouny HMS, Bakry BA, Sadak M S (2015) Effect of arbuscular mycorrhiza and glutamic acid on growth, yield, some chemical composition and nutritional quality of wheat plant grown in newly reclaimed sandy soil. *RJPBCS* 6: 1038-1054.

Abdel-Fattah GM, Asrar AWA (2011) Arbuscular mycorrhizal fungal application to improve growth and tolerance of wheat (*Triticum aestivum* L.) plants grown in saline soil. *Acta Physiol Plant* 34: 267-277.

Abdel-Fattah GM, Asrar AA, Al-Amri SM, Abdel-Salam EM (2014) Influence of arbuscular mycorrhiza and phosphorus fertilization on the gas exchange, growth and phosphatase activity of soybean (*Glycine max L.*) plants. *Photosynthetica* 52: 581-588.

Abdel-Fattah GM (1991) *Some ecological and physiological studies on vesicular arbuscular mycorrhizal fungi*. Ph. D. Thesis, Botany. Department, Faculty of Science. Mansoura University. Egypt. pp. 202.

Abdullahi R, Sheriff HH (2013) Effect of arbuscular mycorrhizal fungi and chemical fertilizer on growth and shoot nutrients content of onion under field condition in Northern Sudan Savanna of Nigeria. *J of Agri and Veter Sci* 5(3): 85-90.

Al-Karaki G, McMichael B, Zak J (2004) Field response of wheat to arbuscular mycorrhizal fungi and drought stress. *Mycorrhiza* 14: 263-269.

Allen SE (1989) *Chemical Analysis of Ecological Materials*. 2nd Ed. Blackwell Sci. Pub. Osney, Oxford, London, 50-84.

Armstrong MJ, Kirkby EA (1979) Estimation of potassium recirculation in tomato plants by comparison of the rates of potassium and calcium accumulation in the tops with their fluxes in the xylem stream. *Plant Physiol* 63: 1143-1148.

Azcón R, Ambrosano E, Charest C (2003) Nutrient acquisition in mycorrhizal lettuce plants under different phosphorus and nitrogen concentration. *Plant Sci* 165:1137-1145.

Barker SJ, Tagu D (2000) The roles of auxins and cytokinins in mycorrhizal symbioses. *J Plant Growth Regul* 19: 144-154.

Baslam M, Garmendia I, Goicoechea N (2013) Enhanced accumulation of vitamins, nutraceuticals and minerals in Lettuces associated with Arbuscular Mycorrhizal Fungi (AMF): A question of interest for both vegetables and humans. *Agriculture* 3: 188-209.

Bavaresco, L, Fogher C (1996) Lime-induced chlorosis of grapevine as affected by root-Stock and root infection with arbuscular mycorrhiza and *Pseudomonas fluorescens. Vitis* 35:119-123.

Bayani R, Saateyi A, Faghani E (2015) Influence of arbuscular mycorrhiza in phosphorus acquisition efficiency and drought- tolerance mechanisms in barley (*Hordeum vulgare* L.). *Int J Biosci.* 7: 86-94.

Beadle CL (1993) Growth analysis. *In: Photosynthesis and Production in a Changing Environment. Springer Science*, Netherlands, pp. 36-46.

Berendsen RL, Pieterse CM, Bakker PA (2012) The rhizosphere microbiome and plant health. *Trends Plant Sci* 17: 478-486.

Bhattacharjee S, DuttaSharma G (2012) Effect of Dual Inoculation of Arbuscular Mycorrhiza and *Rhizobium* on the chlorophyll, nitrogen and phosphorus contents of Pigeon Pea (*Cajanus cajan* L.). *Adv Microbiol* 2: 561-564.

Bouamri R, Dalpé Y, Serrhini M, Bennani A (2006) Arbuscular mycorrhizal fungi species associated with rhizosphere of *Phoenix dactylifera* L. in Morocco. *Afr J Biotechnol* 5: 510-516.

Bucher M (2006) Functional biology of plant phosphate uptake at root and mycorrhiza interfaces. *New Phytologist* 173: 11-26.

Cardon ZG, Whitbeck JL (2007) *The Rhizosphere*, Elsevier Academic Press, Amsterdam, pp. 235.

Celleno L, Tolaini MV, D'Amore A, Perricone NV, Preuss HGA (2007) A dietary supplement containing standardized *Phaseolus vulgaris* extract influences body composition of overweight men and women. *Int J Med Sci* 4: 45-52.

Chapman HD, Pratt PF (1961) Ammonium vandate-molybdate method for determenation of phosphorus. *In: Methods of Analysis for Soils, Plants and Water.* 1st Ed. California Univ., California, pp. 184-203.

CIAT (2007) *Common bean: the nearly perfect food*, Annual Report. pp. 1-4.

CIAT (2004) *Regional bean variety testing in southern Africa.*

De Almeida Costa GE, da Silva Queiroz-Monici K, Pissini Machado Reis SM, de Oliveira AC (2006) Chemicalcomposition, dietary fibre and resistant starch contents of raw and cooked pea, common bean, chick pea and lentil legumes. *Food Chem* 94: 327-330.

El-Shaikh K, Mohammed M (2009) Enhancing fresh and seed yield of okra and reducing chemical phosphorus fertilizer via using VA- mycorrhizal inoculants. *World J Agri Sci* 5: 810-818.

El-Waraky YB (2014) Effect of sowing date, plant density and phosphorus fertilization on seed yield of Okra. *Alexa J Agri* 59: 27-41.

Evelin H, Giri B, Kapoor R (2011) Contribution of *Glomus intraradices* inoculation to nutrient acquisition and mitigation of ionic imbalance in NaCl-stressed *Trigonella foenum-graecum. Mycorrhiza* 22: 203-217.

Ezawa T, Yamamoto K, Yoshida S(2000) Species composition and spore density of indigenous vesicular-arbuscular mycorrhizal fungi under different conditions of P-fertility as revealed by soybean trap culture. *Soil Sci Plant Nutr* 46: 291-297.

FAO (2013) *Food and Agriculture Organization* of United Nations.

Gerdemann JW, Nicolson TH (1963) Spores of mycorrhizal endogone species extracted from soil by wet sieving and decanting. *Trans Br MycolSoc* 46: 235-244.

Giri B, Kapoor R, Mukerji KG (2007) Improved tolerance of *Acacia nilotica* to salt stress by Arbuscular Mycorrhiza, *Glomus fasciculatum* may be partly related to elevated K/Na ratios in root and shoot tissues. *Microbial Ecology-Springer* 54: 753-760.

Grant C, Bittman S, Montreal M, Plenchette C, Morel C (2005) Soil and fertilizer phosphorus: Effects on plant P supply and mycorrhizal development. *Can J Plant Sci* 85: 3-14.

Grover M, Ali SZ, Sandhya V, Rasul A, Venkateswarlu B (2011) Role of microorganisms in adaptation of agriculture crops to abiotic stresses. *World J Micro Biotech* 27: 1231-1240.

Gryndler M, Larsen J, Hršelová H, Řezáčová V, Gryndlerová H, Kubát J (2005) Organic and mineral fertilization, respectively, increase and decrease the development of external mycelium of arbuscular mycorrhizal fungi in a long-term field experiment. *Mycorrhiza* 16:159-166.

Hamel C (2004) Impact of arbuscular mycorrhizal fungi on N and P cycling in the root zone. *Can J Soil Sci* 84: 383-395.

Harborne JB (1984) Photochemical methods: A guide to modern techniques of plant analysis. Chapman and Hall, London. pp 132. *Inter Cent for Tropi Agri* pp. 1-2.

Ishac YZ, Angle JS, El-Borollosy ME, El-Demerdash ME, Mostafa MI, Fares CN (1994) Growth of *Vicia faba* as affected by inoculation with vesicular-arbuscular mycorrhizae and *Rhizobium leguminosarum* bv. *Viceae* into soils. *Biol Fertil Soils* 17: 27-31.

Jia Y (2004) The influence of *Rhizobium* and Arbuscular Mycorrhizal fungi on nitrogen and phosphorus accumulation by *Vicia faba*. *Ann Bot* 94: 251-258.

Johri AK, Oelmüller R, Dua M, Yadav V, Kumar M, Tuteja N, Varma A, Bonfante P, Persson BL, Stroud RM (2015). Fungal association and utilization of phosphate by plants: success, limitations, and future prospects. *Front Microbiol* 6:1-13.

Karthikeyan B, Joe M, Cheruth A (2009) Response of some medicinal plants to vesicular arbuscular mycorrhizal inoculations. *J Sci Res* 1: 381-386.

Khalil AA (2013) Significance of some soil amendments and phosphate dissolving bacteria to enhance the availability of phosphate in calcareous soil. *ISRN Soil Science* 2013:1-7.

Koide RT (1991) Nutrient supply, nutrient demand and plant response to mycorrhizal infection. *New Phytologist* 117: 365-386.

Krishna H, Singh SK, Sharma RR, Khawale RN, Grover M, Patel VB (2005) Biochemical changes in micropropagated grape (*Vitis vinifera* L.) plantlets due to arbuscular-mycorrhizal fungi (AMF) inoculation during exvitro acclimatization. *Sci Horti* 106: 554-567.

López S, Davies DR, Giraldez FJ, Dhanoa M, Dijkstra J, France J (2005) Assessment of nutritive value of cereal and legume straws based on chemical composition and in vitro digestibility. *J Sci Food Agr* 85: 1550-1557.

Menge JA, Johnson ELV, Platt RG (1978) Mycorrhizal dependency of several citrus cultivars under three nutrient regimes. *New Phytologist* 81: 553-559.

Miller RM, Jastrow JD (2000) Mycorrhizal Fungi Influence Soil Structure. *In: Arbuscular Mycorrhizas: Physiology and Function. Springer Science*, pp. 3-18.

Nelson DW, Sommers LE (1973) Determination of total nitrogen in plant material. *Agron* 65: 109-112.

Norhazimah AH (2009) *Effects of using enhanced biofertilizer containing N-fixer bacteria on patchouli growth*. EngD. Thesis. Universiti Malaysia Pahang.

Okon IE (2014) Growth response of okra (*Abelmoschus esculentus* (l.) moench) to arbuscular mycorrhizal fungus inoculation in sterile and non-sterile soil. *IJRHAL* 2: 31-38.

Omotoso SO, Shittu OS (2007) Effect of NPK fertilizer rates and method of application on growth and yield of Okra (*Abelmoschus esculentus* (L.) Moench) at Ado-Ekiti Southwestern, Nigeria. *Int J Agri Res* 2: 614-619.

Parial R, Mohajan S, Hussain MT, Hashem MA, Manisha D, Islam MM (2014) Isolation of Arbuscular Mycorrhizal fungi and evaluation of its effect on plant growth over chemical fertilizers for better human health. *SMU Medi J* 1: 192-206.

Patton W, Sema A, Martí C (2002) Effect of different levels of nitrogen and phosphorus on growth, flowering and yield of okra cv. *Arka Anamika* grown under the foothills. *Hortic J* 15: 81-88.

Peksen E, Artik C (2005) Antibesinsel maddeler ve yemeklik tane baklagillerin besleyici değerleri. *OMÜ Ziraat Fakültesi* dergisi [Nutritive values of antibiotic substances and edible grain legumes. *OMÜ Agricultural Faculty* magazine] 20: 110-120.

Pharudi JA (2010) *Effect of mycorrhizal inoculation and phosphorus levels ongrowth and yield of wheat and maize crops grow n on a phosphorus deficient sandy soil*. Master thesis, University of Stellenbosch, pp. 171.

Porcel R (2004) Arbuscular mycorrhizal influence on leaf water potential, solute accumulation, andoxidative stress in soybean plants subjected to drought stress. *J Exp Bot* 55: 1743-1750.

Queiroz KdS, Oliveira ACd, Helbig E, Reis SMPM, Carraro F (2002) Soaking the common beanina domestic preparation reduced the contents of Raffinose-type oligosaccharides but did not interfere with nutritive value. *J Nutr Sci Vitaminol* 48: 283-289.

Rai N, Ashiya P, Rathore DS (2014) Comparative study of the effect of chemical fertilizers and organic fertilizers on *Eisenia foetida*. IJIRSET 3: 12291-12298.

Richardson AE, Barea J, McNeill AM, Prigent-Combaret C (2009) Acquisition of phosphorus and nitrogen in the rhizosphere and plant growth promotion by microorganisms. *Plant Soil* 321: 305-339.

Romero-Arenas O, Damián-Huato M, Rivera-Tapia J, Báez-Simón A, Huerta-Lara M, Cabrera-Huerta E (2013) The nutritional value of beans (*Phaseolus vulgaris* L.) and its importance for feeding of rural communities in Puebla, Mexico. *Int Res J Biol Sci* 2: 59-65.

Sadek MS (2005) *Studies on seed borne diseases of important leguminous crops in Egypt*. Master Thesis, Botany Department, Faculty of Science, Mansoura University, Egypt, pp. 37-78.

Sadhana B (2014) Arbuscular Mycorrhizal Fungi (AMF) as a biofertilizer. A review. *Int J Curr Microbiol Appl Sci*. 3: 384-400.

Safapour M, Ardakani M, Khaghani S, Rejali F, Zargari K, Changizi M, Teimuri M (2011) Response of yield and yield components of three red bean (*Phaseolus vulgaris* L.) genotypes to co- inoculation with *Glomus intraradices* and *Rhizobium phaseoli*. *Ameri.-Euras. Am J Agric Environ Sci*. 11: 398-405.

Selvaraj T, Nisha MC, Rajeshkumar S (2009) Effect of indigenous arbuscular mycorrhizal fungi on some growth parameters and phytochemical constituents of *Pogostemon patchouli Pellet. Maejo Int J Sci Technol* 3: 222-234.

Sheng M, Lalande R, Hamel C, Ziadi N (2013) Effect of long- term tillage and mineral phosphorus fertilization on arbuscular mycorrhizal fungi in a humid continental zone of Eastern Canada. *Plant Soil* 369: 599-613.

Shukry WM. (2001). Effect of industrial effluents polluting the River Nile on growth, metabolism and productivity of *Triticum aestivum* and *Vicia faba* plants. *Acta Botanica Hungarica* 4: 1153-1159.

Singh K, Kalloo G (2000) Nutrient management in vegetable crops. *Fertiliser News* 45: 77-81.

Smith SE, Jakobsen I, Grønlund M, Smith FA (2011) Roles of arbuscular mycorrhizas in plant phosphorus nutrition: interactions between pathways of phosphorus uptake in arbuscular mycorrhizal roots have important implications for understanding and manipulating plant phosphorus acquisition. *Plant Physiol* 156: 1050-1057.

Soleimanzadeh H (2010) Effect of VA-Mycorrhiza on growth and yield of Sunflower (*Helianthus annuus* L.) at different phosphorus levels. *World Acad Sci Eng Technol* 71: 414-17.

Southgate D (2009) Population growth, increases in agricultural production and trends in food prices. *EJSD* 1: 29-35.

Tanwar A, Aggarwal A, Kadian N, Gupta A (2013) Arbuscular mycorrhizal inoculation and superphosphate application influence plant growth and yield of *Capsicu mannuum*. *J Soil Sci Plant Nutr* 13: 55-66.

Toussaint JP, Smith FA, Smith SE (2007) Arbuscular mycorrhizal fungi can induce the production of phytochemicals in sweet basil irrespective of phosphorus nutrition. *Mycorrhiza* 17: 291-297.

Turkmen O, Sensoy S, Demir S, Erdinc C (2008) Effects of two different AMF species on growth and nutrient content of pepper seedlings grown under moderate salt stress. *Afr J Biotechnol* 7: 392-396.

Utkhede R (2006) Increased growth and yield of hydroponically grown greenhouse Tomato plants inoculated with Arbuscular Mycorrhizal fungi and *Fusarium oxysporum* f. sp. *radicis-lycopersici*. *Biocontrol* 51: 393-400.

Valentine AJ, Osborne BA, Mitchell DT (2001) Interactions between phosphorus supply and total nutrient availability on mycorrhizal colonization, growth and photosynthesis of cucumber. *Sci Hortic* 88: 177-189.

Weisenburger DD (1993) Human health effects of agrichemical use. *Hum Pathol* 24: 571-576.

Wortmann CS (2006) *Phaseolus vulgaris* L. (common bean). Record from PROTA4U. *In: PROTA (Plant Resources of Tropical Africa/Ressources végétalesdel' Afrique tropicale)* (Eds. Brink M, Belay G) Wageningen, Netherlands.

Wu SC, Cao ZH, Li ZG, Cheung KC, Wong MH (2005) Effects of biofertilizer containing N-fixer, P and K solubilizers and AM fungi on maize growth: a green house trial. *Geoderma* 125: 155-166.

Xian-bin Z, Yi-ping C, Fu-suo Z, Zun-quan Z, Chun-jian L, Guo-shun L, De-ping X (2000) Circulation, accumulation and distribution of potassium in flue-cured tobacco with different potassium levels. *JPNF* 6(2): 166-172.

Zaki RN, Radwan TEE (2006) Impact of microorganisms activity on phosphorus availability and its uptake by faba bean plants grown on some newly reclaimed soils in Egypt. *Intl J Agri Biol* 8: 221-225.

ABOUT THE EDITORS

Dr. Rajeshwar P. Sinha is a Professor of Molecular Biology at the Centre of Advanced Study in Botany, Banaras Hindu University, Varanasi, India. He received his PhD in Biotechnology form Banaras Hindu University, Varanasi, India. He is a fellow of the Society for Applied Biotechnology, India and recipient of the prestigious DAAD fellowship. He has visited several countries such as Argentina, Austria, Belgium, Canada, China, Germany, Greece, France, Italy, Japan, Luxembourg, Norway, Poland, Republic of Korea, Spain, Switzerland, The Netherlands and United Kingdom, in the field of academics and research on one or the other scientific assignments. He has over 23 years of research and teaching experience. He has been working on effects of UV radiation on aquatic ecosystems and has concentrated on effects on DNA damage and repair, phycobilliproteins, mycosporine-like amino acids and scytonemin induction in cyanobacteria and algae. He is a life member of various national and international scientific societies and an editorial board member of several national and international scientific journals. He has published over 240 original research papers, reviews and book chapters and edited four books. He has over 7000 citations with an h-index of 43.

Dr. Umesh P. Shrivastava works as Associate Professor at the Department of Botany, Tribhuvan University, Thakur Ram Multiple Campus, Birgunj, Nepal. He obtained his Ph. D. in Biotechnology form Banaras Hindu University, Varanasi, India. He is a recipient of Young Scientist Award by Department of Science & Technology, Government of India, as well as SAARC fellowship by Government of India under Indo-Nepal Cultural Program of Indian Council of Cultural Relation (ICCR). He has over 26 years of research and teaching experience. He has been working on plant microbe interaction. He is a life member of various national and international scientific societies and an editorial board member of several national and international scientific journals. Presently, he is working as the Editor-in-Chief of "International Journal of Applied Sciences and Biotechnology (IJASBT)". He has published over 40 original research papers, reviews and book chapters. He has also published a book.

INDEX

#

A

B

C

H

I

K

L

M

N

O

P

Q

R

S

T

U

V

W

X

Y